C0-FXT-176

Biology of Suspension Feeding

BY

C. BARKER JØRGENSEN
Laboratory of Zoophysiology
University of Copenhagen

PERGAMON PRESS
OXFORD · LONDON · EDINBURGH · NEW YORK
PARIS · FRANKFURT

95510

591.53
J82

Pergamon Press Ltd., Headington Hill Hall, Oxford
4 & 5 Fitzroy Square, London W.1
Pergamon Press (Scotland) Ltd., 2 & 3 Teviot Place, Edinburgh 1
Pergamon Press Inc., 44-01 21st Street, Long Island City, New York 11101
Pergamon Press S.A.R.L., 24 rue des Écoles, Paris 5e
Pergamon Press GmbH, Kaiserstrasse 75, Frankfurt-am-Main

Copyright © 1966
Pergamon Press Ltd.

First edition 1966

Library of Congress Catalog Card No. 64-7812

PRINTED IN GREAT BRITAIN BY THE BAY TREE PRESS

2137/66

CONTENTS

PREFACE

ACCORDING to recent estimates the total primary production in the oceans amounts to about the same as the total production on land (Steemann Nielsen, 1960). However, in the sea and the larger freshwaters by far the greater part of the primary producers are microscopically small, and constitute the phytoplankton. Only within a narrow zone of shallow coastal water may larger plants contribute significantly to the primary production.

It is of both theoretical and practical interest to study the transfer of energy stored in the primary production through the various levels of consumers. The practical interest arises from the fact that upon the number of trophic levels introduced between the primary producers and the organisms used as food by man, depends the proportion of the primary production that is ultimately made available to man. At each level only the smaller fraction of the energy is generally utilized for growth, so that each new link in the "food chain" strongly taxes the energy originally stored by the photosynthetic activity of the phytoplankton.

The primary production in the sea is hardly directly utilized by man, and more intensive utilization is not yet possible due to the difficulties encountered in harvesting the crop. As the single phytoplankton organisms are far too small to be handled individually, large volumes of water would have to be processed, for instance, centrifuged or filtered, in order to concentrate and recover the organisms from the highly dilute suspensions in which they are present in the water.

Mechanisms for extracting on a mass scale small particles suspended in the surrounding water has been developed by numerous aquatic organisms: the suspension feeders. The development of such mechanisms has in fact been a preliminary for the evolution of larger aquatic animals living on phytoplankton cells and other dispersed food particles that are too small to be sensed and seized individually.

The dominance of the suspension feeders amongst the primary consumers is of special importance to man because the suspension feeders reduce the number of links in the food chain that would have been required if the phytoplankton was only fed upon by consumers that were small enough to seize the individual phytoplankton cell, and these consumers by secondary consumers, etc., that were also raptorial feeders.

Suspension feeding tends to develop a high degree of automaticity, characterized by continuous transport at constant rate of a feeding current to structures that retain suspended particles without regard to their value as food. Probably in most suspension feeders the rate of accumulation of food in the feeding organs is not regulated according to the actual need of the individual, but is determined by the fixed rates at which water is transported to the particle-retaining structures, by the concentration of food particles in the water, and by the efficiency with which these particles are retained in the feeding organs. Evidently, the subject of suspension feeding is well suited for a quantitative treatment. Quantitative aspects of suspension feeding have therefore been especially stressed in this treatise. Further, more attention has been given to features that have been found to be of common occurrence amongst suspension feeders, rather than to individual adaptations, an attitude which may in places have led to overgeneralizations.

The present account naturally fell into three parts. Part 1 deals with structure and function of the feeding organs, with special reference to rates of feeding and efficiency of retention of particles of various sizes and kinds in the feeding organs. Part 2 surveys the potential sources of food for the suspension feeders: phytoplankton, organic detritus, bacteria or dissolved organic matter. The methods used in measuring the various fractions of organic and inorganic matter in natural waters are described, and attempts have been made to evaluate the accuracy of the methods. Part 3, finally, discusses the food of suspension feeders in three main habitats: coastal waters, the euphotic zone of oceanic waters, and the deep sea. The discussions and conclusions are largely based on the evidence presented in Parts 1 and

2, and on data from the literature on feeding experiments performed in different representatives of the suspension feeders. Both in Parts 2 and 3 the main emphasis has somewhat arbitrarily been placed on marine habitats. However, the discussions in Part 3 concerning the food of suspension feeders in coastal waters apply to many freshwater biotops, too.

It has been necessary to bring together material belonging to many different subjects that are not generally dealt with together. No attempt has been made exhaustively to cover the literature pertaining to the various subjects. However, I hope that no material has been left out or overlooked that would seriously question the validity of conclusions drawn or problems formulated, or answer questions that have been emphasized as still being unanswered. In providing documentation, mostly recent papers have been selected, in which references to the older literature may be found.

I am greatly indebted to Dr. Gerald Kerkut for asking me to write the book and for reading and correcting manuscript and page proofs; to Mrs. Lis Olesen Larsen for reading the manuscript and for numerous valuable comments and suggestions; and to Miss Gerda Begtrup Hansen for typing the manuscript.

A grant for translation from the Danish Science Foundation has made it possible to include some of the rich marine biological literature in Russian pertaining to the questions dealt with, especially concerning feeding in planktonic suspension feeders and food problems in the deep sea.

ACKNOWLEDGEMENTS

My thanks are due to authors, publishers and societies for permission to reproduce figures from the following publications:

J. Mar. biol. Ass. U.K.	Figs. 1.4, 1.6, 1.7, 1.29, 1.30, 1.49, 1.50.
Trans. Roy. Soc. Edinburgh	Figs. 1.8, 1.9, 1.10, 1.11, 1.12.
J. exp. Biol.	Figs. 1.17, 1.18.
Phil. Trans. Roy. Soc. B.	Figs. 1.21, 1.37.
Nature Lond.	Fig. 1.22.
Proc. Zool. Soc. Lond.	Fig. 1.23.
Quart. J. micr. Sci.	Figs. 1.28, 1.33, 1.35, 1.36, 1.41.
Biol. Rev.	Fig. 1.34
Verh. Deutsch. Zool. Ges.	Figs. 1.1, 1.14.
Z. vergl. Physiol.	Fig. 1.2.
Z. Zellforsch.	Fig. 1.3.
Zool. Anz.	Fig. 1.5.
Wiss. Meeresunters, Abt. Kiel	Fig. 1.19.
Z. Fisherei u. deren Hilfswissensch.	Fig. 1.20.
Intern. Rev. Ges. Hydrobiol.	Figs. 1.25, 1.26, 1.27, 1.42.
Z. Morph. u. Ökol.	Figs. 1.45, 1.46.
Pubbl. Staz. Zool. Napoli	Fig. 1.43.
Zool. Bidr. Uppsala	Figs. 1.15, 1.16.
Biol. Bull.	Fig. 1.13.
J. Morph.	Figs. 1.31, 1.32.
Fish and Wildlife Serv. Fish. Bull.	Figs. 1.38, 1.39.
J. E. Morton: Molluscs. Hutchinson and Co., Lond. 1958	Fig. 1.24.
W. Kükenthal and T. Krumbach, Handb. der Zool.	Fig. 1.44.

INTRODUCTION

THE term "suspension-feeders" was used by Hunt (1925) to designate benthic animals that "fed by selecting from the surrounding water the suspended micro-organisms and detritus". It has later been extensively applied to indicate feeding-type of not only benthic, but also planktonic and nectonic organisms. The introduction of the concept of suspension feeding has proved most useful, but the use of the term "selecting" in the original definition is unfortunate, since most suspension feeders seem to be non-selective in the sense that retention of suspended matter from the water passing the feeding organs depends more on physical properties of the suspended particles than on their value as food.

Often the term "filter-feeder" is used synonymously with "suspension-feeder"; however, in the strict sense of the word many suspension feeders are not feeding by filtering the water. It is therefore reasonable to restrict filter feeding to denote feeding by passing the surrounding water through structures that retain particles mainly according to size and shape. In many groups of suspension feeders the particulate food is obtained by other methods, as expounded in Part 1 (see p. 131).

The suspension feeders generally constitute a well-defined group; difficulties are encountered in deliminating the group from other feeding groups mainly within very small forms and within forms feeding in the transition zone between the free water masses and the bottom. In the first case, the difficulty may arise when the difference in size between the consumer and its food particles gets too small. The difficulty is especially conspicuous amongst ecologically very important groups of the zooplankton, e.g. of the copepods. The phytoplankton algae range in size roughly from 1 to 100 microns. To a zooplankton organism about 1 millimetre in length the smallest phytoplankton cells are, therefore, too small to be consumed individually, whereas a phytoplankton organism 100 μ in diameter on the other hand would represent an object exclusively for raptorial feeding.

Suspension feeders amongst the benthic epifauna obtain their feeding current from water masses that are often only little contaminated with re-suspended bottom materials. The food conditions for such forms are therefore comparable to those of pelagic suspension feeders. However, many forms that live more or less buried in the bottom may take their food from the water immediately above the bottom. They may still be suspension feeders, but the suspended matter often originates from stirred up bottom materials, so the food conditions of these forms are not directly comparable with those of pelagic or typical epifaunal suspension feeders. Moreover, some infaunal suspension feeders may actively stir up already deposited bottom materials and thus form a transition to the true deposit feeders, which feed on particulate matter on or in the bottom. Cloudsley-Thompson (1960) has suggested that deposit feeding may sometimes be secondary and have evolved from suspension feeding in ancestral forms, but instances are also known where the opposite development seems to have occurred (p. 66).

Considering the dominant role of the phytoplankton in the primary production of natural waters, it is understandable that suspension feeding is of widespread occurrence; suspension feeders are found in almost all animal classes represented in fresh waters or in the ocean. Sponges, ascidians, brachiopods and most bivalves are suspension feeders, and also many crustaceans, gastropods, worms, etc. However, our knowledge about feeding in the various types varies; some few forms have been thoroughly investigated, in others knowledge does not go beyond main features of the anatomy of the feeding organs. In this review, a detailed treatment will only be given of forms in which the activities of the feeding organs have been experimentally investigated, especially activities as rates of water transport, efficiency of particle retention, and others needed for quantitative considerations. Forms in which little is known about the function of the feeding organs have been summarily dealt with; or omitted, e.g. Entoprocta, Polyzoa (Bryozoa), Phoronida, the hemichordate

Pterobranchia (van Gansen, 1960), and the recently discovered Pogonophora, which may even not be suspension feeders (Southward and Southward, 1963).

Only invertebrate suspension feeders have been considered in the present review. Suspension feeding is also of scattered occurrence within all of the vertebrate classes, e.g. in the ammocoete larva of the lamprey (Schroll, 1959); in the basking shark (Matthews and Parker, 1950; Parker and Boeseman, 1955); in a number of teleosts (Whitehead, 1959; Yasuda, 1960); in anuran tadpoles (Dodd, 1950; Savage, 1950); in flamingoes (Jenkin, 1957). However, the larger vertebrate suspension feeders, which are filter feeders, mostly do not utilize the primary production directly, as their filters are so coarse that they only efficiently retain larger food particles, for instance, zooplankton organisms.

Reviews on feeding in suspension feeders have previously been given by von Buddenbrock (1956), van Gansen (1960), Jørgensen (1955), Yonge (1928).

The methods used in determining rates of water transport through the feeding organs, or the efficiencies with which particles of various sizes and kinds are retained are either direct or indirect. For technical reasons the direct methods are only applicable to a restricted number of suitably shaped suspension feeders, and sometimes the results obtained may not be representative of normal undisturbed feeding behaviour. In the indirect methods that have been most promising, and which are theoretically applicable to all suspension-feeding organisms, the rates of water transport to the feeding organs are calculated from the rates at which feeding animals remove suspended particles from the surrounding water.

In suspension feeders that filter the water, the volume of water cleared of suspended particles, per unit time, in the following termed "clearance", is equal to the volume transported through the filters of the feeding organs when the suspended particles are so large that they cannot pass the pores of the filters. When part of the suspended particles pass the filters, the volume of water cleared is smaller than the volume transported, or filtered. From measurements of clearances in suspensions of particles of various size,

INTERNATIONAL SERIES OF MONOGRAPHS IN PURE AND APPLIED BIOLOGY

Division: **ZOOLOGY**

GENERAL EDITOR: G. A. KERKUT

VOLUME 27

BIOLOGY OF SUSPENSION FEEDING

INTERNATIONAL SERIES OF MONOGRAPHS IN PURE AND APPLIED BIOLOGY

Division: **ZOOLOGY**

GENERAL EDITOR: G. A. KERKUT

VOLUME 27

BIOLOGY OF SUSPENSION FEEDING

it is therefore possible to determine the porosity of the filters and, thus, to estimate how efficiently the experimental animal may retain particulate matter of various kinds suspended in its natural surroundings.

Also in non-filtering suspension feeders, measurements of clearances are important or indispensable in order to evaluate the availability to the animal of the different organic fractions in the normal environment.

Rates of water transport through the feeding organs of different suspension feeders are generally comparable only when expressed relative to the biomass in terms of, for instance, dry weight or nitrogen content of the soft parts of the body. However, in the literature often only length, volume or total weight of the animals under experiment are stated, so that direct comparisons are not possible if the species differ too much in organisation. In some instances conversion factors are known that have permitted calculation of the organic matter from length, total weight, etc., of the animals. In the following, water transport is expressed, when possible, per unit dry weight or nitrogen content of the soft parts. For specific purposes the water transport has also been expressed in relation to oxygen consumption (see p. 137).

PART 1

STRUCTURE AND FUNCTION OF FEEDING ORGANS

1.1. PORIFERA

The feeding mechanisms are rather similar in all sponges, calcareous as well as siliceous, but characteristic differences exist between primitive and more differentiated forms. The highly developed freshwater Spongillidae form the most thoroughly investigated group. They are therefore used as an example of feeding in the sponges.

Water, with its content of suspended food particles, enters through the pores of the surface epithelium, flows through the incurrent canals, passes the flagellated chambers, which are the water-transporting organs, and leaves the sponge body through the osculum via the excurrent canals and the cloaca (Fig. 1.1). Only small particles can pass through the pores of the epithelium, which in *Spongilla* are less than about 80 μ in diameter (van Trigt, 1919). Larger particles can, however, be phagocytized directly by the surface epithelium (van Weel, 1949; Kilian, 1952). The epithelium of the incurrent canals also ingests particles, especially those which are too big to enter the flagellated chamber. Phagocytosis in the incurrent canals is facilitated by means of protoplasmic bridges arranged as web-like structures across the canals, thus filtering the water that flows through the canals (van Weel, 1949). Only the smallest particles can pass the prosopyles of the flagellated chamber (Fig. 1.2). In *Spongilla* and other siliceous sponges the prosopyles are normally 3–4 μ, seldom as large as 7–8 μ, but somewhat wider in the calcareous sponges *Sycon* and *Grantia* (Bidder, 1920; van Trigt, 1919). The choanocytes of the flagellated chambers remove the particles that come into contact with their cell body or collar.

B

Until recently it was widely believed that the collar is a thin homogeneous cytoplasmatic membrane. However, electron microscopical investigations have shown that the collar is built of coarser fibrils connected by very fine transversal fibrils (Fig. 1.3). The distance between the fibrils in Spongillidae is about 0·2 μ (Fjerdingstad, 1961a; Rasmont, 1959). Most likely, therefore, flagellar movements will draw water inside the collar through the pores between the fibrils, so that water is filtered by the collars. This would explain the observation of van Trigt that small particles entering the flagellate chamber, when reaching the level of the collar, suddenly deviate and stick to the collar.* By a mechanism not yet understood, the adhering particles are transported downward to the base of the collar, taken up by the cell body and quickly transported down to the base of the cells. Here the particles are forwarded to the wandering phagocytes, archaeocytes, in which digestion is carried out (van Weel, 1949; Pourbaix, 1933a).

In primitive sponges, both calcareous, e.g. *Grantia* and *Sycon,* and siliceous, e.g. *Oscarella* and *Chondrosia,* with large choanocytes and flagellated chambers, uptake and digestion of particles is mainly performed by the choanocytes. In forms with medium-sized flagellated chambers which constitute transitional stages to the Spongillidae and other highly developed sponges, many particles are ingested before they can reach the choanocytes (Pourbaix, 1931, 1933b).

Apparently it is common to all sponges to ingest suspended particles without discriminating between matter with and without food value. Carmine particles and other inert material are as

* The collar of the choano flagellates has been shown to be similar in structure to that of the choanocytes (Fjerdingstad, 1961b). Furthermore, the flagellum of the sponge has been found to be typical of that of the flagellated algae: the flagellum possesses two rows of fine so-called flimmer hairs standing perpendicular to the flagellar axis. This type of flagellum has not been found in other metazoa. Flimmer flagella, possibly of the type characteristic of choano flagellates, have been found in both siliceous and calcareous sponges. The hypothesis is thus supported that the sponges are descendants of the choano flagellates (Afzelius, 1961). It is therefore an interesting possibility that filter feeding mechanisms that developed in unicellular organisms have been maintained with little change in even large multicellular organisms.

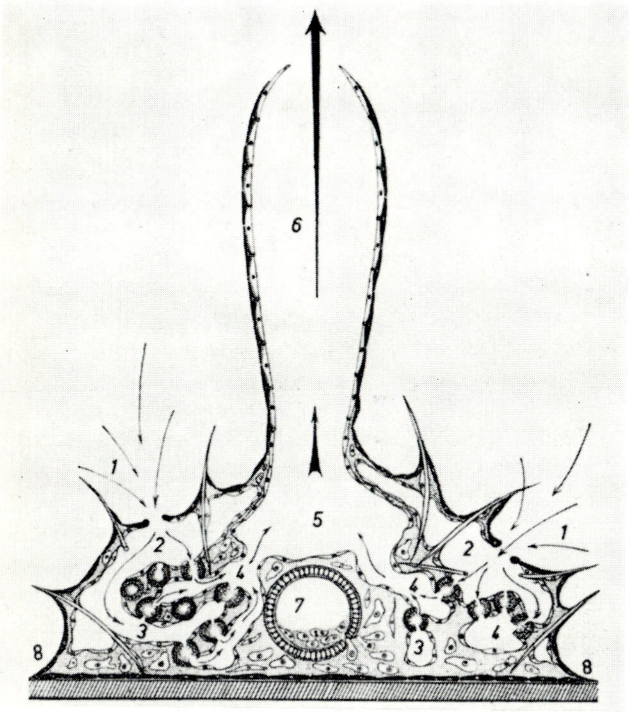

FIG. 1.1. Schematic section through a young sponge, *Ephydatia fluviatilis*.

1, Water entering through the pores; 2, subdermal space; 3, incurrent canals; 4, excurrent canals; 5 and 6, cloaca; 7, empty gemmule. (After Ankel, 1948.)

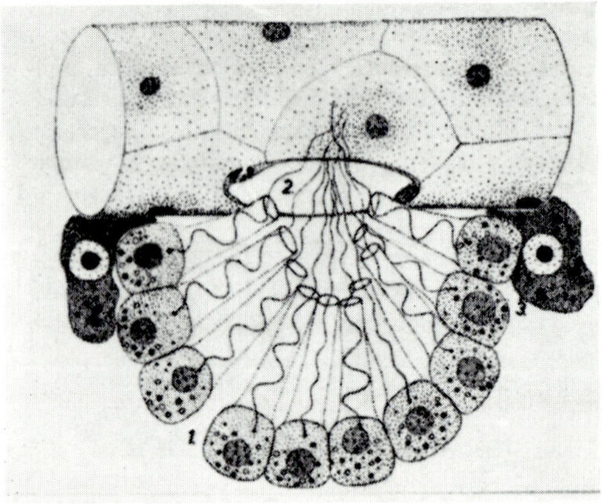

FIG. 1.2. Schematic section through a flagellated chamber of *Ephydatia fluviatilis*.
1, Prosopyle; 2, apopyle opening into an excurrent canal; 3, archaeocytes. (After Kilian, 1952.)

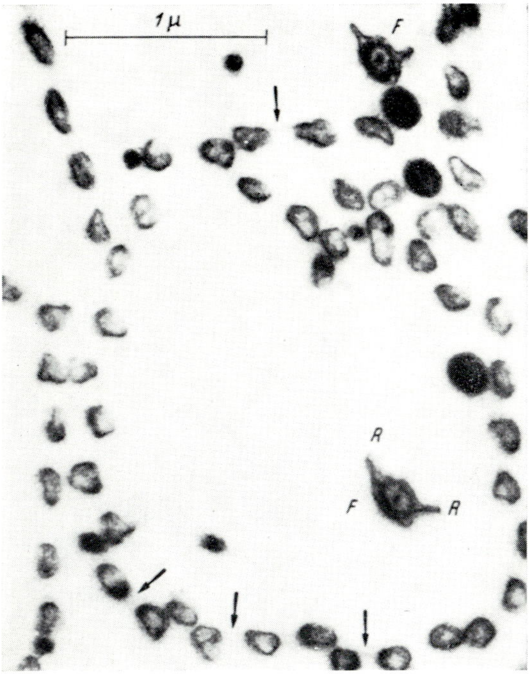

FIG. 1.3. Cross section through collar of a choanocyte in *Spongilla lacustris*. Parts of adjoining collars are seen.
F, flagella; R, projecting ridges of the flagellum. At the arrows some material may be suspected between the collar fibrils. (After Fjerdingstad, 1961.)

effectively ingested as food particles such as algae or bacteria (Bidder, 1937; Kilian, 1952). Particles without food value are, however, rapidly carried by phagocytes to the excurrent canals into which they are ejected.

In most sponges the water leaves the osculum with great speed, and is carried far away from the sponge so that the inspired water is not mixed with food-depleted water. Bidder (1923) showed that in finger-shaped sponges, such as, for example, *Leucandra aspera,* the osculum is adjusted in size to the rate of water transport in such a way that maximum length of the jet of water through the osculum is achieved. He also described other means by which mixing of inspired and expired water is avoided. In many sponges, a stalk lifts the body from the bottom so that the angle of supply is increased. Some sponges, especially in the deep sea, take the shape of discs or fans with one surface for intake and one for outflow of water. It has been shown experimentally that water currents may influence the form of sponges. Warburton (1960) found that the cells of *Microciona prolifera* squeezed through bolting silk would rebuild small sponges with central oscular chimneys perpendicular to the substrate in still water, but with excentric chimneys pointing downstream in flowing water; the chance of re-circulating the expired water is thus reduced.

The Porifera possess mechanisms which protect them against overfeeding as well as against noxious substances in water. When excessive amounts of particles are added to the water, the flagellated chambers can be seen to contract so that the water currents stop. The same effect can also be produced by the addition of chemicals to the water, by heat or light, or by mechanical disturbances. When normal conditions are re-established the sponge soon starts functioning again. These reversible reactions have been observed by Ankel (1948, 1950) in young gemmules of *Spongilla* and *Ephydatia* growing between microscope slide and cover- glass.

EFFICIENCY OF PARTICLE RETENTION

In living *Spongilla,* van Trigt (1919) observed that 0·5–1 μ particles seldom escape through the flagellated chambers. The distance between the collars of the choanocytes at their apex is about 1 μ. I have found that graphite particles of about 1 μ are effectively retained by the calcareous sponges *Sycon* and *Grantia* and the siliceous sponge *Halichondria* (Jørgensen, 1949b). The collar filter presumably retains particles down to 0·2 μ in diameter (see above).

RATE OF WATER TRANSPORT

In a piece of the siliceous sponge *Spinosella*, 10cm in length and 4cm in diameter, Parker (1914) measured a through-current of about 3l./hr. Calculations were made from observations of the rate at which particles travelled in a glass tube inserted in the osculum, and from the width of the tube. The hydrostatic pressure in the osculum varied in different sponges from 1·3 to 2·9mm water, under laboratory conditions. Under natural conditions in the sea it was, however, observed that water might still flow from oscula raised 4mm above the water-level. Therefore, the values of water transport obtained in the experiments are probably too low. Similar conclusions were reached by Bidder (1923), who found that the oscular velocities always decreased with time in sponges brought into the laboratory. In a freshly captured calcareous sponge, *Leucandra aspera,* 10cm in length, the flow rate amounted to 0·9l./hr. The dry weight of the protoplasm was about 0·2g. Assuming a nitrogen content of about 10 per cent, the water transport was thus about 45ml/hr/mgN.

From measurements of the speed at which colloidal graphite was removed from the water by the filtering activity of the calcareous sponges *Grantia compressa* and *Sycon coronatum;* a flow rate of about 170ml/hr/mgN has been calculated, and 65ml/hr/mgN in the siliceous sponge *Halichondria panicea*; the temperature was 18–19°C (Jørgensen, 1949b).

1.2. COELENTERATA

Most coelenterates are carnivorous, and live on large prey, but suspension feeding has been adopted both by members of the Scyphozoa and Anthozoa.

Suspension feeding has been described in the common jellyfish *Aurelia aurita* by Southward (1955b). Particles in the water are caught by mucus on both umbrellar surfaces and passed by ciliary currents to the marginal food groove (Fig. 1.4). In the groove ciliary tracts transport the material towards eight food pouches in which the masses of mucus and particles accumulate. The masses are then picked up by the tips of the oral arms and carried along grooves in the arms to the gastro-intestinal grooves,

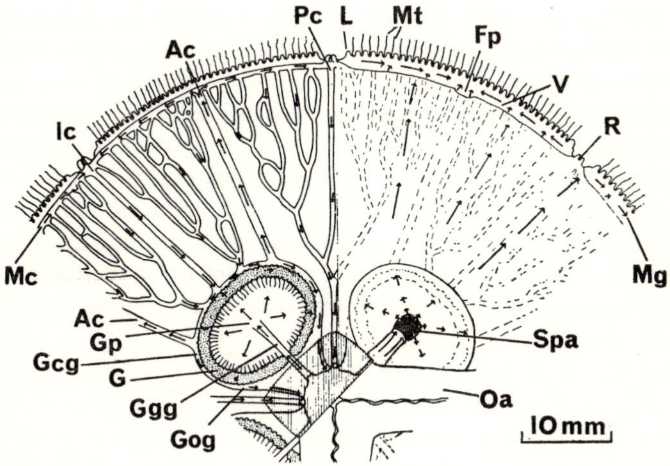

FIG. 1.4. *Aurelia aurita*. Part of the disc seen from the sub-umbrellar surface, with two of the oral arms removed (the cut bases hatched). The right side shows the currents on the sub-umbrellar surface and at the margin, while the left side shows the major currents in the canals and in one of the gastric pouches.
Ac, adradial canal; Fp, food pouch; G, gonad; Gcg, gastro-circular groove; Gog, gastro-oral arm groove; Gp, gastric pouch; Ic, interradial canal; L, lappet; Mc, marginal canal; Mg, marginal groove; Mt, marginal tentacle; Oa, oral arm; Pc, parradial canal; R, rhopalium; Spa, aperture of sub-genital pit; V, velarium. (After Southward, 1955b.)

and farther on to the gastric pouches. Examination of the mucus masses in the food pouches of freshly captured specimens usually showed suspended matter typical of the place of capture, for instance copepods, diatoms, ciliates, flagellates, or detritus. Feeding experiments showed rapid accumulation of the diatom *Nitzschia* or the copepod *Tigriopus*, first in the food pouches, later in the gastric pouches. Particles up to some millimetres in size can thus be accumulated, or larger particles than in most other suspension feeders (p. 133); but food particles above 5mm in diameter were always dropped or rejected from the margin or oral grooves.

Aurelia possess several mechanisms for selection of the collected material according to its value as food. Thus, the position of the numerous tentacles along the marginal groove determines the fate of the material collected on the umbrellar surfaces: when the tentacles are relaxed, they serve as rejection tracts, but when they are contracted, the material has access to the marginal groove. The oral arms constitute the major selective device outside the gastric pouches; they respond to proteins and other nitrogenous substances, but not to carbohydrates or inorganic particles. If the material accumulated in the food pouches contains much inert material, it may, therefore, not be picked up by the arms, or it may be dropped from the arms before it reaches the gastric pouch. Selection also takes place in the gastric pouches. Similar ciliary feeding mechanisms for handling fine particles have been described in some other medusae.

Suspension feeding in an anthozoan has been demonstrated recently by Roushdy and Hansen (1961), who found that the octocoral *Alcyonium digitatum* would feed on suspensions of the diatom *Sceletonema costatum* and other algae, labelled with carbon-14. The uptake of carbon-14 in the tissues of the *Alcyonium* colonies showed that the ingested diatoms were also assimilated; this coral is thus not only able to digest animal food.

Apparently, the suspension feeding coelenterates are not filter feeders, but they use the large mucus secreting ciliated external surfaces for directly catching particles from the water which is kept continuously streaming over the surfaces.

1.3. ROTATORIA

The Rotatoria are heterogeneous with respect to their feeding habits; some forms are exclusively carnivorous, others may feed both on large food and on suspended particles, still others seem to be exclusively suspension feeders (Myers, 1941). Feeding on fine particles is by means of cilia on the wheel-organ (Fig. 1.5), but the feeding mechanisms may vary between the species, as appears from the work of Gossler (1950), who studied the function of the wheel-organ with the stroboscope. Even when location and function of the ciliary tracts on the wheel-organ vary between species, their main function always seems to be to create and direct water currents with the suspended particles towards

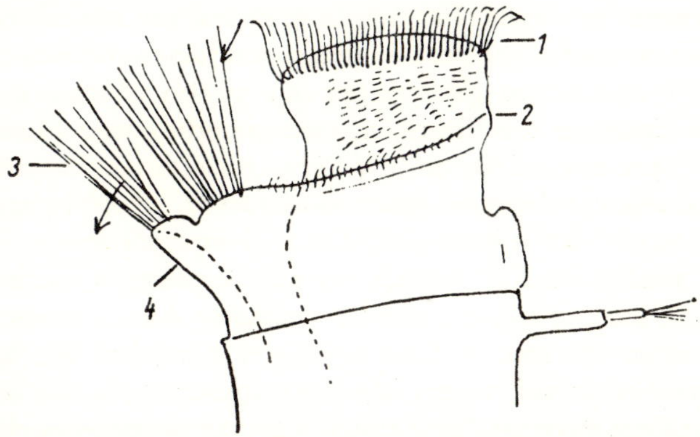

FIG. 1.5. *Philodina* sp. Lateral view of anterior end. 1, Trochus of the wheel-organ; 2, cingulum; 3, filtering setae on the lower lip (4). The arrows indicate the direction of the feeding current. (After Remane, 1932.)

the oral funnel and the mouth where fine particles are retained and large ones rejected. True filters in which particles are strained from the water to be carried by special mechanisms to the mouth do not seem to be present.

EFFICIENCY OF PARTICLE RETENTION

The Rotatoria are small animals of a body length mostly less than 1mm; only small particles are therefore able to pass the mouth opening. In a number of suspension feeding species, Gossler (1950) found that the maximal size of particles ingested was about 10 μ; Pourriot (1957) found 20 μ in *Brachionus*. Minimum size is not stated, but as it does not seem to depend on porosity of filters, it is probably not well defined.

RATE OF WATER TRANSPORT

Erman (1956) measured the rates at which the benthic *Philodina roseola* and the planktonic *Brachionus urceolaris* cleared suspensions of *Scenedesmus* colonies or *Chlorella* cells. The experiments were made on single individuals. In *Philodina,* clearances ranged from

0·4 to 1·9mm^3/hr, average *ca.* 1mm^3; they were independent of the concentration of algae, which was varied from about 10^4 to 10^7 cells or colonies per ml. At the highest concentrations the animals ingested several cells per second or about 10 times their own weight per hour. In *Brachionus,* clearances varied with the concentration of algae; maximum clearances were measured at about 10^5 *Scenedesmus* colonies per ml. At this concentration the average clearance was 7mm^3/hr, corresponding to a food ingestion of 5 per cent of the body weight. At high concentrations of algae, clearances dropped again, down to *ca.* 1mm^3/hr at a concentration of $7 \cdot 10^6$ colonies/ml. The high concentrations depressed the vitality of the animals, so the decrease in the rates of clearing concentrated suspensions probably does not imply that in *Brachionus* the rate of food uptake is regulated by the food concentration in the water. More surprising is the finding that the clearances decreased also at low algal concentrations, down to 230 colonies/ml. Erman suggests that the decrease is due to starvation and consequently reduced vitality. Similar results were obtained by Nauwerck (1959) who found that mixed populations of Rotatoria cleared about 1mm^3 lake water (0·1 to 2mm^3) of nannoplankton per hour. The plankton was labelled with carbon-14.

1.4. BRACHIOPODA

All investigated Brachiopoda are suspension feeders, which use the lophophore as a feeding organ. The lophophore varies extremely in shape between species, and is mostly of very elaborate structure. However, its function is basically similar throughout the Brachiopoda, and a simple type as met in *Megathyris detruncata* (Atkins, 1960) may be used to illustrate the principles of function. During feeding, the lophophore is expanded and the marginal filaments protrude beyond the opening of the shell (Fig. 1.6). A current of water is passing between the filaments produced by lateral ciliary tracts on both sides of the filaments (Fig. 1.7). In *Megathyris* the lophophore carries only a single series of filaments; in most other Brachiopoda two series are present. The filaments are very flexible, and when a brachiopod is in the feeding position they are arranged in such a way that the mantle cavity is divided in an inhalant and exhalant chamber (Atkins and Rudwick, 1962). The efficiency with which the lophophore retains particles from the passing water is not known, but Atkins (1960) observed in *Megathyris* that many peridineans or graphite particles offered to a feeding animal passed out between the filaments all around the lophophore. Apparently, the lophophore is not a

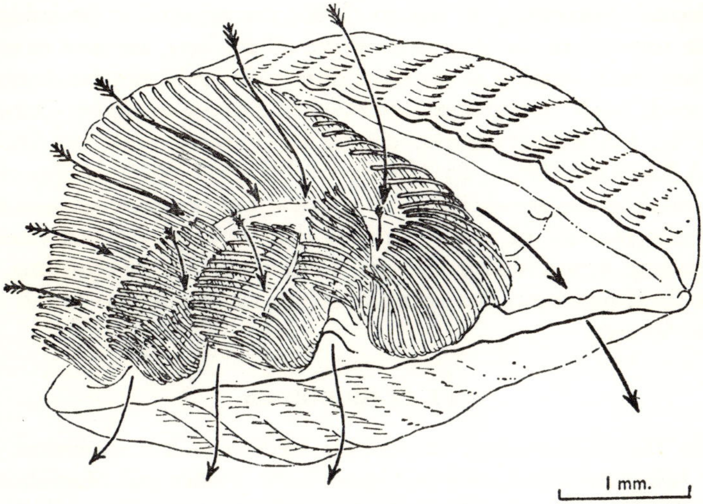

FIG. 1.6. *Megathyris detruncata.* Obliquely lateral view of a feeding animal to show the usual inhalant (feathered arrows) and exhalant (plain arrows) currents. Currents are reversed when strong cleansing action is needed. (After Atkins, 1960.)

filtering structure, but similar in functional type to, for instance, the polychaete gill crown, which also functions by keeping water continually streaming over a ciliated surface on which suspended particles can be retained by a mechanism that is probably not yet finally understood. On the filaments of brachiopods, mucus cells are present, and Orton (1914) ascribes the mucus spread out in a thin layer an important role in capturing and retaining food particles. However, according to Chuang (1956), mucus plays only a minor role in the transport of food particles.

The frontal ciliary tracts carry particles down to the base of the filaments where the tracts join other tracts in the food grooves of the lophophores, which transport the collected material to the mouth (Fig. 1.7).

The frontal ciliary tracts beat towards the base of the filaments only when small particles in not too large amounts are present on the tracts. In heavy suspensions or in suspensions of large particles, the direction of current of the tract is reversed, and the accumulated particles are carried to the tip of the filaments and rejected. According to Atkins (1956) the reversal is probably a true reversal of the direction of the ciliary beat, because frontal ciliary tracts beating in opposite directions could not be demonstrated. Such tracts were described in *Lingula unguis* by Chuang (1956), but their existence denied in *Crania anomala* by Atkins and Rudwick (1962). In *Megathyris,* Atkins (1960) also observed reversal of the lateral cilia. This reversal was probably responsible for the reversal of water currents that took place when strong cleansing action was called for, for instance on overloading of the food grooves. The reversal of the water currents lasted for 3 to 10 secs.

1.5. ECHIUROIDEA

In the Echiuroidea, suspension feeding is the exception, so far being described only in *Urechis caupo* (Fisher and MacGinitie, 1928). This worm dwells in U-shaped burrows in the sandy mud of shallow waters. The mode of feeding has been observed in conveniently shaped glass tubes. When feeding, the animal secretes a mucus net by a girdle of glands behind the oral chaetae. The net is attached to the walls of the burrow, generally close to the entrance. Water is drawn through the mucus net by peristaltic movements of the body, which thus acts in a piston-like fashion. Peristalsis becomes more energetic as the net wall fills. When loaded with particles the net is slipped "over the head" and picked up by the pharynx. Larger particles which happen to have been caught in the filter are rejected as the net is swallowed.

EFFICIENCY OF PARTICLE RETENTION

MacGinitie (1945) investigated the size of the mesh openings in the mucus feeding nets of *Urechis* and the polychaete *Chaetopterus* by adding to the water proteins of different molecular weight.

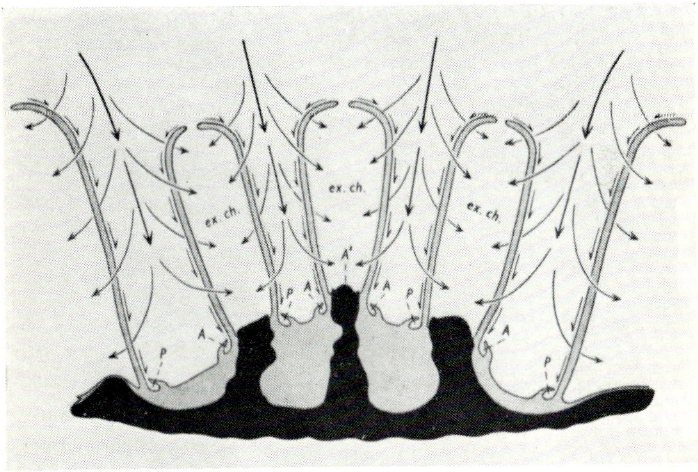

FIG. 1.7. *Megathyris detruncata.* Diagram of a section through the lobes of the lophophore to show the direction of the inhalant currents and of the frontal currents on the filaments of the feeding animal.

A, anteriorly and P, posteriorly directed currents in the food grooves. A^1, anteriorly directed current along the crest of the median septum. ex.ch. excurrent channel. The shell with the septa shown black. (From Atkins, 1960.)

The proteins were stained with Evans Blue so that retention could be estimated from the colouring of the net. He found that haemocyanin (mol. wt. *ca.* 450,000) was completely retained, whereas serum globulin (mol. wt. 176,000) was only partially stopped, and ovalbumin (mol. wt. 44,000) passed through quantitatively. The pore size thus seemed to be around 0·004 μ. The experiments further show that adsorption to the mucus net was of negligible importance for the retention of the proteins. The *p*H of sea water is on the alkaline side of the isoelectric point of ovalbumin and most other proteins, which therefore behave as anions in sea water.

RATE OF WATER TRANSPORT

Hall (1931) found that during feeding a 60g *Urechis caupo* pumped on an average 30ml/hr/g fresh weight. The pumping rate was measured on animals kept in glass U-tubes of the same dimensions as those of the natural burrows. The water transported by the worms was collected in a measuring cylinder via a siphon tube and a flask which was so adjusted that, when filled, water overflowed at the same level as that in the aquarium. Much higher pumping rates were found by Jørgensen (1955). Direct observations and measurements on a feeding *Urechis* (21g) in a glass tube of 25mm internal diameter yielded the following results. The peristaltic waves passing down the worm travelled over a distance of about 10cm from behind the head to the end of the animal. The diameter of the contracted parts of the body on both sides of the inflated piston-like portion was about 13mm. Thus, the volume of water driven in front of one complete peristaltic wave was $10\ (\pi.1{\cdot}25^2-\pi.0{\cdot}65^2)=36$ml. Eight to nine waves passed per minute, viz. about nine immediately after the feeding net was built, and about eight prior to the eating of the net. Some 300ml were thus filtered per minute, or 900ml/hr/g fresh weight. The temperature was 20°C, as against about 17°C in Hall's experiments. The difference in temperature thus does not explain the great discrepancy between the feeding rates measured by Hall and calculated by me. In the calculations a correction could be introduced for back-flow of water between the inflated parts of the body and the walls of the tube. This correction is probably insignificant, since the body is pressed tightly against the glass over a distance of about 1cm. As far as can be seen the only difference in experimental conditions that might have something to do with the great difference in pumping rates is that the worms in Hall's experiments had to transport water through long siphon tubes, whereas my observations were made on worms where the living–tube opened directly into the aquarium. Hall does not state the dimensions of the siphon tubes used, so it is not possible to estimate the resistance they might have offered to the flow of water.

LENGTH OF FEEDING TIME

Urechis caupo feeds discontinuously, and two types of resting periods are recorded. The first type consists of intervals of rest of 1–5 min which regularly alternate with periods of activity of 5–9 min. The second type is a continuous rest for an hour or more (Fisher and MacGinitie, 1928). These long periods of quiescence are also stated to occur under natural conditions (MacGinitie and MacGinitie, 1949). In the experiments of Hall (1931), *Urechis* was found to feed on an average 13min/hr, or only about one-fifth of the time.

1.6. POLYCHAETA

Within the Polychaeta, feeding on suspended matter is characteristic of the tube-dwelling Serpulimorpha, including such families as Serpulidae and Sabellidae. They feed by means of the crown of so-called gills. The gill filaments are equipped with cilia driving water through the gill crown and carrying particles towards the mouth (Johansson, 1927; Nicol, 1931; MacGinitie and MacGinitie, 1949). Similar feeding mechanisms are also described outside the Serpulimorpha, for instance by Dales (1957a) in *Owenia fusiformis*. The fullest account of feeding has been provided by Nicol (1931), in *Sabella pavonina*.

When the worm is feeding the gill-crown is fully expanded (Figs. 1.8 and 1.9). The filaments forming the branchial crown stand out stiffly at regular intervals and are kept in position by the membraneous web at their base. The pinnules on each filament are arranged in two rows, which make an angle with each other rather greater than 90°. Water currents are produced by the abfrontal and the latero-frontal cilia of the pinnules (Figs. 1.10

FIG. 1.8. *Sabella pavonina*. Terminal view of the branchial crown with the filaments fully spread, to show the collecting tracts of the filaments and lips. The arrows indicate the path taken by particles passing to the mouth or the ventral sacs. (After Nicol, 1931.)

and 1.11). Apparently, there is no true filtering of the water entering the branchial funnel, but eddies forming in front of the pinnules keep fresh water flushing over the frontal ciliary tracts. Particles suspended in the water are collected on the tract and transported by the cilia to the base, where they pass between the basal expansions of one pinnule and the next and enter the longitudinal groove of the gill-axis, down which they are carried by cilia to the gill folds (Fig. 1.12A, B). The gill folds are arranged in pairs with their inner faces almost touching; they carry three parallel horizontal tracts, one along the edge of each fold, one along the middle and one along the base. Particles are sorted into three grades. Large particles cannot enter between the gill folds;

they are forced upwards by the cilia of the edges and carried along the edges to rejection tracts on the lips (Fig. 1.9). Medium sized particles pass between the folds along the middle groove and are transferred to tracts on the lips leading to the ventral sac

FIG. 1.9. *Sabella pavonina*. Terminal view of the branchial crown with the filaments fully spread, to show the rejection tracts of the filaments, palps and lips. The arrows indicate the path taken by the particles passing to the tips of the palps. (After Nicol, 1931.)

in which accumulates building material for tube formation (Fig. 1.8). Only very fine particles pass straight down between the folds in the basal grooves, which are continuous with labial tracts that carry the fine particles directly into the mouth (Fig. 1.8). High loads of particles stimulate a copious secretion of mucus from the filaments and bulky strings of mucus with entangled particles are produced. These strings are carried either along the middle groove or along the rejection tracts on the edges of the folds; only loose, fine particles can be transported along the food-collecting tracts. No selection of particles according to food value has been described. Dales (1957b) found that various serpulimorphs ingested much material when they were placed in suspensions of

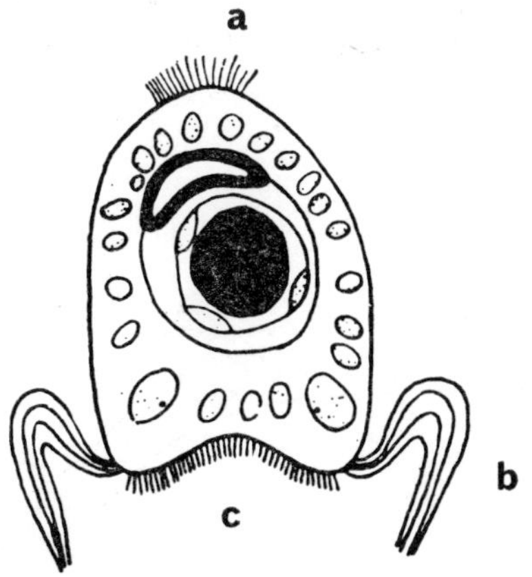

FIG. 1.10. *Sabella pavonina.* Transverse section of a pinnule to show the ciliation.
a., Abfrontal cilia; b., latero-frontal cilia; c., frontal cilia. (After Nicol, 1931.)

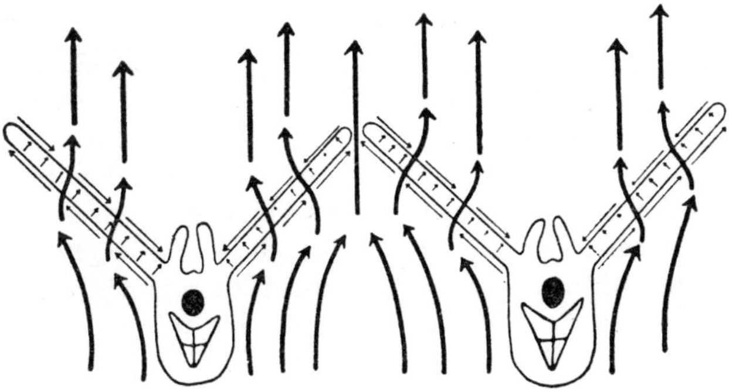

FIG. 1.11. *Sabella pavonina.* Diagrammatic section of two gill-filaments, to show the direction of the flow of water entering the branchial funnel, and the direction of beat of the cilia which cause the current. The small arrows indicate the direction of beat of the cilia, the large arrows the direction of flow of water. (After Nicol, 1931.)

colloidal graphite; dense graphite suspensions were even better tolerated than dense suspensions of various algae.

Feeding by filtering through mucus nets has been developed in at least two polychaetes, viz. *Chaetopterus variopedatus* (MacGinitie, 1939a) and the unrelated *Nereis diversicolor* (Harley, 1950). *Chaetopterus* lives buried in the bottom in a leathery U-shaped

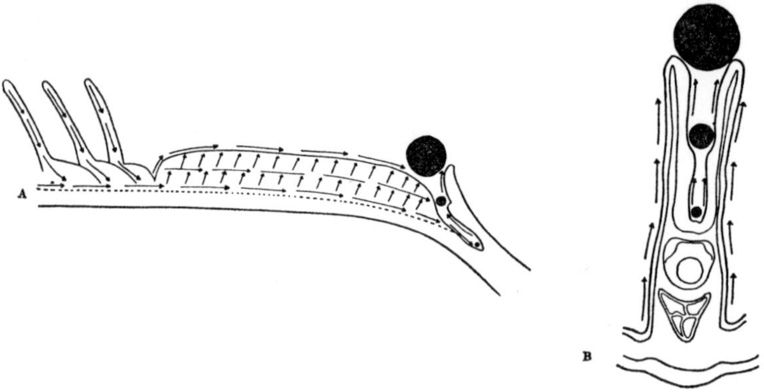

FIG. 1.12. *Sabella pavonina*. Diagrammatic representations of the base of a filament to illustrate the sorting mechanisms.
A, Diagrammatic representation of the lateral lip and the base of a gill-filament with one basal fold removed to show the tracts of the lower lip, with the relative sizes of particles passing along them, and the direction of beat of cilia on the basal fold. B, Diagrammatic transverse section through a pair of basal folds to show the position of the longitudinal tracts and the relative sizes of particles passing along them. The arrows indicate the direction of beat of the cilia. (After Nicol, 1931.)

tube which is constricted at both ends (Fig. 1.13). The two extremities of the tube usually protrude several centimetres above the bottom (Enders, 1909). When feeding, the worm secretes a mucus net by the aliform notopodia of the 12th segment (Fig. 1.13). Water is drawn through the mucus net by means of fan-shaped structures on the 14th–16th segments. The feeding net is continuously produced, the posterior end being rolled up into a ball within the dorsal cupule. At about 15-min intervals, the secretion of mucus is interrupted, and the mucus ball with entrapped particles is carried along the dorsal ciliated groove towards the mouth for ingestion.

Previous investigators of feeding in *Chaetopterus* (Enders,

1909; Faulkner, 1931) did not notice the mucus-net feeding. They state that suspended food particles in the incurrent water are directly caught by various ciliary tracts, for instance, around the mouth in the dorsal groove, on the aliform notopodia and in the dorsal cupule. The particles are embedded in mucus and either carried to the mouth for ingestion or rejected.

Further investigations are needed in order to find out whether

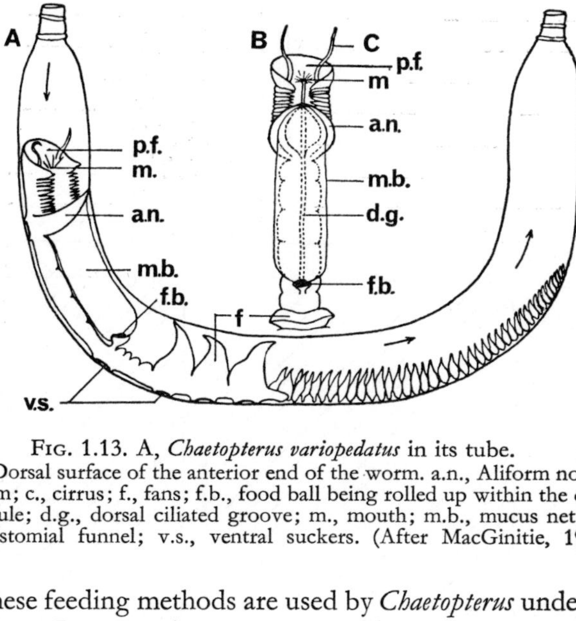

FIG. 1.13. A, *Chaetopterus variopedatus* in its tube. B, Dorsal surface of the anterior end of the worm. a.n., Aliform notopodium; c., cirrus; f., fans; f.b., food ball being rolled up within the dorsal cupule; d.g., dorsal ciliated groove; m., mouth; m.b., mucus net; p.f., peristomial funnel; v.s., ventral suckers. (After MacGinitie, 1939a.)

both these feeding methods are used by *Chaetopterus* under natural conditions. It seems, however, more likely that only suspension feeding by means of a mucus net is normal. It is difficult to imagine how the ciliary tracts described can effectively strain suspended particles from the current passing through the tube. Further, it looks as if food ingestion is a discontinuous process. Normally, the dorsal groove ends blindly behind the dorsal lip of the mouth. But when a food bolus travels up the groove, the dorsal lip turns backwards forming a temporary ciliary pathway which leads directly from the dorsal groove into the mouth (Faulkner, 1931). This reaction of the dorsal lip can be released by stimulating the upper parts of the dorsal groove mechanically with a stiff hair.

Such a mechanism suggests that in worms feeding normally food is ingested discontinuously, as it is bound to be during mucus-net feeding, though not if food particles were caught only by the different ciliary tracts.

Chaetopterus is able to get rid of unwanted material by reversal

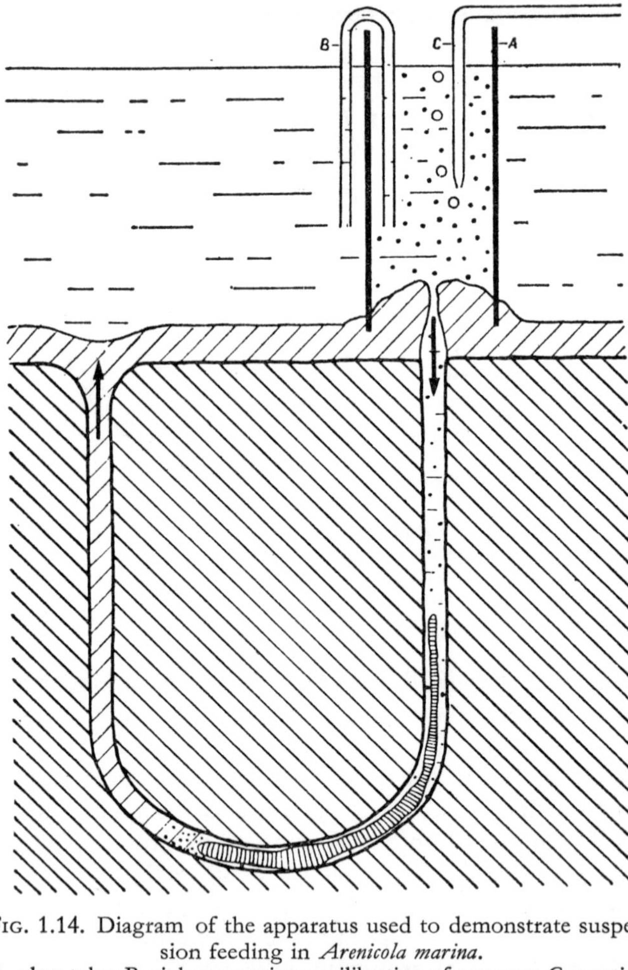

FIG. 1.14. Diagram of the apparatus used to demonstrate suspension feeding in *Arenicola marina.*
A, glass tube; B, siphon securing equilibration of pressure; C, aeration. The black dots in the tube represent the particles which are retained by the sand immediately in front of the worm. The arrows indicate the direction of the through-current. (After Krüger, 1959.)

of the direction of beat of the fan-shaped parapodia. Werner (1959) observed in young worms that when the mucus net contained too much unwanted matter, for instance stirred up bottom material, the net was not rolled up and eaten, but washed out of the tube by the reversed water current.

Whereas *Chaetopterus* feeds exclusively on suspended matter, *Nereis* apparently feeds by means of mucus nets only occasionally, and, as it seems, in response to the presence of particulate matter in the surrounding water. Thus, Harley (1950) found that when a suspension of carmine was added to the water, *Nereis* sometimes moved to one end of the glass tube in which it was being kept fixing the ends of fine mucus threads, secreted by the parapodial glands, to the sides of the glass tube. In retreating down the tube these threads were made into a mucus funnel through which water was drawn by means of undulating movements of the more distal parts of the body. From time to time the particle-loaded feeding net was then swallowed, and the whole process might be repeated.

Until recently, the lugworm *Arenicola marina* was believed to obtain its food from the organic matter normally present in the sand swallowed by the worm. However, Krüger (1959) found that the content of organic nitrogen was higher in the faeces (*ca.* 0·17mg/g) than in the surface sand (0·11mg/g); even higher values were found in the content of the anterior part of the intestinal tract (*ca.* 1mg/g). Experiments suggest that *Arenicola* enrich the sand by driving water through the column of sand in front of the head. Diatoms and other particles in the water are filtered off and eaten with the sand (Fig. 1.14).

Efficiency of particle retention in the filter feeding *Chaetopterus* has been dealt with above (p. 10). In the non-filter feeding serpulid *Myxicola infundibulum,* Dales (1957b) found that suspended graphite particles, varying in size from 1 to 100 μ, were removed at the same rate, indicating that particles within this size range are retained in the food collecting organs with equal efficiency.

RATE OF WATER TRANSPORT

Dales (1957b, 1961) measured the rates at which suspensions of colloidal graphite were cleared by various serpulimorph polychaetes and by *Chaetopterus variopedatus.* The results are shown in

Table 1.1. In *Sabellidae* and *Serpulidae* the clearances decreased from 2·09 to 0·07l./g/hr, when body weight increased from 0·1mg to 2·7g.

TABLE 1.1. RATES OF CLEARING SUSPENSIONS OF COLLOIDAL GRAPHITE PARTICLES IN POLYCHAETES (After Dales, 1957b, 1961.)

Species	Average body weight in mg	Clearance l./g/hr
Sabellidae:		
Myxicola infundibulum	2,700	0·10
Schizobranchia insignis	1,000	0·07
Sabella pavonina	190	0·39
Serpulidae:		
Pomatoceros triqueter	19	1·40
Hydroides norvegica	12	0·90
Spirorbis borealis	0·2	0·95
Salmacina dysteri	0·1	2·09
Chaetopteridae:		
Chaetopterus variopedatus	6g	0·05

LENGTH OF FEEDING TIME

Dales (1961) states that undisturbed *Schizobranchia* feed most of the time, and the same is probably true of the other suspension feeding sabellids and serpulids. It is not known how much time the intermittently feeding *Chaetopterus variopedatus* spends on mucus-net feeding under natural conditions (Wells and Dales, 1951).

1.7. CRUSTACEA

Suspension feeding is common throughout the Crustacea, especially amongst the small forms. It has evolved independently in several groups, and the feeding mechanisms often differ fundamentally from group to group (Tiegs and Manton, 1958). Even within smaller groups, suspension feeding may be of polyphyletic origin; for instance, in the Ostracoda the two filter feeding forms *Asterope* and *Cytherella* apply functionally similar feeding mechanisms, but the analogue processes are carried out by different limbs (Cannon, 1933b). Generally some of the head or trunk limbs have developed into filtering organs, the filtration being performed by plumose setae. Very often suspension feeding is only one of more feeding methods adopted by one species.

The feeding mechanisms have been extensively studied in several of the suspension feeding crustaceans, especially in species belonging to Phyllopoda, Cladocera, Ostracoda (Cannon, 1933b), Copepoda and Cirripedia, and to the malacostracan Euphausiacea, Mysidacea (Cannon and Manton, 1927), Anomostraca (Cannon and Manton, 1928) and Anomura (Nicol, 1932). Smallness, elaborate structure and complicated and rapid movements, etc., of the parts of the feeding organs may make it extremely difficult to analyse the normal feeding processes; the various investigators, therefore, often disagree in their descriptions and interpretations.

Quantitative aspects of suspension feeding have mainly been studied on Copepoda, Cladocera, Cirripedia and, less extensively, in Euphausiacea; the feeding mechanisms in these groups have therefore been dealt with in some detail. The pelagic phyllopods, cladocerans, copepods and euphausians include some of the most important zooplankton organisms, cladocerans mainly in freshwaters, copepods both in freshwaters and in all parts of the sea down to great depths, and euphausians mainly in the open ocean; the pelagic groups will be treated together, and the sessile barnacles separately.

1.71. ZOOPLANKTON CRUSTACEA

PHYLLOPODA AND CLADOCERA

The feeding mechanisms of the primitive suspension feeding Phyllopoda and the more diversified Cladocera have been described by Storch (1926), Cannon (1933a), Lowndes (1933), Eriksson (1934), Lochhead (1936), and others. The following description is mainly based on the accounts given by Cannon and Eriksson who on certain important points differ in opinion from Storch and Lowndes, but also Cannon and Eriksson disagree in certain points (Cannon, 1935).

Suspension feeding in the Phyllopoda probably developed from the locomotory movements of the thoracal limbs of primitive crustaceans. In *Lepidurus* which uses the thoracal limbs for locomotion and not for feeding, Eriksson observed water currents capable of carrying larger particles towards the mouth region.

The currents were similar to the feeding currents in the suspension feeding phyllopods. In the Branchiopodidate the limbs serve both swimming and feeding: currents created by the movements of the 11 similarly shaped pairs of thoracic limbs are partly drawn through filters formed by special setae on the limbs (Fig. 1.15).

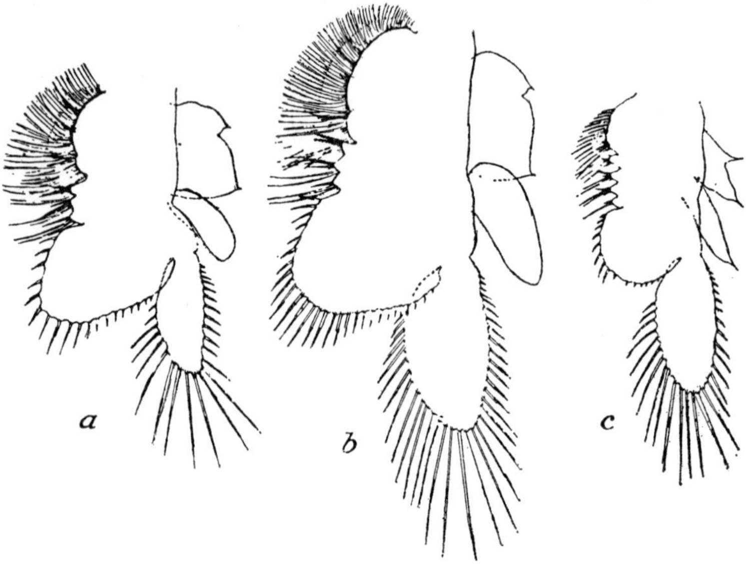

FIG. 1.15. *Tanymastix stagnalis*, 1., 6., and 11. limb. The endites at the base of the limbs are furnished with closely spaced filtering setae. (After Eriksson, 1934.)

The endites of the limbs are furnished with setae in two rows, of which the long, plumose ultimal setae act as filters. The exopodite and distal part of the limb are very flexible. The position and movements of the limbs and their parts, and the filtration process are shown diagrammatically in Fig. 1.16A, B and C. Each pair of the limbs move synchronously, but a little out of phase with the pairs in front and behind, so that a metachronal wave results moving from behind forwards. During the abductory phase of the limb movement, the exopodite and distal parts of the limbs passively flap backwards and make contact laterally and ventrally with the following limb. Thus, a small closed chamber is formed only communicating medially with the space between the two rows of limbs (Fig. 1.16A, B). During abduction, when the limb forming the anterior wall of the small lateral chamber is moving away from the limb forming the posterior wall, the hydrostatic

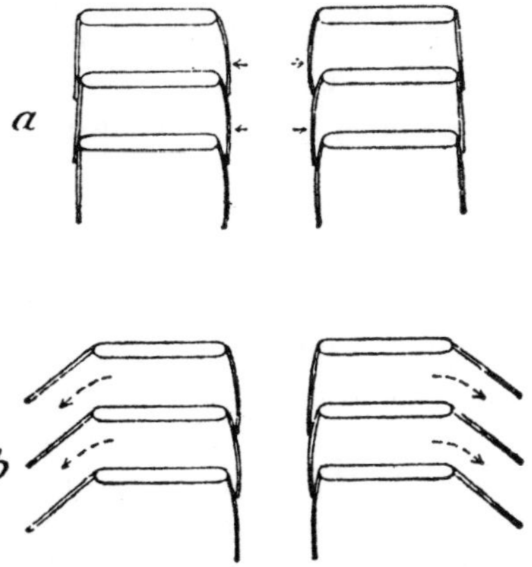

FIG. 16A

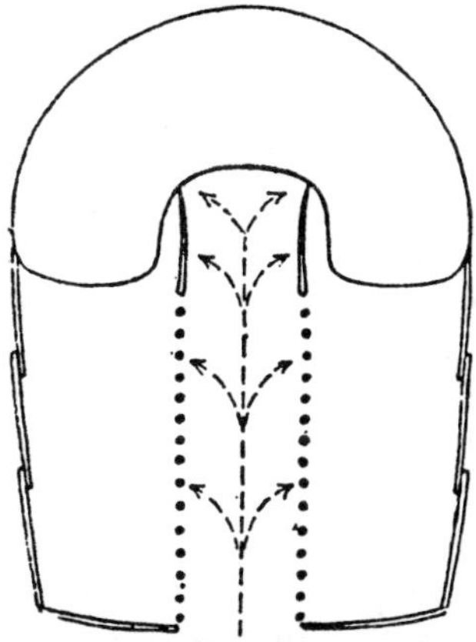

FIG. 16B

FIG. 1.16A. *Tanymastix stagnalis.* Diagrammatic horizontal section through two pairs of suction chambers.
a, in the abductory phase; b, in the adductory phase. (After Eriksson, 1934.)

FIG. 1.16B. *Tanymastix stagnalis.* Diagrammatic transversal section through a pair of suction chambers. The arrows indicate the direction of the feeding currents. (After Eriksson, 1934.)

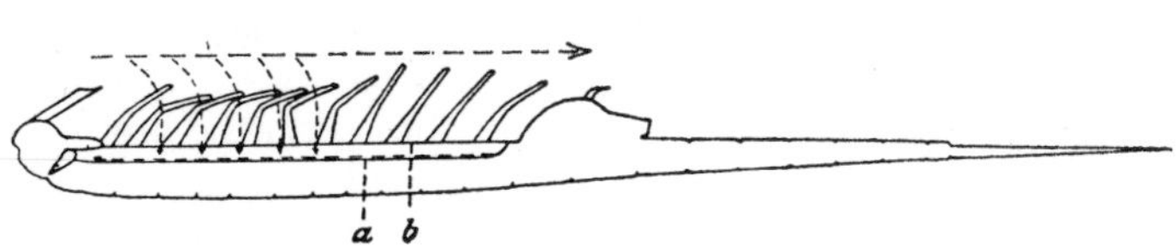

FIG. 16C

FIG. 1.16C. *Tanymastix stagnalis.* Diagrammatic median section.
a, Dorsal, and b, ventral boundary of the ventral groove. The arrows indicate the direction of the water currents connected with the food uptake. (After Eriksson, 1934.)

pressure is decreased and water consequently sucked into the chamber from the median space between the rows of the limbs, with maximum suction at their base. Particulate matter too coarse for passing the filters formed by the endites will be left behind and eventually accumulate in the ventral groove at the base of the limbs. During the adductory phase, pressure in the lateral chambers rises again; the chambers open laterally and ventrally, through which openings the water leaves. A small backflow of water between the bases of the limbs seems to create a current towards the mouth region of the concentrated suspension of particles. Forward transport of the suspension may also be assisted by special setae on the 1. endites, which move forward and sweep the ventral groove during the adductory phase. The limbs generally make a few hundred cycles per minute, and a complete wave can extend over about 10 pairs of limbs. Apparently, the metachronal character of the movements of the locomotory thoracal limbs in primitive crustaceans was a preliminary for the development of filter feeding by means of suction chambers. The ventral groove is continuous with the so-called atrium enclosed by the mandibles, 1. and 2. maxillae and other mouth parts. Material in the ventral groove is stated to be sucked into the atrium, suction being exerted by lateral movements of the paragnaths. Finally, the material is pushed between the masticatory surfaces of the mandibles by the 1. maxillae. No true mastication takes place, merely transfer of the food into the oesophagus.

In the Cladocera the feeding mechanisms are much more varied and specialized than in the Phyllopoda, but apparently the mechanisms can be derived from the phyllopod type just described, with the shell-bearing phyllopod group Conchostraca representing intermediary forms. In Conchostraca and Cladocera, the shells help forming the suction chambers. During abduction the limbs move laterally so that the exopodites are pressed against the shells, which thus close the suction chambers laterally; during adduction the limbs move a little medially.

In the primitive cladocerans, the Ctenopodia the number of thoracal limbs are reduced to six pairs. Of the ctenopods, *Sida crystallina* and *Diaphanosoma brachyarum* have been used in experiments on feeding rates, etc. In the Daphnidae like the Ctenopoda mainly pelagic or semi-pelagic, only five pairs of thoracic limbs are present, but all participate in the formation of suction chambers. In other groups of cladocerans only the fifth pair of the limbs are used exclusively for filter feeding, the anterior limbs may partly or totally have been adapted to other purposes, for instance crawling or collection of larger food masses on the bottom. However, in all instances where filter feeding does persist in the Cladocera the main features in the feeding process are unchanged: water is sucked from the median space into the lateral chambers, and fine particles are filtered off by the setae of the endites. It is also characteristic of cladocerans that the inhalant and exhalant currents separate; water is taken in between the shells in the anterior end and passed out behind, thus securing that already filtered water does not mix with the food-containing water.

The role played by glandular secretions in feeding in Phyllopoda and Cladocera has been a matter of discussion. Recent findings by Fryer (1962) in the cladoceran *Eurycercus lamellatus* show the importance of secretions produced by large glands at the base of the fourth trunk limbs. The secretion is passed towards the food groove and forward, carrying with it filtered-off particles. Glands adjacent to the food groove have also been described in Phyllopoda. Finally secretions from labral glands around the mouth is mixed with the food particles before ingestion.

The feeding organs of phyllopods and cladocerans are specialized for handling of small particles; large particles are excluded by special means. In the shell-less phyllopods, the setae on the 3.–5. endites overbridge the median space, and thus prevent large particles from entering the feeding organs. In the Cladocera the same function is exerted by setae on the highly specialized 1. endit. In this group the largest particles have already been sorted by not being able to pass between the shells. If despite these protective mechanisms the feeding organs should become clogged up, for instance if the surrounding water contain too many fine particles, Cladocera and Conchostraca can clean the filtering organs by means of a special structure on the abdomen. This very flexible structure carries claws, which are used to scrape off and expel unwanted material from between the shells.

All filter feeding phyllopods and cladocerans are further able to get rid of material that already has reached the mouth region. If this material is not ingested, it may be expelled from the atrium by means of the 1. pair of thoracal limbs. In order that the limb can have access to the atrium the upper lip is lifted.

COPEPODA

Whereas filter feeding in Cladocera is apparently the primitive method of feeding, raptorial feeding is probably primary in the Copepoda, and filter feeding is a specialized condition especially developed within the Calanoidia. However, few, if any, copepods rely exclusively on filter feeding; even specialized filter feeders can also feed by other methods, for instance by scraping or by

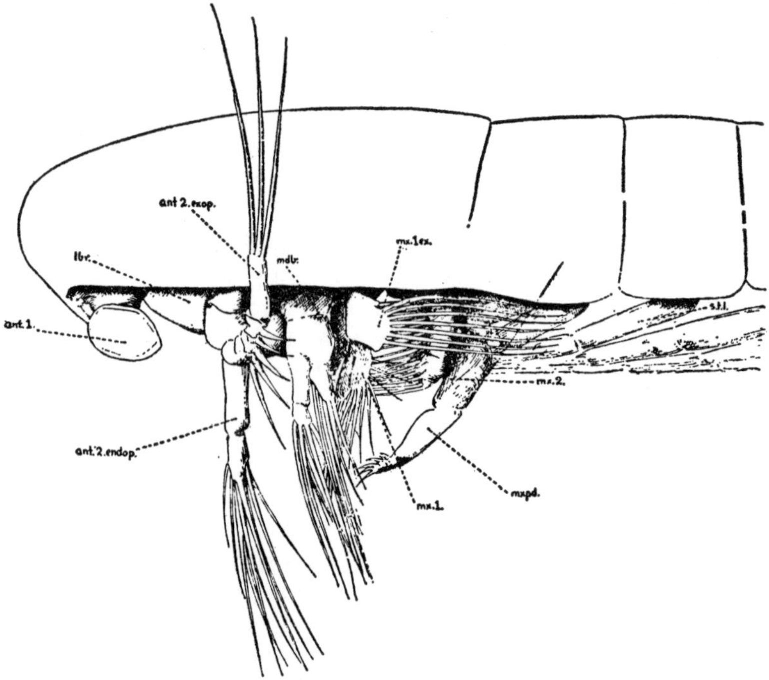

FIG. 1.17. *Calanus finmarchicus,* lateral view. The 1. antenna has been cut off close to the body.
ant. 1., 1.Antenna; ant. 2. endop., endopodite of 2.antenna; ant. 2. exop., exopodite of 2.antenna; lbr., labrum; mdb., mandible; mx.1, 1.maxilla; mx.2., 2.maxilla; mxpd., maxilliped; s.t.l., swimming trunk limb. (After Cannon, 1929.)

seizing larger food, living or dead (Beklemishev, 1954; Conover, 1960; Fryer, 1957; Lowndes, 1935; Petipa, 1959, 1960).

In most suspension feeding copepods, as the marine *Calanus finmarchicus* and freshwater *Diaptomus* species, a filter chamber is enclosed between the ventral body wall and the 2. maxillae which project ventro-anteriorly (Figs. 1.17 and 1.18). The 2. maxillae carry long plumose setae, extending antero-medially towards the mouth, and forming the lateral walls of the filter chamber. The feeding currents are produced chiefly by rapid vibratory movements of the antennae. The rate at which the antennae vibrate varies. In *Diaptomus gracilis* the minimum rate was 600 beats per minute, and the maximum 1,620 beats; similar or even higher values have been measured in other copepods (Lowndes,

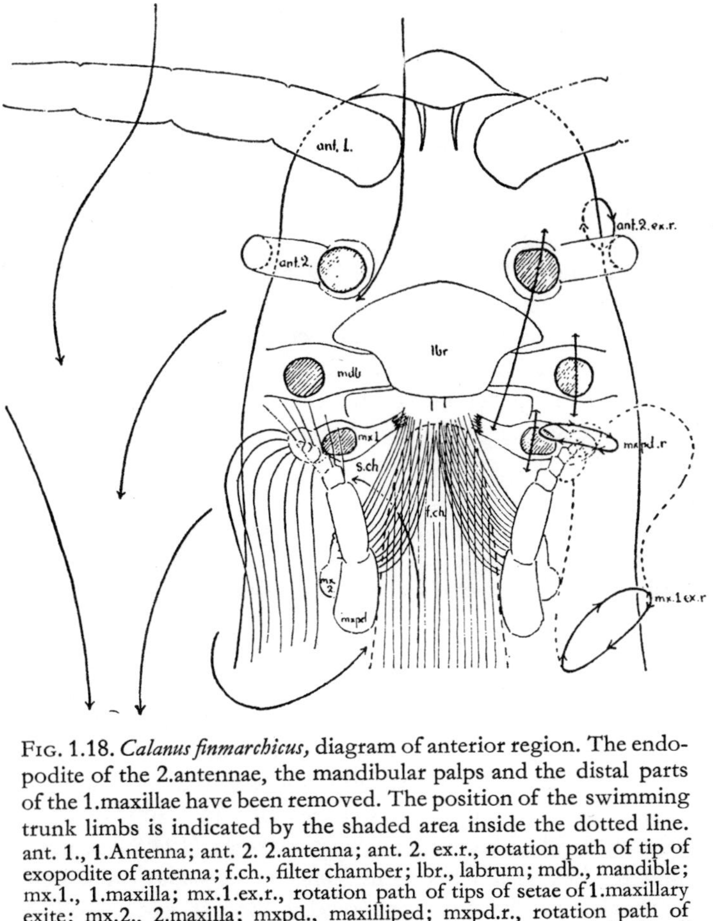

FIG. 1.18. *Calanus finmarchicus,* diagram of anterior region. The endopodite of the 2.antennae, the mandibular palps and the distal parts of the 1.maxillae have been removed. The position of the swimming trunk limbs is indicated by the shaded area inside the dotted line. ant. 1., 1.Antenna; ant. 2. 2.antenna; ant. 2. ex.r., rotation path of tip of exopodite of antenna; f.ch., filter chamber; lbr., labrum; mdb., mandible; mx.1., 1.maxilla; mx.1.ex.r., rotation path of tips of setae of 1.maxillary exite; mx.2., 2.maxilla; mxpd., maxilliped; mxpd.r., rotation path of the tip of maxilliped; s.ch., suction chamber. Arrows on the left side of the figure indicate direction of water currents, on the right, the limb movements (After Cannon, 1929.)

1935). Also mandibles, 1. maxillae, and probably maxillipeds, are of importance in the production of the feeding current and in directing the current through the filter chamber, as has been demonstrated by analyses of the limb movements by means of the stroboscope and ultra-rapid photography. In some species, *Acartia tonsa* and *A. clausi,* the 2. maxillae seem not to act only as passive filters, but the setae of the maxillae are alternately spread apart to form an open basket-like structure and quickly drawn together (Conover, 1956). Particles retained by the bristles of the setae are scraped off by the endites of the 1. maxillae and specialized long setae of the maxillipedes, and passed forward to the mandibles and mouth.

The complicated feeding processes can vary from species to species, and have furthermore been a matter of discussion. Detailed descriptions and discussion of varying statements should not be given here, but reference is made to papers of Cannon (1929), Kozhova (1956), Lowndes (1935), Storch (1929) and Storch and Pfisterer (1926).

The fact that so many copepods combine suspension feeding with raptorial feeding is perhaps partly a consequence of suspension feeding being secondary within the group, the filters are relatively coarse (see p. 32); but it may also be of importance that many phytoplankton are too large for uptake by filtration (Fig. 1.19). Algal cells larger than some 30–50 μ will thus often be available to copepods only by direct seizing. An advantage of combining the two types of feeding methods is presumably that phytoplankton can be utilized as food both when it is dominated by large and by smaller cells.

EUPHAUSIA

In the euphausians the six long limbs of the cephalothorax form the so-called basket which can function as a filtering structure, but which is apparently also used for seizing larger individual prey, e.g. copepods, especially in deep sea species (Einarsson, 1945; Mauchline, 1959). Barkley (1940) investigated the filter feeding mechanisms in *Euphausia superba,* on which the larger whales in the southern oceans almost exclusively feed. Most of the information was obtained from studies of the building of the limbs in specimens 1–6 cm long, and from

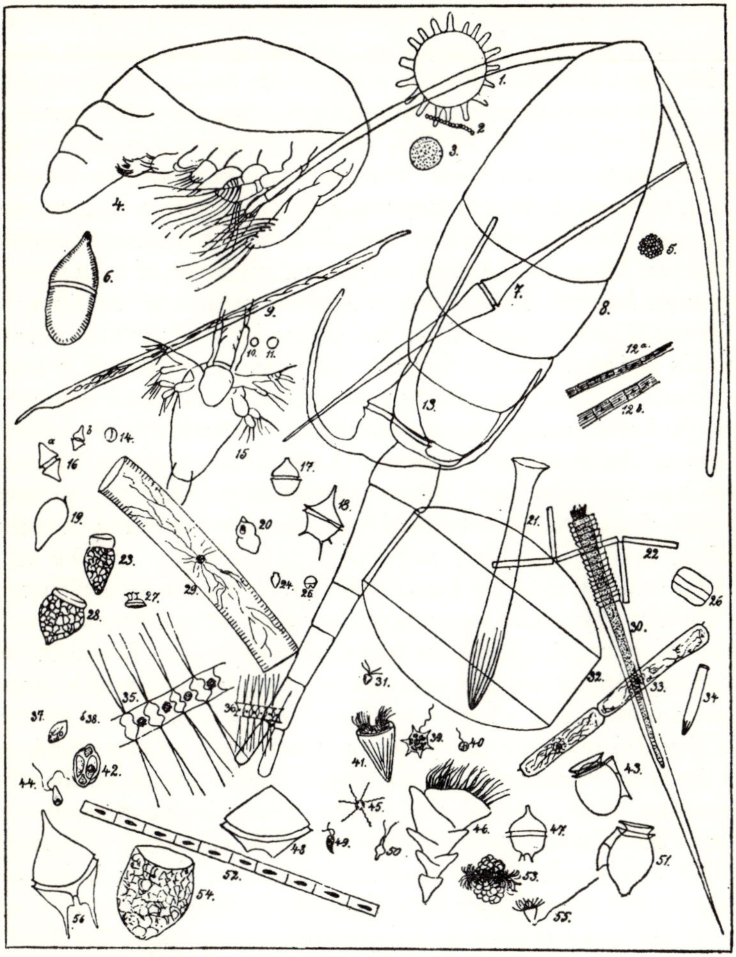

FIG. 1.19. Relative sizes of planktonic copepods and various phytoplankton, etc.
8, *Oithona similis,* adult; 15, *O. similis,* nauplius; 4, *Temora longicornis,* nauplius. (After Lohmann, 1908.)

observations on living animals of how the various parts of the limbs were orientated when functioning (Fig. 1.20). The basket is closed on all sides. Ventrally and anteriorly the distal segment of the thoracopods on one side touch the corresponding ones of the opposite side; the gap posteriorly between the 6. pairs of thoracopods is closed by long setae on the limbs. The various segments of the thoracopods are furnished with plumose setae which form the true filtering walls of the basket, overbridging the spaces between neighbouring limbs. The distances between setae and bristles are smallest on the proximal segments, and largest on the most distal. The distances between the bristles which determine their functioning in the filter is 7 μ on the segments closest to the body wall, 21 μ on the median segments and 35 to 230 μ on the distal segments, on some of which they are even lacking. Presumably the whole structure functions as follows: When feeding, the animal swims forward with the thoracopods pointing in the direction of swimming. Water is consequently pressed through the coarse filter of the distal segments into the basket and out again laterally and posteriorly through the filters of the proximal segments. Accordingly, the smaller zooplankton organisms and largest phytoplankton cells or colonies should, therefore, be prevented from passively entering into the basket, and the smallest suspended particles that can be efficiently retained must be larger than 7 μ. Extensive investigations by Barkley of stomach contents of *Euphausia* are in agreement with this deduction. Amongst the identifiable organisms found in the stomach content smaller diatoms of smooth and regular shape always dominated, whereas large and spiny forms were absent. Barkley did not investigate how the particulate matter retained on the filter is transferred to the mouth, but he points out that the structures of the ventral side of the cephalothorax are similar to that of the related *Hemimysis* investigated by Cannon and Manton (1927), and similar function can therefore be expected. Special setae on the thoracopods may comb the particulate matter from the filters into the food groove between the bases of the limbs; transport towards the mouth-parts of the content of the groove seems to be by means of sweeping movements of setae on the coxopodits at the base of the limbs.

1.711. QUALITATIVE SELECTION OF FOOD

Zooplankton crustaceans that besides suspension feeding also feed by seizing large particles individually, apparently select these particles according to their food value. Kozhova (1953, 1956), for instance, observed the calanoid copepod *Epischura baicalensis* to

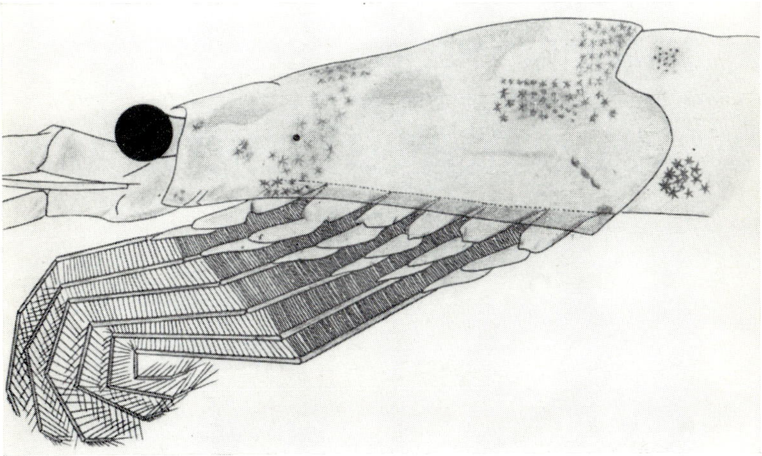

FIG. 1.20. *Euphausia superba*. Schematic drawing of cephalothorax and thoracopods in the feeding position. (After Barkley, 1940.)

seize living, but not dead, algal cells or detritus particles. The copepod even showed a pronounced preference for the diatom *Cyclotella baicalensis*, which might dominate the content of the gut even when other species dominated in the surrounding water. It is, however, not finally settled whether or to what extent qualitative selection is also possible of particles passively retained in the filtering organs. At any rate, selection of food particles to the exclusion of inorganic or other particles without food value does not occur. As in suspension feeders generally, much indigestible material of suitable particle size which happens to be present in the water will also be filtered off and ingested by zooplankton crustaceans. Thus, copepods from waters with much suspended inorganic material also contained many inorganic particles in the intestinal tract (Beklemishev, 1954). In the copepod *Calanus finmarchicus*, Marshall and Orr (1955a) found little or no selection of food. Material collected in the filter chamber of copepods can be discarded *en masse* instead of being ingested, but mechanisms that might allow sorting and selection of the individual particles have not been described.

According to Gaevskaya (1949) qualitative selection takes place in the feeding organs of *Daphnia magna* and other daphnids. Thus, in mixtures of fine glass particles and algal cells, the material reaching the mouth region was sorted so that the material closest to the mouth was free of glass particles. The possible sorting mechanisms are unknown. It is not clear to what extent sedimentation during the experiments of heavy particles in the suspensions may have influenced the results. In another cladoceran *Penilia avirostris* from the Black Sea, Pavlova (1959) found no selection in various suspensions of digestible and non-digestible materials, for instance of carmine particles and bacteria, or of asbestos and algae.

1.712. EFFICIENCY OF PARTICLE RETENTION

In suspension feeding zooplankton crustaceans it is important to know not only the minimum size of particles that are efficiently retained but also the maximum size that can be handled by the filtering organs, since the maximum size always seems to fall within the size range of the phytoplankton algae.

The Cladocera are apparently adapted for feeding on very small particles. Sushchenya (1959a), thus found that in mixtures of large or colonial algae, as *Asterionella, Fragilaria* and *Ceratium,* and small algae, as *Scenedesmus* and *Chlorella,* species of *Daphnia, Diaphanosoma* and *Bosmina* removed the small algae at about equal rates from the suspension, whereas no significant uptake occurred of the large organisms. Sushchenya further found that in six species of Cladocera the rate of removal of algal cells from mixtures increased with decreasing cell size; the mixtures contained four species of algae of sizes 3·5, 15, 20 and 70 μ. Sushchenya assumes that the increased rates of uptake with decreasing cell size express preference for small cells. However, possibly the faster uptake of small cells is not due to real qualitative selection, but to structural properties of the filtering organs, which are especially adapted for removing small particles from a suspension; in *Penilia avirostris* the maximal size of algal cells taken up is stated to be only 8 μ (Pavlova, 1959).

The minimum size of particles that are efficiently retained in the filters of cladocerans is not accurately known, but feeding experiments indicate that particles down to bacterial sizes can be efficiently retained. Thus, Monakov and Sorokin (1961) found that *Daphnia longispina* cleared suspension of 1 μ bacterial cells at rates only a little below the rates at which suspensions of 6 μ *Chlorococcus* cells were cleared. Clearances of bacteria in various cladocerans measured by Manuilova (1958) are probably not significantly lower than clearances of small algal cells measured by several other investigators (Table 1.3). Also phyllopods seem able to retain bacteria efficiently, as indicated by the finding that *Artemia salina* could be reared to sexual maturity on a diet of bacteria. But autolyzed yeast and algae, or dead dissolving algae could not be utilized as food until contaminated with bacteria. Therefore, organic colloids or dissolved materials from decaying cells are presumably not to any great extent retained in the filtering organs (Bond, 1933). Similar results are obtained in other feeding experiments in cladocerans (see pp. 244, 258).

The filtering organs of adult Copepoda generally select larger particles than the filtering organs of Cladocera; both maximal and minimal size of particles taken up by filtration are within the size range of the phytoplankton algae. In the filter feeding freshwater copepod *Diaptomus*, Rylov (1930) found rapid uptake of 5–20 μ particles, slow uptake of 40–50 μ particles and no uptake of particles larger than 50 μ.

The minimal size of particles that can be efficiently retained has been determined in several copepods, both from the relationship between the size and the rate of clearing of the particles and from measurements of the mesh openings in the filters. Clearances of algae less than 10 μ were found to be lower than clearances of larger algae (Marshall and Orr, 1955, 1956, 1961). In females of *Acartia, Calanus, Centropages* and *Temora* the maximal clearances of small flagellates varied from 1 to 7 per cent of the maximal clearances of large flagellates. Thus, if the rates of water transport were the same in the experiments with large and small algae, and if the large algae were unable to pass the filters, only 1–7 per cent of the small algae appear to have been strained from the water passing the filters. Only in *Pseudocalanus minutus* the clearance of small algae was relatively high, viz. 41 per cent of the clearance of large flagellates. In a mixture of algae of various sizes, Petipa (1959) found that *Acartia clausi* removed 6–8 μ flagellates slowly, whereas 16 μ *Gymnodinium* cells and 39 μ *Prorocentrum* cells were cleared at high and about equal rates. In *Diaptomus,* Malovitskaya and Sorokin (1961a) found very little uptake of suspended bacteria, but effective assimilation of aggregated bacteria.

These results are in accordance with measurements of the distances between bristles of the setae on the maxillulae, which function as the filters in copepods. Ussing (1938) obtained the following values for the smallest distances in a number of copepods of different stages from Greenland waters: adult females of *Metridia longa* 8·2 μ, *Calanus finmarchicus* 5·7 μ, *C. hyperboreus* 4·7 μ and *Pseudocalanus minutus* 1·6 μ; copepodid stage V of *Calanus finmarchicus* 3·8 μ, *C. hyperboreus* 3·3 μ and *Pseudocalanus minutus* 1·5 μ, stage III of *Calanus finmarchicus* 3·2 μ. In *C. finmarchicus* from British waters Marshall and Orr (1956) found a smaller minimum distance between bristles than did Ussing in the larger specimens from the Arctic. They further found that in adult *C. helgolandicus* the bristles are wider apart than in *C. finmarchicus.* Their measurements of distances in the various stages of the two forms of *Calanus* are shown in Table 1.2. In *Acartia clausi,* Petipa (1959) found the distances between bristles to vary from 6 to 16 μ (average 10 μ) in the large form and from 3·3 to 6·6 μ (average

5 μ) in the small form. Similar results were obtained by Conover (1956). It thus appears that generally small species and copepodid stages are able to strain finer particles from the water than larger species and adults. But in no instance is it to be expected that food particles smaller than a few to several microns can effectively be filtered from the water by older stages or adult copepods. However, nauplial stages of both filter feeding copepods, *Diaptomus*, and non-filter feeding copepods, Cyclopidae, have been found to be capable of ingesting and assimilating bacteria dispersed in the surrounding water. Even natural concentrations of bacteria supported growth of the nauplii, which, therefore, presumably efficiently retain particles as small as bacteria.

In *Euphausia superba,* Barkley (1940) found, as mentioned, that down to 7 μ large particles were efficiently retained in the filters, whereas particles above 35 μ might be prevented from entering the filtering basket. The porosity of the filters may thus be about the same as in the much smaller copepods.

1.713. RATE OF WATER TRANSPORT

A number of determinations is available on the rates at which cladocerans and copepods clear water of suspended particles (Tables 1·3 and 1·4), but it is difficult to evaluate how representative these widely varying values are of the volumes of water filtered by the undisturbed animals in nature.

Cladocera

As already mentioned (p. 32), clearances in Cladocera decrease only little, if at all, with decreasing size of suspended particles down to bacterial sizes; even clearances of bacteria may, therefore, express total volumes of water passed through the filtering organs. However, the concentration of suspended matter has repeatedly been found to influence the rates at which the suspensions are cleared (Monakov and Sorokin, 1961a, b; Pavlova, 1959; Ryther, 1954; Sushchenya, 1958a, b). It might, therefore, seem as if Cladocera differ from most other filter feeders in which rates of water transport through the filters are independent of particle concentration in the water below critical values. In most experiments on cladocerans, clearances have been found to decrease with increasing concentration of algae throughout the

Table 1.2. Distances Between Barbs in *Calanus*, in μ (from Marshall and Orr, 1956)

	Stage	Coxa			Basipod		Endopod			
		External seta	Endite 1	Endite 2	Endite 3	Endite 4	Segment 1 (Endite 5)	Segments 2–4	Segment 5 "Feather"	Segment 5 Others
C. finmarchicus	♂	—	5–12	4–8	6–9	5–11	4–16	5–21	4–5	6–17
C. helgolandicus	♂	2	5–9	8–9	8–10	8–11	12–17	11–19	7–9	13–22
C. finmarchicus	♀	2	2–6	3–6	3–7	3–8	4–7	5–14	3–5	5–14
C. helgolandicus	♀	2	5–9	6–9	4–9	6–12	7–13	11–19	3–9	12–20
C. finmarchicus	V	1–3	2–4	3–4	2–4	2–4	3–6	4–8	2–5	4–10
C. helgolandicus	V	1–3	3–9	4–7	3–8	4–10	3–15	5–17	3	7–20
	IV*	2	3	3–4	3–6	3–5	3–5	4–7	3–4	5–9
	III*	2	2–4	3–4	3–4	2–5	4–7	4–8	3–4	4–11
	II*	—	2–3	3–4	2–4	3–4	3–5	3–7	3–4	3–7
	I*	—	2–4	2–4	2–5	2–5	3–6	4–7	2–4	4–8

* *C. finmarchicus* and *C. helgolandicus* cannot be distinguished until Copepodite V.

range applied. However, in some experiments, for instance by Ryther (1954) in *Daphnia magna* (Table 1.3) even the lowest concentrations used seem to be high, so it still remains possible that filtration rates might be independent of concentration at more normal concentrations of algae. In the experiments of Monakov and Sorokin (1961a, b) the lowest concentrations used of chlorococcae and bacteria were about 0·3 to 1mg living weight per litre of suspension, or well below maximal concentrations in nature; the concentrations of algae in the Kuibyshev Reservoir at Volga from which the *Daphnia* were taken increased in periods to 3–7mg/l. The natural concentrations might, therefore, increase to concentrations that would strongly influence rates observed of water transport through the filtering organs. Also in the experiments of Sushchenya (1958a, b) low concentrations of algae were used, down to about 0·3mg/l. wet weight. The clearances obtained at these low "normal" concentrations were higher than any recorded in freshwater cladocerans by other investigators. Only *Daphnia magna* also filtered at low rates in the low concentrations of algae (Table 1.3). But again clearances decreased regularly with increasing concentration of algae throughout the range investigated. However, both in the experiments of Monakov and Sorokin and of Sushchenya, simultaneously with the increase in algal concentration the volume of water available per animal was decreased, a fact that may be, at least partially, responsible for the decreasing clearances. In copepods at any rate, it has been shown that clearances decrease with decreasing volumes of water below certain values (see below). The very low rate at which *Daphnia magna* cleared algae even at low concentrations may also be due to the small volume of water available, namely 2ml per 1mg body weight of the animal against 280–630ml per 1mg body weight in the experiments with *Diaphanosoma, Simocephalus* and *Bosmina*. It is, therefore, still possible that below certain concentrations of particles in the water, also Cladocera will filter at rates that are independent of the concentration of particles.

It further remains to be settled whether the decreasing filtration rates with increasing concentration of algae express true adaptation of feeding rate to food requirement, as is often implied, or whether in dense suspensions of particles the large amounts of material accumulating in the filters mechanically interfere with the activity of the filtering organs. High algal concentrations may also have toxic effects.

It seems that variations in concentration of suspended matter and volume of water available to the experimental animals can explain most of the variation found in the clearances in cladocerans from various types of freshwater localities: rivers, lakes, reservoirs, or laboratory cultures. It is, therefore, indicated (Table 1.3) that rates of water filtration in various groups of freshwater Cladocera at low or average concentration of food in the water amounts to about 1–2l./day/mg dry weight. Pavlova (1959) found much higher values in the marine cladoceran *Penilia avirostris* from the Black Sea; the species is a coastal form of wide distribution. It filtered on an average 26·5l./day/mg dry weight, a very high value compared with filtration rates measured in other marine suspension feeders. Pavlova used low concentrations of algae and large volumes of water taken directly from the sea.

Marine copepods

Even greater variation in the rates of clearing water of suspended particles have been found in marine copepods than in cladocerans. Measurements or estimates of daily volumes cleared by adult *Calanus finmarchicus* thus vary from about 5ml to some litres. Much of the variation is apparently due to the varying efficiency of retaining the particles used. As mentioned above later copepodid stages and adult copepods mostly only inefficiently retain particles smaller than 5–10 μ. Only clearances of larger particles are, therefore, likely to express total volumes of water passed through the filtering organs. However, even when only clearances of algae that cannot pass the filters are considered, the scatter is large. According to Marshall and Orr (1962), "in an experiment there are usually some individuals which do not feed at all or filter only a very small volume," so that the lower end of the range of clearances obtained is often near zero. They therefore consider average volumes misleading. But if the copepods in an experiment were filtering at rates as under natural conditions, the average rates of filtration would of course be of greater significance than the maximum values,

TABLE 1.3. RATES OF CLEARING SUSPENSIONS OF ALGAE OR BACTERIA IN CLADOCERANS

Species	Micro-organism in suspension	Clearance* ml/24hr		Locality	Concentration of algae in mg dry wt./l.	References
		per Animal	per mg dry wt.			
Freshwater species						
Sididae:						
Diaphanosoma brachyurum	*Chlorella pyrenoidosa*	10	2,340	Lake Naroch, U.S.S.R.	0·06	Sushchenya, 1958a, b
Diaphanosoma brachyurum	Nannoplankton	1	—	Lake Erken, Sweden	Natural conc.	Nauwerck, 1959
Daphnidae:						
Daphnia magna	Green algae or diatoms	25–73	200–600	Laboratory culture	*ca.* 5×10^4 cells/ml	Ryther, 1954
Daphnia magna	*Chlorella pyrenoidosa*	8	—	—	70	Lefèvre, 1942
Daphnia magna	*Chlorella pyrenoidosa*	7	72	Lake Naroch, U.S.S.R.	2	Sushchenya, 1958a, b
Daphnia longispina	*Chlorococcus* sp.	14	1,100	Kuibyshev reservoir at Volga, U.S.S.R.	0·2	Monakov and Sorokin, 1961a, b
Daphnia longispina	Bacteria	5	440		0·4	

Daphnia longispina	Bacteria	25	—	Tributary to Volga	0·2	Manuilova, 1958
Daphnia longispina	Nannoplankton	3	—	Lake Erken, Sweden	Natural conc.	Nauwerck, 1959
Daphnia cucullata	Bacteria	14	220	Tributary to Volga	0·2	Manuilova, 1958
Daphnia pulex	*Chlamydomonas reinhardi*	—	200–300	Laboratory culture	6–25	Richman, 1958
Daphnia pulex	*Chlorococcus* sp.	64	2,000	Volga, U.S.S.R.	0·06	Monakov and Sorokin, 1961b
Simocephalus vetulus	*Chlorella pyrenoidosa*	133	1,850	Lake Naroch, U.S.S.R.	0·05	Sushchenya, 1958a, b
Simocephalus vetulus	Bacteria	26	370	Tributary to Volga	0·2	Manuilova, 1958
Bosminidae:						
Bosmina longirostris	*Chlorella pyrenoidosa*	2·6	1,620	Lake Naroch, U.S.S.R.	0·15	Sushchenya, 1958a, b
Bosmina coregoni	Bacteria	10	—	Tributary to Volga	0·2	Manuilova, 1958
Bosmina coregoni	Nannoplankton	1	—	Lake Erken, Sweden	Natural conc.	Nauwerck, 1959
Marine species						
Sididae:						
Penilia avirostris	Various flagellates	106	26,500	Black Sea, U.S.S.R.	Natural conc.	Pavlova, 1959

* Clearances as measured at lowest concentrations of micro-organisms in suspension.

because it is the average rates which determine the daily amounts of food taken up. Presumably, Marshall and Orr therefore imply that the maximum values measured in the laboratory are more representative of natural filtration rates than are the average values. But even the maximum filtration rates obtained may perhaps be lower than under natural conditions, since marine copepods do not seem to adapt easily to experimental conditions; they are generally difficult to maintain or rear in the laboratory.

In order to feed maximally under laboratory conditions, copepods have been found to require a certain volume of water, but the various authors do not agree upon how large volumes are needed. According to Cushing (1958a), in *Temora longicornis* rates of clearing of *Skeletonema* increased from 5·7 to 150ml/day/animal when the total volume of suspension available to two animals was increased from 150 to 500ml. However, Berner (1962) found that the clearance values only increased with the volume at volumes less than 15ml per copepod. At volumes ranging from 15 to 625ml, clearances of *Skeletonema* averaged 8ml/day/animal, that is, much below the values obtained by Cushing. In *Calanus finmarchicus,* Marshall and Orr (1962) found that the volume filtered depended upon the volume available up to 70–100ml.

Clearances have been found not to vary with the concentration of algae (Harvey, 1937; Gauld, 1951, Fuller, 1937; Marshall and Orr, 1955; Conover, 1956).

In Table 1.4 are shown rates at which various copepods have been found to clear about 10 μ or larger algae from suspensions. In *Calanus finmarchicus,* on which by far the greater number of experiments have been made, clearances mostly vary from 50 to 100ml/day/animal. Only in suspensions of the diatom *Ditylium,* clearances amounted to about 200ml (Harvey, 1937). The *Ditylium* cells are, however, so large (50 × 100 μ) that they are probably mostly seized individually, and the volumes cleared of *Ditylium* cells, therefore, probably do not express volumes of water filtered. [Raymont and Gross (1942) obtained very low clearances of *Ditylium*.] Harvey further found that spiny, chain-forming *Chaetoceros* cells were not removed from the suspensions under conditions where *Calanus* cleared 70 ml/day of 30 μ *Lauderia* cells. This experiment, therefore, also strongly indicates that the large phytoplankton algae are not taken up by copepods by filtration, but only by direct seizing, and not automatically when the cells are encountered. However, the 50–100ml cleared per day of 10–30 μ large algal cells presumably represent volumes of water passed through the filtering organs. Thus, Gauld (1951) calculated from his experiments that one animal collected from four to more than 30 cells per second, depending on the concentration, speeds that strongly indicate automatic uptake of the cells by filtration.

The very large clearance values of up to several litres per day obtained by Cushing and Vucetic (1963) are not directly measured, but calculated from changes with time in the concentration of diatoms in a natural population in the North Sea that was assumed to be controlled by grazing by the stock of copepods, mainly *Calanus finmarchicus*. Cushing (1959a) has suggested that the diatoms are not extracted from the water by filtration, but by direct seizing.

Also Cowey and Corner (1963) have calculated clearances. They measured the loss of aminoacids from starving *Calanus helgolandicus* and the concentrations of the aminoacids in the particulate matter in the water off Plymouth, in which the animals had been living. The volumes that should be cleared of various essential aminoacids present in the surrounding water in particulate form in order to cover the losses were found to vary from 16 to 50 (average 30) ml per *Calanus* per day in winter and from 10 to 75 (average 50) ml in summer. However, the aminoacids in particulate form were obtained by filtering the water through membrane filters with pore size 0·5–1·0 μ. The clearances calculated, therefore, represent minimum values for the volumes of water to be filtered, and should be corrected for the amounts of particulate aminoacids not retained by the filters of the copepods. Moreover, the clearances are for maintenance of starving animals, higher values being needed for growing or reproducing copepods.

Freshwater copepods

Diaptomus was found by Nauwerck (1959) to clear 100–300ml of suspended phytoplankton daily per mg dry weight, Malovitskaya and Sorokin (1961b) obtained slightly higher maximum

TABLE 1.4. RATES OF CLEARING SUSPENSIONS OF ALGAE IN COPEPODS

Species	Algae in suspension	Clearance, ml/24hr/animal	References
Marine species			
Calanus finmarchicus, V + VI	*Ditylium*, 50 × 100 μ	170–240	
	Lauderia, 30 μ	50–100	Harvey, 1937
	Chaetoceros	0	
Calanus finmarchicus, V		100	
Calanus finmarchicus, IV	*Chlamydomonas*, 7 × 12 μ	50	Gauld, 1951
Calanus finmarchicus, III		29	
Calanus finmarchicus and *C. helgolandicus*, ♀	>10 μ flagellates	Max. 84	Marshall and Orr, 1961a, b
	>10 μ diatoms	Max. 43	
C. finmarchicus, Nauplius III	>10 μ algae	Max. 1·0	
C. finmarchicus, Nauplius IV		Max. 1·2	
C. finmarchicus, Nauplius V		Max. 1·7	
C. finmarchicus, Nauplius VI		Max. 2·0	Marshall and Orr, 1956
C. finmarchicus, Copepodite I		Max. 2·8	
C. finmarchicus, Copepodite II		Max. 6·4	
C. finmarchicus, Copepodite III		Max. 9·2	
Calanus finmarchicus, V		128–7,018, av. 2,850*	Cushing and Vucetic, 1963
Calanus finmarchicus, V	Natural phytoplankton	60	Cushing, 1958a

C. helgolandicus		16–50, av. 30; winter† 10–75, av. 50; summer	Cowey and Corner, 1963
Pseudocalanus minutus, ♀	>10 μ flagellates	9	Marshall and Orr, 1962
Centropages hamatus, VI	*Chlamydomonas*, 7 × 12 μ	13	Gauld, 1951
Centropages hamatus, ♀	>10 μ flagellates	Max. 15	Marshall and Orr, 1962
	>10 μ diatoms	Max. 2·7	
Temora longicornis, VI	*Chlamydomonas*, 7 × 12 μ	8	Gauld, 1951
Temora longicornis, ♀	>10 μ flagellates	Max. 16	Marshall and Orr, 1962
	>10 μ diatoms	Max. 18	
Temora longicornis, ♀	*Skeletonema costatum* chains	Max. 27, av. 6·5	Berner, 1962
Temora longicornis, ♀	*Skeletonema costatum* chains	150	Cushing, 1958a
Acartia clausi ♀	>10 μ flagellates	Max. 9	Marshall and Orr, 1962
	>10 μ diatoms	Max. 2·7	
Metridia lucens, ♀	>10 μ flagellates	Max. 1·3	
	>10 μ diatoms	Max. 0·4	
Oithona similis, ♀	>10 μ flagellates	Max. 0·02	
	>10 μ diatoms	Max. 0	
Freshwater species			
Diaptomus gracilis and *D. graciloides*	*Chlorococcus*, *ca.* 6 μ	4·1 (370ml/mg dry wt.)	Malovitskaya and Sorokin, 1961
Diaptomus graciloides	Natural phytoplankton	1–3 (200ml/mg dry wt.)	Nauwerck, 1959

* Estimates from field observations.

† Calculated as volumes of water that should be cleared of various essential aminoacids in particulate form in order to cover requirements of starving copepods.

values (Table 1.4). In both instances clearances were measured by means of carbon-14 labelled algae. Nauwerck found the clearances to be independent of the concentration of algae within the normal range, but Malavitskaya and Sorokin obtained values that decreased with increasing concentration of suspended food. However, simultaneously with the increase in concentration, the volume of suspension available per animal decreased. As small volumes are known to depress filtration rates in marine copepods it appears necessary to investigate the effect of volume of suspension on clearances in *Diaptomus* before definite conclusions can be made as to the importance of particle concentration.

Euphausia

Lasker (*cit.* Holmes, 1958a) found the oceanic *Euphausia pasifica* to clear 6ml/day/mg dry weight of the flagellate *Dunaliella primolecta* from a suspension containing the very high amount of 5–14mg organic matter per litre. More experiments are needed to decide whether these values express filtration rates under natural conditions.

LENGTH OF FEEDING TIME

As mentioned, in laboratory experiments copepods have been found to feed discontinuously (Berner, 1962; Gauld, 1951; Lowndes, 1935; Marshall and Orr, 1962). Fuller found that *Calanus* sometimes ceased feeding for 24 hr or more, but indications of a diurnal rhythm in the feeding were also obtained in *Calanus*. In order to investigate whether feeding is also discontinuous in nature, Gauld (1953) investigated the diurnal feeding activity in stage V and adult females of *Calanus* by recording the gut content of freshly caught animals taken in the sea at 4-hr. intervals through 24 hr. Hauls were taken simultaneously at depths of 2–3, 35–40 and 75–80m. At the surface 80–100 per cent of the *Calanus* contained plenty of food in the gut at all hours of the day, but the percentage of animals with food decreased with increasing depth parallel to a decrease in the amount of phytoplankton in the water. Thus, under natural conditions *C. finmarchicus* feeds continuously, its daily feeding rate depending upon the length of time it spends in the food-containing layers of the water. Diurnal vertical migrations were observed on some occasions, but not all. Continuous feeding in nature has also been found in *Acartia clausi* (Petipa, 1959), and in the freshwater copepods *Epischura* (Kozhova, 1956) and *Diaptomus* (Nauwerck, 1959). However, feeding intensities may vary. *Acartia* shoved two diurnal maxima, one at night and one at day; *Diaptomus* fed more in light than in darkness, but the differences in filtration rates were small. It therefore seems that filter feeding copepods, and probably also other zooplankton crustaceans, feed more or less continuously, with filtration rates that perhaps do not vary greatly diurnally.

1.72. CIRRIPEDIA

The account of feeding in barnacles is based on the studies of Crisp and Southward (1961). Barnacles feed by means of six pairs of cirri on the head and thorax (Fig. 1.21). The mouth and mouthparts are placed between the bases of the first and second

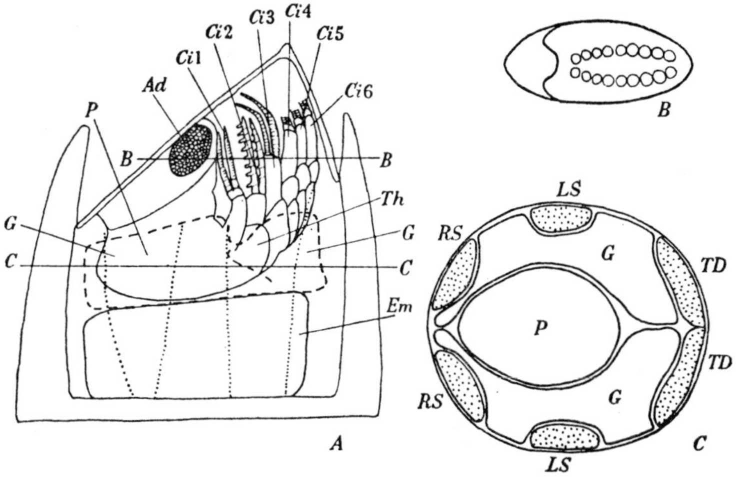

FIG. 1.21. Diagram of large balanoid (based on *Balanus hameri*) to show the arrangement of parts and spaces inside the shell.
A, section of shell in carino-rostral plane, with depressor muscles and branchia of one side removed to display prosoma and thorax. The distal segments of the large cirri (4–6) are not drawn. B, horizontal section across line B to B; C, section across line C to C. Ad, adductor muscle; Em, mass of developing eggs or larvae; Ci 1 to 6, individual cirri; G, branchiae; LS, lateral scutal depressor muscle; P, prosoma; RS, rostral scutal depressor muscle; TD, tergal depressor muscle; Th, thorax. (After Crisp and Southward, 1961.)

cirri. The first two (in chthamalids) or three (in balanoids) of the pairs of cirri are short and stout, the remaining four or three are long and thin and project outside the shell when active. The diversified cirral activity varies with habitat as well as from group to group. Rhythmic extention and withdrawal of the long cirri, formed into a sweep net, has often been considered typical of feeding in sessile barnacles, but has not been universally demonstrated. In most Balanidae, however, the rhythmic beat represents an important feeding type. The extension of the cirri during the beat is slow and due to filling of the cirri by body fluid forced into the cirri by muscular activity. The contraction of the cirri is rapid, performed by means of flexor muscles that roll up the cirri inside the mantle cavity. Particles caught by the cirri are scraped off by special setae on neighbouring cirri closer to the head end, and finally passed to the mouth. In many species the frequency of beat may increase, with lack of full withdrawal of the cirri; this feeding type seems to represent an adaptation to feeding in still water. Finally, the long cirri may remain extended for longer or shorter periods before being withdrawn; the cirri then function as a passive net, held across the direction of the water current. This type of cirral activity is found in forms living in moving water; it is the only or main type of activity of the long cirri in the stalked barnacles (Batham, 1945; Barnes and Reese, 1959), and also in some balanoids (Crisp and Southward, 1961). However, all three types of activity of the long cirri can often be exhibited by one species, especially within the balanoids which are the most diversified with respect to feeding activities; the type of feeding adopted will depend upon environmental factors as concentration and kind of food particles and rate of water flow.

The long cirri are mainly used in feeding on larger particles such as small zooplankton organisms, large phytoplankton cells, etc.; small phytoplankton cells easily pass through the cirral net. Particles too small to be efficiently retained by the long cirri can, however, be filtered off by the short cirri which during feeding are held spread out into a fan across the entrance to the mantle cavity. The cirri are furnished with closely spaced bristles, the spaces between individual bristles may be as small as 1 or 2 μ. Even small naked flagellates or bacteria should, therefore, be filtered off from the water passing this cirral net. The water currents passing the short cirri are produced partly by the beat of the long cirri, partly by the pumping action arising from the movements of the operculum; in opening to the mantle cavity, the operculum is raised, the volume of the mantle cavity increased and hence water sucked in, passing over the short cirri; in closing, the operculum is lowered, and water is expelled through a narrow aperture between the thorax and terga. The unidirectional flow of water and its jet-like expulsion secure separation of the water in the inhalant and exhalant currents. Feeding by filtration of fine particles in the short cirri occurs simultaneously with capture of larger food particles in the long cirri, but opercular movements can maintain a water current to the short cirri and thus secure filter feeding also in the absence of activity of the long cirri.

The relative significance of the short and long cirri in feeding and of small and large particles as food for the barnacles probably varies with the species as well as with the size and kind of particles in the surrounding water, but little definite is known. Southward (1955a) and Crisp and Southward (1961) measured the rates at which balanoids cleared suspensions of organisms of various sizes. In *Balanus perforatus,* clearances of 750–1,000 μ long nauplii varied from 4 to 15ml/hr, of 30 μ *Peridinium* 2–9ml and of 10 μ *Gymnodinium* 2–3ml. It thus appears that even 10 μ large particles are rather efficiently retained. Also 2 μ long rod-shaped bacteria were filtered off, since they were found in the faeces, but their clearance was not measured. Efficient retention of particles down to bacterial sizes is suggested by the finding of significant amounts of 1–2 μ particles in the stomach content of various barnacles. In *B. nubilis,* 2–$<$1 μ particles amounted to 38 per cent of the dry weight of total particulate matter in the stomach, in *B. cariosus* to 10 per cent (Barnes, 1959). However, it is not clear how large a part of these small particles were directly ingested, and how large a part originated in the digestive tract through the breakdown of larger particles.

Barnacles seem to be largely non-selective feeders; Barnes (1959) found that the stomach content of various barnacles reflected the material available in the surrounding water. However, some selection of food can take place. Crisp and Southward (1961) found special structures on the first pair of cirri by means of which unwanted matter retained by the cirri was rejected. Also material collected in the extended long cirri was observed sometimes to be rejected; the cirral net was rotated on its axis through 90–180° and the current was allowed to sweep away the accumulated material.

1.73. NAUPLII

Gauld (1959) has compared the feeding mechanisms in the nauplial stages of various crustaceans, *Artemia, Balanus,* some copepods and euphausians, and others. Euphausians and some copepods did not feed in the nauplial stages, but in the others food particles were observed to be collected by a sort of sweep net formed by setae of the antennae or mandibles. In *Artemia* the setae of the antennae sweep the particles into the oral region, where they are carried to the mouth by spines on the reduced mandibles. In *Balanus* and in most planktonic copepods the mandibles both sweep particles into the oral region by the long ventral setae and pass the particles to the mouth by spines like those of *Artemia.*

1.8. INSECTA

Filter feeding has evolved independently in several aquatic insects, but its occurrence is sporadic. Setae or nets built of threads secreted by salivary glands act as filters.

Nymphs of the ephemeropteran *Isonychia* sp., which lives in streaming waters, bear two rows of long stiff setae along the anterior side of the forelegs. The rows of setae are set at an angle with each other, opening forward, upstream. Sufficiently large suspended particles in the water streaming through this filter are retained and passed to the mouth by the mouth parts (Needham *et al.,* 1935). Also a lake-dwelling ephemeropteran nymph, *Povilla adysta* (Fig. 1.22), has been found to possess pinnate setae on forelegs and mandibles; but as it lives in still water it must create its own feeding current: during feeding the gills on the abdominal segments beat rhythmically for a few seconds with short intervals, producing an intermittent current of water from the anterior end. Particles caught on the filtering brushes are removed by the palps and transferred to the mouth (Hartland-Rowe, 1953).

Further, some filter feeding trichopteran nymphs are known; they are rheophilous forms that passively filter the streaming water; however, the filtering mechanisms vary (Wesenberg-Lund, 1943). The nymphs of *Oligoplectrum maculatum* use the setae of the second and third pairs of legs as filters. The nymphs orient

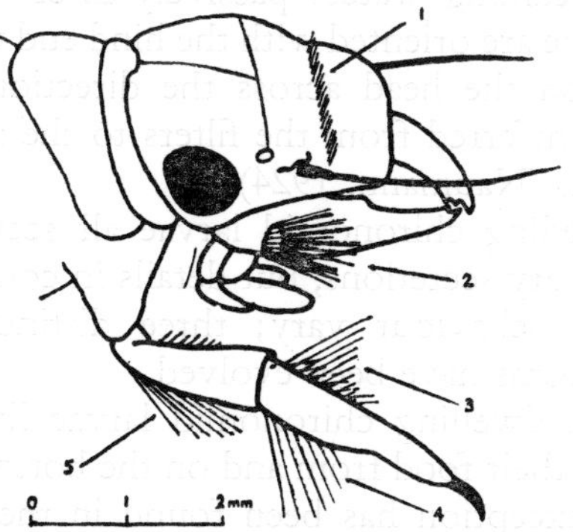

FIG. 1.22. Positions of filtering setae in *Povilla adusta*.
1, Frontal setae; 2, mandibular setae; 3, dorsal tibial setae; 4, ventral tibial setae; 5, femoral setae. (After Hartland-Rowe, 1953.)

themselves with the head directed against the stream and the legs, with the filtering setae stretched out into the water, perpendicular to the body. At intervals, one of the legs in turn is bent towards the head, and caught particles are removed by the fore-legs, formed into a ball and eaten (Nielsen, 1943). The tube-dwelling nymphs of *Hydropsyche* are stimulated by streaming water to spin nets in front of their tubes. The nets are hung up between pebbles, small twigs and other protruding structures in the bottom of the stream. The meshes of the nets are exceptionally regular, square or rectangular in shape. The *Hydropsyche* nymphs are omnivorous, and besides eating the matter passively caught in the nets, they are voracious carnivores that will actively attack even large prey (Kaiser, 1962; Sattler, 1958). The related *Macronema,* living in streams in the Amazon area, secretes a complicated tube encrusted with sand grains. The tube is constructed in the shape of a Pitot-tube with the tip directed against the streaming water. The feeding net, also with regular rectangular meshes, is spun across the lumen at the bend of the tube (Sattler and Kracht, 1963).

Within the Diptera, filter feeding has been described in larvae belonging to the Culicidae, the Simulidae and the Chironomidae. Culicide and *Simulium* larvae have developed filtering setae on the head. The culicide larvae live in stagnant waters and are active filters, feeding when swimming, whereas the *Simulium* larvae from streaming waters passively filter the water: when feeding the larvae are oriented with the hind end upstream and the filtering setae on the head across the direction of the stream. Particles are transferred from the filters to the mouth by means of the mandibles (Naumann, 1924).

The filter feeding chironomid larvae all seem to build nets made from salivary secretions, but details in construction of nets and in feeding behaviour vary; three distinct types of filter feeding mechanisms have been evolved.

Most bottom dwelling chironomid larvae living in tubes in the mud obtain their food from and on the bottom (Thienemann, 1954), but an exception has been found in the species *Tendipes plumosus,* which spins a conic net across the lumen of its U-shaped tube anterior to the body. Violent undulations of the body produce a current of water through the tube, and particles suspended in the water are caught by the net. At intervals of $1\frac{1}{2}$–2 min. the net with particles is eaten and a new net built. About half the time is spent in undulating and the other half in making the net (Walshe, 1947). As none of closely related species spin nets across their tubes, it is clear that filter feeding in *T. plumosus* larvae is a recent, isolated acquisition (Walshe, 1951).

Chironomid larvae mining in aquatic plants also spin a net across the tube and drive water through the tube by movements of the body, but after having spun the net they turn round so that the net is behind the body. The net is much deeper than that of *T. plumosus* (Fig. 1.23) (Leathers, 1922; Burtt, 1940; Walshe, 1951). Filter feeding in leaf miners has been described in three genera, *Glyptotendipes, Endochironomus* and *Pentapedilum,* which are not all closely related to one another. More closely related genera are not leaf miners or filter feeders. Filter feeding may, therefore, have evolved independently more than once within the leaf mining chironomids and thus represent a striking example of parallelism in evolution imposed by similar modes of life.

The leaf-mining chironomid larvae are not obligatory filter

feeders, but may also feed on the tissue of the host plant and other plant materials; for instance when phytoplankton or other suspended food particles in the water are scarce.

The third type of filter feeding in chironomid larvae is found in *Tanytarsus* from fast streams. The larvae build cases which bear spreading arms at their anterior end. When feeding, the

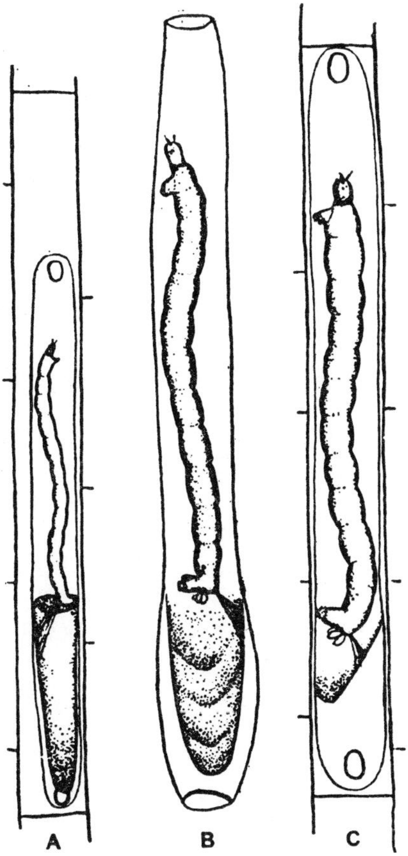

FIG. 1.23. Appearance of the filtratory salivary nets of leaf-mining chironimid larvae. *Pentapedilum* (A) and *Glyptotendipes* (C) are shown in their burrows within the tissue of the plant, *Endochironimus* (B) in a salivary tube on the surface of the leaf. All three larvae are in the position which they assume when irrigating their tubes to fill the salivary nets. (After Walshe, 1951.)

larvae spin a net between the arms. The net is placed across the direction of the stream and acts as a passive filter. All filter feeding insect larvae living in streaming water thus seem to be passive filterers making use of the natural flow of a stream to bring them the suspended food particles. At intervals the *Tanytarsus* larvae nip the net, or part of it, free of the arms. The net with particles is formed into a ball by the mouth parts and anterior legs and eaten, or used for the construction of the case (Walshe, 1951).

The filter feeding insect larvae appear to be mainly non-selective feeders. In culicide and *Simulium* larvae using setae as filters, Naumann (1924) observed no selection according to food value of the suspended particles; and neither did Burtt (1940) nor Walshe (1947) in various net-spinning chironomid larvae. In the *Tanytarsus* larvae some selection is indicated since these larvae reject large particles that may have become attached to the net (Walshe, 1950). Whether or not particles suspended in the water passing the filters are caught and ingested by the larvae probably mainly depend upon the size of the particles.

POROSITY OF FILTERS

Apparently the filters of insect larvae are mostly rather coarse, but measurements of porosities are only available in some instances. The mesh diameters in the nets spun by *Glyptotendipes glaucus* vary from 5 to 40 μ (Burtt, 1940); and in *Tendipes plumosus,* Walshe (1947) found that all particles above 17 μ and most above 12 μ were retained by the net. Nielsen (1943) found that in the intestinal content of young *Oligoplectrum* nymphs diatoms dominated, but they were scarce in the intestine of old nymphs even when abundant in the surrounding water. Presumably, therefore, the filters become coarser with increasing size of the nymphs. Also the net spinning nymphs of the trichopteran *Hydropsyche* produce highly porous filters with meshes about 50 to 100 μ in diameter (Kaiser, 1962), whereas the nets of the related *Macronema* have much finer meshes, measuring 3×19 μ (Sattler and Kracht, 1963).

The porosity of the filters of *Simulium* larvae has not been directly measured, but the larvae have been reared on a diet only consisting of suspended bacteria in natural concentrations

(Fredeen, 1960). It is therefore indicated that the filtering setae of *Simulium* larvae are much less porous than the filters of most other insect larvae.

RATE OF WATER TRANSPORT

Little is known about feeding rates and their regulation in filter feeding insect larvae. Senior-White (1928) calculated the rates at which suspensions of various algae were filtered by larvae of *Anopheles rossi*. Values varied from *ca.* 10 to *ca.* 900mm^3/day, increasing with size of the larvae. In leaf-mining chironomids, Burtt (1940) and Walshe (1951) found the concentration of particles in the water to have little or no influence on the irrigation time, which is thus not determined by the amount of particles accumulated in the net. However, if few or no suspended particles were present in the water or if the entrance to the tube was blocked, the larvae stopped net building and irrigation, and explored and reopened the entrance. They are, therefore, in some way able to register when the food supply is running short.

1.9. GASTROPODA

In Gastropoda, feeding on suspended matter is confined to some pectinibranch Prosobranchia and to the pelagic thecosomatous Pteropoda.

1.91. PROSOBRANCHIA

Suspension feeding is of special interest in the prosobranchs since filtering mechanisms have evolved independently within several families; many striking examples are known of convergent as well as divergent lines of evolution. Suspension feeding is apparently confined to the Mesogastropoda; it is unknown amongst the primitive Archaeogastropoda and the most advanced prosobranchs, the Neogastropoda. It is adopted both by freshwater and marine forms. In freshwater, representatives are found in the families Paludinidae (*Viviparus contectus;* Cook, 1949; Fretter and Graham, 1962), Hydrobiidae (*Bithynia tentaculata;* Schäfer, 1953) and Valvatidae (*Valvata piscinalis;* Tsikhon-Lukanina, 1961a, b). In the sea, suspension feeding is known to occur within several families: Turritellidae (Graham, 1938), Siliquaridae (Morton, 1951b), Vermetidae (Boettger, 1930; MacGinitie and MacGinitie, 1949; Morton, 1951a; Yonge, 1932), Calyptraeidae

(Orton, 1914; Werner, 1953a, b, 1959), Capulidae (Yonge, 1938), and Struthiolariidae (Morton, 1951c). More than half of the discoveries of suspension feeding in prosobranchs have been published since 1949, so the numbers of prosobranchs recognized as suspension feeders may still increase.

The Mesogastropoda are distinguished from the Archaeogastropoda by the possession of a monopectinate gill, which generally seems to have been a preliminary for the establishment of suspension feeding in prosobranchs.* It is the well-developed cleansing mechanisms of the gill and of the mantle cavity that have repeatedly been modified and taken into use as food-collecting and -transporting mechanisms (Yonge, 1938).

Only minor modifications of the monopectinate gill are necessary to change it into an efficient filtering and particle-collecting organ. The unmodified gill consists of an axis from which a single row of triangular gill filaments extends (Fig. 1.24) (Morton, 1958). The axis is fused with the mantle wall. The gill partly divides the mantle cavity into inhalant and exhalant chambers. The respiratory water-flow through the gill is produced by the lateral cilia of the gill filaments. Particulate material in the inspired water retained in the gill is entangled in mucus secreted by numerous gland cells, and carried along frontal and abfrontal ciliary tracts towards the tips of the filaments. From here the particle masses are deposited on the floor of the mantle cavity, further consolidated by mucus, and carried by the cilia of rejection tracts towards the exterior. In the filter-feeding prosobranchs, part of the rejection system has been converted into food tracts or grooves. The content of these grooves is conveyed towards the head region where it can be grasped by the radula and ingested instead of being discarded. In addition, the gill has increased in size, completely dividing the mantle cavity into an inhalant and an exhalant chamber. All the inspired water, therefore, has to pass the greatly enlarged gill filter, which, moreover, retains food particles more efficiently than does the unmodified gill (Werner, 1953b; Yonge, 1938).

* *Valvata* seems to represent an exception by being a mesogastropod capable of suspension feeding, but possessing a bipectinate gill (Cleland, 1954; Tsikhon-Lukanina, 1961a, b). However, the suspension feeding mechanism has not been described.

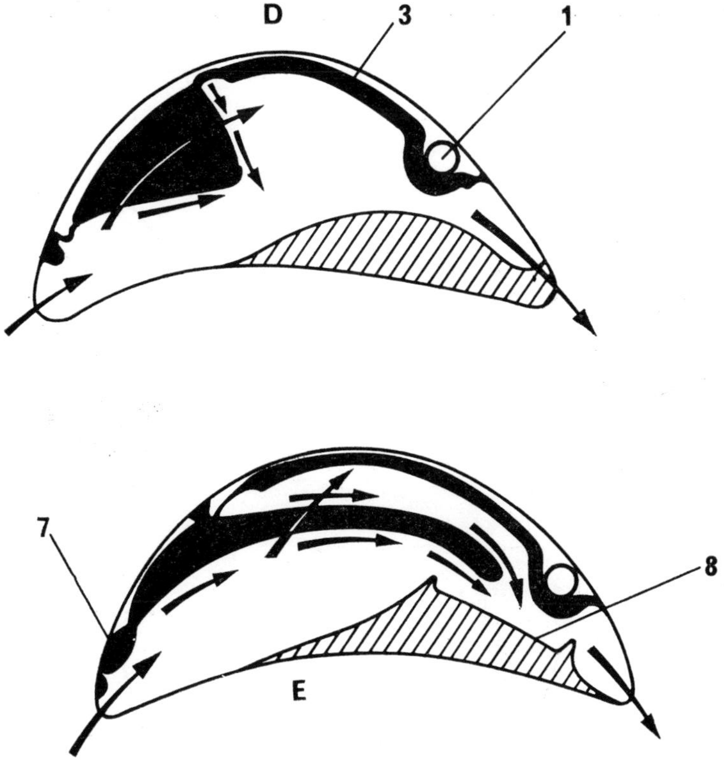

FIG. 1.24. Diagrammatic section through the mantle cavity of two prosobranchs with a single monopectinate gill.
D, *Buccinum,* a non-suspension feeder; E, *Struthiolaria,* a suspension feeder; note the elongated gill filaments. 1, Rectum; 3, hypobranchial gland; 7, "endostyle"; 8, food groove. Arrows indicate directions of water currents and the beat of cilia. (After Morton, 1958.)

As an example of feeding in a filter-feeding prosobranch, *Crepidula fornicata* will be used; it represents an advanced stage in filter feeding and is one of the best-investigated forms.

Crepidula belongs to the family Calyptraeidae. It is sessile, and utilizes oyster shells as a substratum. The feeding mechanisms were first described by Orton (1914) and recently reinvestigated by Werner (1951, 1953a), who added important new observations. It is possible directly to watch the feeding behaviour in *Crepidula* that are allowed to attach themselves to a glass plate. Through the glass, parts of the feeding processes in the mantle cavity are freely observable.

In *Crepidula*, Werner found that the water current produced by the activity of the lateral cilia of the gill filaments is not only filtered at the level of the gills, but also at the entrance to the mantle cavity (Fig. 1.25). This mantle filter is coarse and mainly

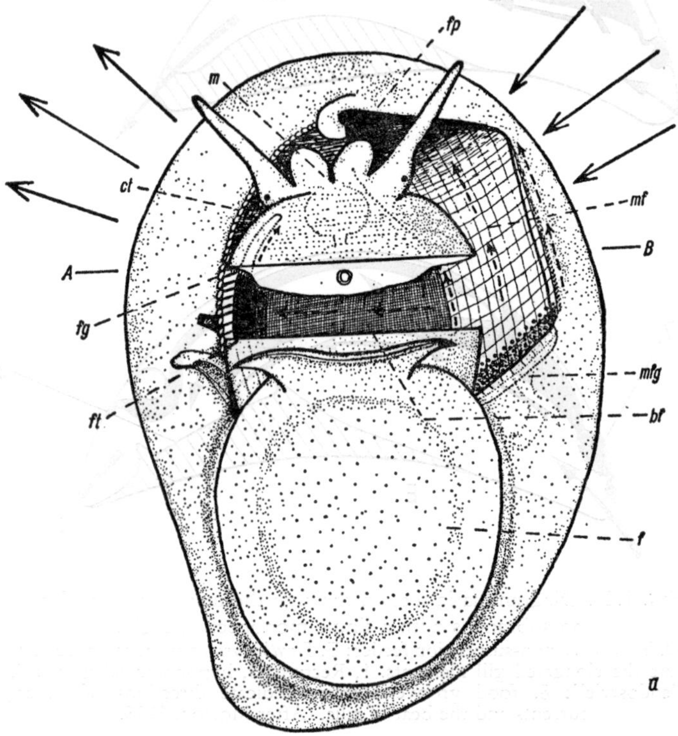

FIG. 1.25. *Crepidula fornicata,* seen from the ventral side. bf, Branchial mucus filter; cg, ciliated grooves; ct, gill; EC, exhalant chamber; en, endostyle; f, foot; fg, food groove; fp, food pouch; ft, food string; gf, gill filament; IC, inhalant chamber; m, mouth; mf, mantle mucus filter; mfg, mantle filter gland. (After Werner, 1959.)

strains large particles from the through-current, but also many small particles were observed to be caught in the mucus of the filter. The filter is built of regularly arranged longitudinal and transversal threads. It is secreted continuously and carried forward across the entrance to the mantle cavity by two lateral ciliary tracts (Fig. 1.26). The particle-loaded mucus piles up in a food pouch in front and to the left of the mouth; it may be ingested by means of the radula or discarded as pseudo-faeces. The position

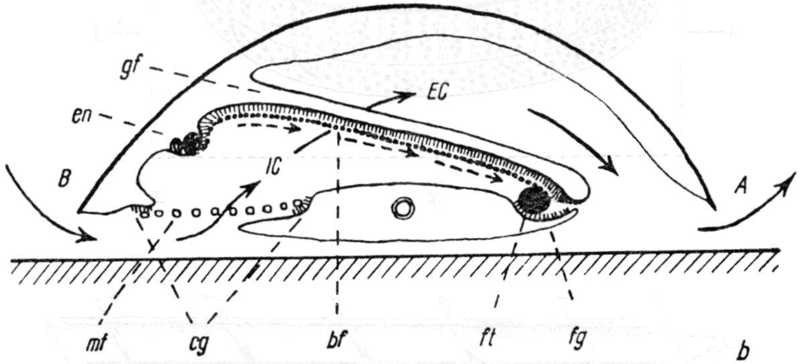

FIG. 1.26. *Crepidula fornicata,* schematic transversal section across line A to B in Fig. 1.25. For explanation of letterings, see text to Fig. 1.25. (After Werner, 1959.)

of the gland producing the filter is shown in Fig. 1.25 and its fine structure in Fig. 1.27. The gland consists of an epithelium containing mucus producing cells and ciliated cells. In the posterior part of the gland these elements are randomly mixed, but in the anterior part they are arranged in alternating bands. The longitudinal threads of the filter are secreted by the bands of mucus cells in the anterior part of the gland, and the transversal threads in the posterior part. The continuously secreted mucus of this part is transported forwards by the cilia scattered between the mucus cells, and the mucus preliminarily accumulates as a band between the posterior and anterior parts of the gland. The cilia of the anterior part beat metachronically, the wave moving from left to right. Each wave draws off material from the accumulated mucus and thus forms the transversal threads of the filter. The cilia of the cuneiform part of the gland (Fig. 1.27a and c, cp) seem to be of special significance in the separation of mucus, and thus the formation of distinct transversal threads. When the mucus threads pass the anterior part of the gland, they adhere to the secretion of the mucus producing cell bands; and as the transversal threads move further on the secretion of each cell band is drawn out into the longitudinal threads of the filter.

The gill filter is secreted by the so-called endostyle, a large

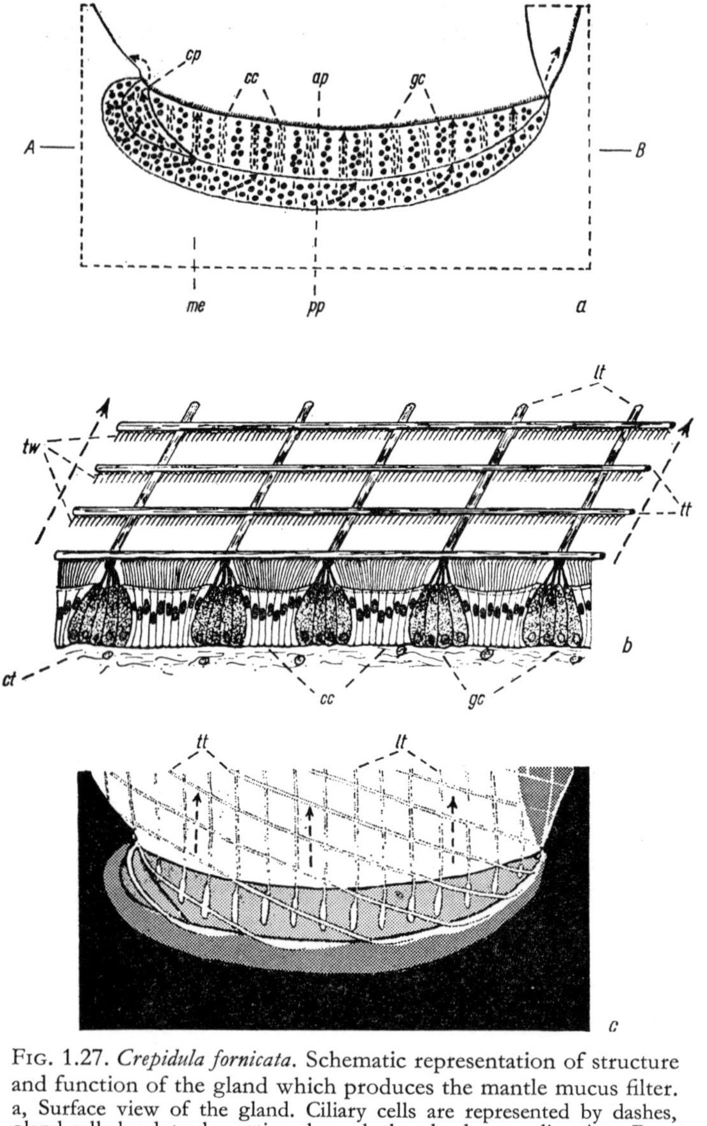

FIG. 1.27. *Crepidula fornicata.* Schematic representation of structure and function of the gland which produces the mantle mucus filter. a, Surface view of the gland. Ciliary cells are represented by dashes, gland cells by dots; b, section through the gland across line A to B; c, illustration of the formation and shape of the mucus filter. ap, Anterior part; cc, ciliated cells; cp, cuneiform part; ct, connective tissue; gc, glandular cells; lt, longitudinal threads; me, mantle epithelium; pp, posterior part; tt, transversal threads; tw, transverse waves of ciliary beating. (After Werner, 1959.)

mucus gland situated along the gill axis. The structure of the endostyle is very similar to that of the mantle filter gland, but the elements secreting the threads are much smaller. The gill filter is, therefore, finer and less porous than the mantle filter. The filter is produced continuously and is carried to the edge of the gill by means of the frontal cilia on the gill filaments. At the free margin of the gill, the filter with enclosed particles is rolled up like a rope and deposited.

Like *Crepidula, Capulus hungaricus,* of the related family Capulidae, is frequently found attached to the shells of bivalves. Its mode of feeding is similar to that of *Crepidula,* but whereas the *Crepidula* possesses linear gill filaments, those of *Capulus* are of less modified triangular shape (Yonge, 1938). Orton (1948) observed specimens of *Capulus* overlying the shell margin of the bivalves *Modiolus* and *Monia*, at places where the food grooves of the demibranchs were within reach of the proboscis. According to Orton there can be little doubt that in such cases *Capulus* takes food collected in the food groove or recurrent mantle path of the bivalves. This assumption has been strongly supported by Sharman (1956) who observed the insertion of the proboscis of *Capulus* between the valves of *Chlamys opercularis*. Many other Capulidae are commensals.

An interesting line of development is found in the sessile worm-like Vermetidae. In a species such as *Vermetus novae-hollandiae,* living in the rough water of the oceanic sides of the coral reefs, feeding is by means of the gill (Yonge, 1932). In the New Zealand species *Serpulorbis zelandicus*, which prefers a moderate amount of wave surge, the gill is more primitive than in *Vermetus novae-hollandiae* and does not completely divide the mantle cavity. No endostyle is present. Still, the gill functions as a food collecting organ, but, in addition, another mode of feeding has been adopted. From the well-developed pedal gland, mucus threads, several at a time, are produced. These threads float freely in the water, catching animal plankton organisms which adhere to the mucus. From time to time the feeding threads are withdrawn and ingested (Morton, 1951a). This development has proceeded further in vermetids from still water, where the gill is even more reduced and of almost unmodified appearance. In these forms, e.g. *V. gigas,* feeding by means of mucus threads is completely dominant

(Boettger, 1930; Yonge and Iles, 1939). In the vermetid *Aletes squamigerus* the feeding mucus is described as a triangular sheet floating in the water (MacGinitie and MacGinitie, 1949).

In the Siliquaridae, which were formerly included in the family Vermetidae but are now considered a separate family, no pedal gland is present, and the gill is the only feeding organ (Morton, 1951b). The mode of feeding in the highly specialized siliquarid *Stephopoma roseum* has been described by Morton. Gill filaments are long, rod-shaped and equipped with longitudinal muscles. The feeding mucus of the gill is exclusively produced by the endostyle, and mucus cells are absent from the filaments. During feeding the filaments bend downwards towards the food groove situated in the floor of the mantle cavity. However, if undesirable material such as carmine or carborundum powder is added to the inspired water, the gill is lifted free of the food groove. The mucus-entangled particles that accumulate at the tips of the filaments are then carried forward along the edge of the gill by apical cilia and ejected.

A peculiar mode of feeding was observed in undisturbed *Stephopoma*. The anterior quarter of the gill is stretched outside the mantle cavity. The gill filaments, covered by mucus from the endostyle, are spread widely apart and swept through the water, thus catching suspended material. The lateral cilia stop beating, so that the mucus is not drawn away from the frontal surface of the filaments. The mucus sheet with entangled food particles is carried by the frontal cilia towards the apex of the filaments. Here the apical cilia, which are motionless when the gill is in the feeding position inside the mantle cavity, carry the mucus posteriorly towards the food groove at the floor of the mantle cavity.

The above-mentioned suspension feeding prosobranchs were all sessile organisms. Recently, however, filter feeding has also been described in a few free-living families. Thus, Graham (1938) found that *Turritella communis,* which lives buried in fine sand or mud, can feed like typical prosobranch filter feeders. An endostyle is present which produces most of the feeding mucus of the gills. Further, the opening to the mantle cavity is guarded by a curtain of tentacles that prevents larger particles from entering with the inhalant current. Yonge (1946b) states that *Turritella* is exclusively a suspension feeder.

A similar mode of feeding was described in the Struthiolariidae by Morton (1951c). During feeding, *Struthiolaria* likewise lies buried in the sand. Water enters and leaves the animal through mucus-lined siphonal tubes made by the extensile and very flexible proboscis (Fig. 1.28). It is worth mentioning that the

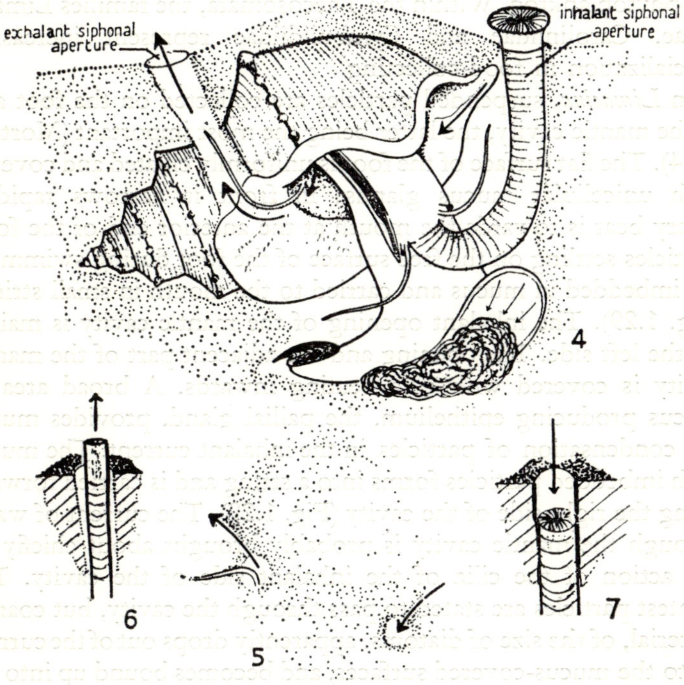

FIG. 1.28. *Struthiolaria papulosa.* 4, in feeding position buried below the surface; 5, surface view of the substratum, inhalant aperture to the right, exhalant to the left; 6 and 7, diagrams showing the action of the proboscis in forming the siphonal tubes. (After Morton, 1951c.)

closely related *Aporrhais* constructs similar mucus-lined tubes when buried beneath the surface (Yonge, 1937). But whereas *Aporrhais* is a deposit feeder, *Struthiolaria* feeds by filtering the current of water that enters from the inhalant tube. The mantle cavity is typically developed, with large, strongly ciliated gill, endostyle and food groove.

1.92. OPISTHOBRANCHIA

In Opisthobranchia, suspension feeding seems to be restricted to the pelagic shelled pteropods, or Thecosomata. Some species, as *Limacina retroversa,* may be important constituents of oceanic zooplankton and thus be of ecological significance as grazers on the phytoplankton. Within the Thecosomata, the families Limacinidae, Caroliniidae and Cymbulidiidae represent increasing specialization in food-collecting.

In *Limacina,* suspended particles are collected on the foot and in the mantle cavity, the latter being the most important (Morton, 1954). The flat surface of the foot is uniformly ciliated and covered with unicellular mucus glands; in front it narrows rapidly; ciliary beat is towards the mouth at the anterior end of the foot. Particles settling on the free surface of the foot during swimming are imbedded in mucus and carried to the mouth in small strings (Fig. 1.29). The inhalant opening of the mantle cavity is mainly on the left side. The opening and the adjacent part of the mantle cavity is covered with cilia beating inwards. A broad area of mucus producing epithelium, the pallial gland, provides mucus for condensation of particles in the inhalant current. The mucus with imbedded particles forms into a string and is carried forward along the right side of the cavity (Fig. 1.29). The current of water through the mantle cavity is probably brought about chiefly by the action of the cilia of the inhalant side of the cavity. The lightest particles are stated to pass through the cavity, but coarser material, of the size of diatoms, apparently drops out of the current on to the mucus-covered surfaces, and becomes bound up into the string which constantly moves forward on the right side. Ciliary tracts carry the string to the mouth where it is ingested by means of the tiny chitinous teeth of the radula. Thus, in contrast to suspension feeding Prosobranchia, the Thecosomata are not filter feeders; gills are absent. Material collected around the mouth, for instance larger particles breaking away from the food string, may also be transported away by cilia on rejection tracts under the surfaces of the wings, and discarded. Morton states that food chiefly consists of dinoflagellates and suitably shaped diatoms.

Feeding in the more advanced families, Cavoliniidae and

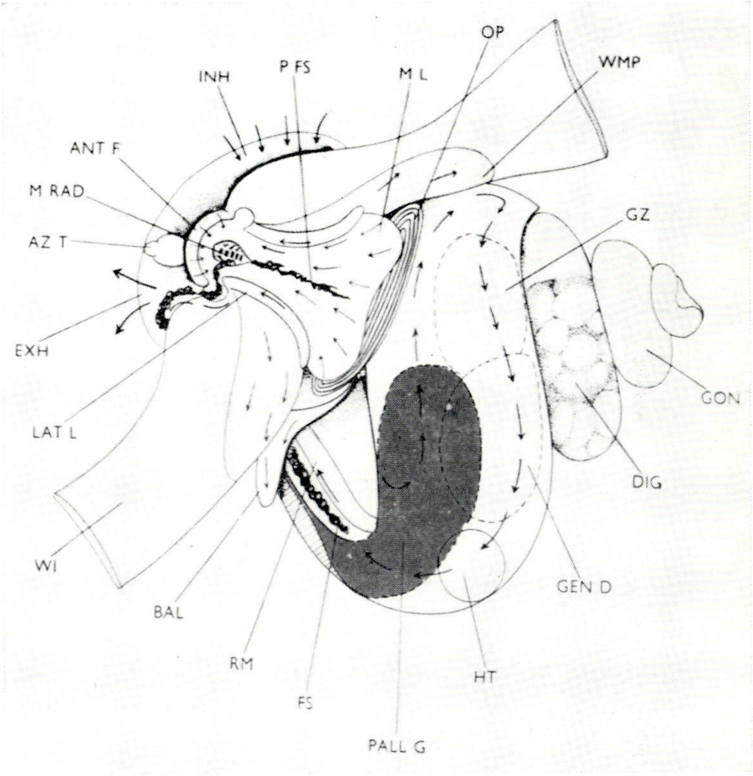

FIG. 1.29. *Limacina retroversa,* ventral view. Diagram showing the ciliary currents employed in food collecting. The surface of the foot and the ciliated portions of the wings are fully expanded, and the distal parts of the wings are omitted. The contents of the pallial cavity are shown as if transparent.

ANT F, Anterior horseshoe-shaped fold; AZ T, azygous tentacle; BAL, balancer; DIG, digestive gland; EXH, exhalant side of pallial cavity; FS, food string within the pallial cavity; GON, gonad; GEN D, genital duct; GZ, gizzard; HT, position of the heart; INH, inhalant side of the pallial cavity; LAT L, lateral lobe of the foot; M L, median lobe of the foot; M RAD, mouth, with the radula; OP, operculum; P FS, food strings formed on the median lobe of the foot; PALL G, pallial mucous gland; RM, rectum; WI, wing; WMP, ciliated "Wimperfeld" at base of wing. (After Morton, 1954.)

Cymbuliidae, was described by Yonge (1926b). In these forms no food collecting takes place in the mantle cavity. In *Cavolinia* and *Creseis,* food collection is by means of the foot and by ciliated fields on the wings; in *Limacina* these fields seem mainly to be rejectory. In *Cymbulia* and *Gleba* the ciliary action of the wing fields is lost and food particles are collected only by the foot. In the advanced Thecosomata there is a reduction of the buccal mass to the complete disappearance of jaws, radula and salivary glands in *Gleba.* The external surfaces become loaded with particles during swimming.

1.93 VELIGERS

A large number of the marine gastropods pass through a pelagic so-called veliger stage, which is feeding on suspended particles. The velum, which characterizes the stage, functions both as a locomotory and feeding organ; when the larvae are swimming they are at the same time automatically feeding. The feeding mechanisms seem to be very similar in both prosobranch and opisthobranch veligers (Lebour, 1931; Werner, 1955; Thompson, 1959). The activity of the long cilia of the edge of the velum produces water currents which cause the swimming of the larva and which also bring suspended particles in contact with food collecting tracts situated between the long marginal cilia and a subvelar row of shorter cilia (Fig. 1.30). Particles are caught by the short cilia in the food groove, imbedded in mucus and rapidly carried to the mouth. Here the material may be sorted; in nudibranch veligers only small particles are ingested and if a too heavy load of suspended matter is carried to the mouth, the surplus material is rejected.

1.94. EFFICIENCY OF PARTICLE RETENTION

Werner (1953) estimated the porosity of the two mucus filters in *Crepidula* by measuring the distance between the elements of the glands forming the longitudinal threads of the filtering nets. The distances between the ciliary bands on the anterior part of the gland secreting the mantle filter was about 40 μ in an animal about 10mm long; in 21 to 53mm long individuals the averages ranged from 60 to 90 μ, apparently independently of the size of the individuals. In the gill filter, the distances between ciliary bands

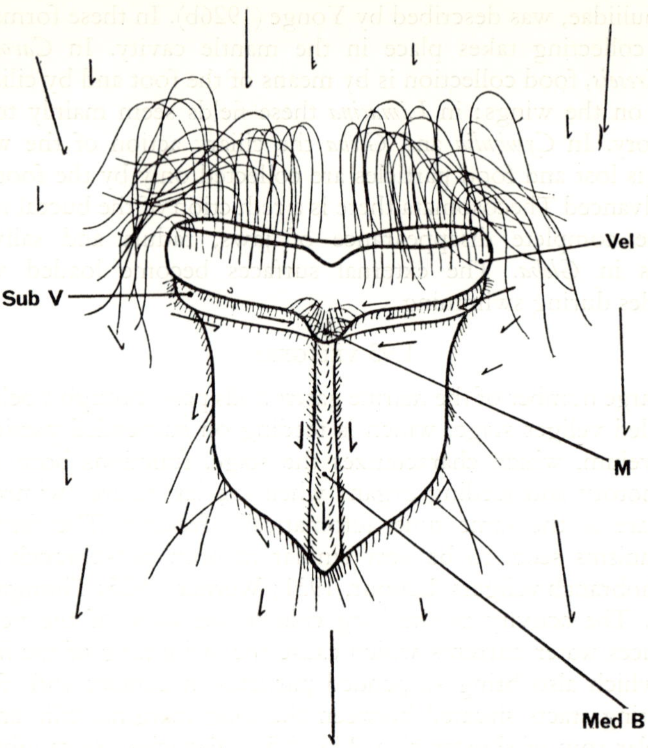

FIG. 1.30. Cephalopedal ciliary apparatus of larval *Archidoris pseudoargus*: semi-diagrammatic view from ventral aspect. The arrows indicate the direction in which particles are impelled. M., mouth; Med.B., median pedal band; Sub.V., subvelar ridge; Vel., velum. (After Thompson, 1959.)

averaged 8 to 14 μ in the various individuals, independently of their size. It is, however, not possible to estimate accurately the minimum size of particles that are efficiently retained by the filters, since the thickness of the mucus threads are not known. In the freshwater *Bithynia* and *Lymnaea,* Rodina (1948b) obtained growth on diets of bacteria fed in amounts corresponding to normal contents in eutrophic and mesotrophic lakes. This strongly indicates that the gastropods efficiently retain particles of bacterial sizes.

In the suspension feeding gastropod veligers, which are not filter feeders, the size range of particles that can be utilized would seem mainly to depend upon the width of the oesophagus. Thompson (1959) found that only particles smaller than *ca.* 20 μ were ingested by *Crepidula* veligers. These maximum sizes of particles ingested are in good agreement with the theoretical maxima estimated by Thorson (1946) from measurements of the width of the oesophagus in various gastropod veligers. The estimates varied from 5 to 45 μ, dependent upon size and species of the veligers.

1.95. RATE OF WATER TRANSPORT

The rates at which the surrounding water is cleared of particles seem not to have been determined in marine suspension feeding gastropods, but only in some freshwater prosobranchs as *Viviparus, Bithynia* and *Valvata*. Schäfer (1953) measured the rates at which *Bithynia tentaculata* and *Viviparus fasciatus* cleared suspension of charcoal (Tierkohle). Calculations show that *Bithynia* cleared about 27ml per animal and hour, and *Viviparus* about 130ml. The weight of *Viviparus* is not stated, but the average weight of *Bithynia* was 135mg. The initial concentrations of charcoal were 250mg/l. in the experiments on *Bithynia* and 500–1,000mg/l. in the experiments on *Viviparus*. In the natural habitats of the snails, the River Rhein and River Lahn, the concentrations of suspended matter vary strongly, but were mostly below 20mg/l. The concentrations of suspended charcoal were, therefore, very high, but as the effect of concentration of suspended matter on clearances was not investigated, it is not known whether the measured clearances are representative of those pertaining in nature. Tsikhon-Lukanina (1961b) determined clearances of suspensions of algae or yeast by *Bithynia tentaculata* and *Valvata piscinalis*. In concentrations of *Scenedesmus* ranging from 10,000 to 100,000 cells per ml *Bithynia* cleared the suspension at an average rate of 25ml/hr, whereas the rate of clearing was about 60ml/hr at concentrations of algae ranging from 500 to 5,000 cells per ml. A similar decrease in rates of clearing with increasing concentrations of suspended cells was also found in *Valvata*. Tsikhon-Lukanina, therefore, concludes that *Bithynia* and *Valvata* to some degree regulate the amount of food consumed depending on the concentration of food particles. However, there is no proof that the decrease in clearance with increasing particle concentrations represents an active adaptation. The size of *Bithynia* and *Valvata* used are not stated so that the clearances cannot be expressed in terms of body weight.

It is noteworthy that none of the freshwater suspension feeding prosobranchs seems to be obligatory suspension feeders. Schäfer observed that in clean water, *Bithynia* will move around in the aquarium and rasp off material adhering to the glass, but when suspended material was added to the water, the animals became sessile as long as suspended material was present. *Bithynia,* therefore, seems to be able to feed exclusively on suspended matter and to prefer this type of food. The fact that charcoal also induces filter feeding indicates that the quality of the suspended matter is not decisive in determining the type of feeding adopted by the snails.

Also *Viviparus* and *Valvata* were found capable of feeding both on suspended matter and by scraping off loose particulate matter from the substratum (Cook, 1949; Cleland, 1954).

1.10. BIVALVIA

Most bivalve molluscs feed on suspended matter in the water. More work has been done on the various aspects of feeding in bivalves than in any other group of suspension feeders.

Suspension feeding seems to be secondary within the bivalves, and, as in prosobranch gastropods, to have developed by adaptations of the water transporting and cleansing mechanisms of gills and mantle cavity. Suspension feeding bivalves of the present day may have originated from ancestors similar, in appearance and habits, to the Protobranchia (Yonge, 1939), but a palp proboscis may not have been present (Stasek, 1961).

According to most investigators Protobranchia are essentially deposit feeders; Nuculidae and Ledidae feed by means of their large mouth palps and palp proboscides, which may be extended for a considerable distance outside the shell and be moved over the bottom. The ciliary currents on the surface of the palps catch and transport particles towards the mouth at the base of the palps (Yonge, 1939). Besides, suspended particulate matter is inhaled with the respiratory current of water. In forms such as *Nucula* and *Acila* that live buried in the substratum, stirred-up bottom material may presumably add significantly to the naturally suspended matter. The suspended particles entering the mantle cavity may be caught directly on the palps; or be carried along ciliary cleansing currents on the surface of the gills and transferred along special ciliary tracts to the palps, to be finally ingested (Stasek, 1961). Suspended food was believed to be important by Caspers (1940) who showed that *Nucula* freely suspended in water by a string attached to the shell would ingest carmine and other particles added to the water. From these and other experiments he concluded that *Nucula* normally obtains its food with the respiratory currents. However, the experiments are inconclusive as to the importance of suspended matter as a normal food source of *Nucula*; the smallness of the gills in relation to the size of the palps strongly indicates that suspended material retained by the gills plays a minor role in the normal nutrition of Nuculidae.

The third family of Protobranchia, the Solenomyidae, are also predominantly deposit feeders, but the gills seem here to be of great importance in the collection of the food. The Solenomyidae, which live in muddy bottom, bring large quantities of bottom material into the mantle cavity by means of muscular activity of the mantle and possibly of the foot. The material accumulates on the relatively large gills from where it is eventually transferred to the relatively small palps and further on to the mouth (Owen, 1961). This division of labour between gills and palps is characteristic of the lamellibranch bivalves.

The present-day concept of structure and function of the feeding organs in typical suspension feeding bivalves is based on works of Ridewood (1903), Wallengren (1905), Kellogg (1915), Yonge (1926a), and others, and was greatly extended by the detailed studies of Atkins (1936, 1937a, b, 1938 and 1943) on some 90 species. More recently, several comparative studies have been made of feeding mechanisms and their adaptations within families or larger groups of bivalves, thus in the minute, largely commensal, forms belonging to the families Erycinidae, Montacutidae and Galommatidae (Popham, 1940; Oldfield, 1955, 1961); the reefboring Tridacnidae (Yonge, 1936, 1953; Purchon, 1955b; Stasek, 1962); the woodboring Teredinidae and Xylophaginidae (Purchon, 1941); the rockboring Pholadidae (Purchon, 1955c); the Lucinacea (Allen, 1958a); the Veneridae (Ansell, 1961); the

Solenidae (Yonge, 1951b), and the Lyonsidae (Yonge, 1951c). MacGinitie (1941) compared feeding mechanisms in forms from four different habitats, viz. *Schizothaerus, Pecten, Ostrea* and *Mytilus*. A number of papers deal with single species of suspension feeding bivalves. Quite naturally, more work has been done on the commercially important species of the oysters, specially on the European *Ostrea edulis* and the North American *Crassostrea virginica,* than on any other bivalve (Yonge, 1960). Feeding mechanisms in *O. edulis* were studied by Yonge (1926a) and in *C. virginica* by Nelson (1923, 1938, 1960), and others.

The organs of feeding in suspension feeding bivalves are the mantle, gills and mouth palps. The mantle edges form the apertures through which the food current passes, and the mantle proper lines the inhalant and exhalant chambers, which are separated by the gills. The mantle does not participate in collection and sorting of particles intended for ingestion; its ciliary currents are for cleansing and for rejection of deposited particles. The mantle edge is typically divided into three folds. The outer fold is concerned with shell formation; the middle fold is sensory in function and may carry tentacles and eyes. The sensory function is especially important around the inhalant region of the mantle, since it provides the first contact between the animal and the incoming water, which may thus be tested before being accepted. The tentacles may also function as filters. In water with much suspended coarse material, the tentacles of oysters may cross to form a screen which strains the larger particles from the inhalant current (Nelson, 1938). The inner fold is muscular and controls water movements into and out of the mantle cavity (Hopkins, 1933; Nelson, 1938). The shape of the mantle edges vary, especially with the habitat. It is least specialized in bivalves belonging to the epifauna or living on or in the uppermost layer of the bottom, but deeper-burrowing forms mostly possess siphons through which a food current can be drawn in from the water masses above and food depleted water expelled.* The

* In *Loripes lucinalis* and other Lucinidae, which are highly specialized burrowing forms, no inhalant siphon reaches the surface of the bottom, but a special inhalant tube is formed by mucus secreted by the foot. This tube is comparable to that built, for instance, by the suspension feeding prosobranchs, Struthiolariidae (see p. 61) (Allen, 1958).

formation and types of siphons within the various groups of bivalves have been treated by Yonge (1948, 1957).

In active suspension feeding bivalves, the inhalant aperture is large and may in some species correspond to the greater part of the opening between the mantle edges, whereas water is expelled through a relatively small opening as a jet which carries the food-depleted water far away so that recirculation of the water is avoided, as already observed by Wallengren (1905). He further noticed that the direction of the inhalant current is at a large angle with the exhalant current, which also minimizes the chance of recirculation of water. In *Venerupis pullastre*, Quayle (personal communication) calculated the speed of the current in the

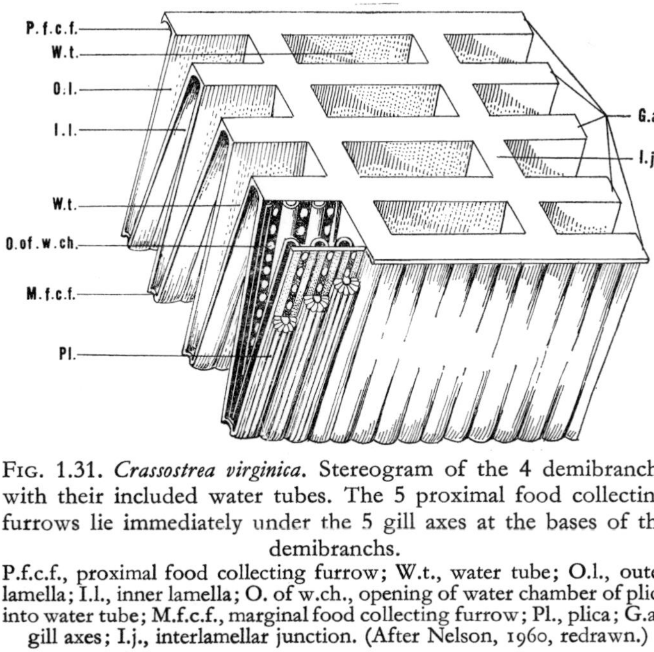

FIG. 1.31. *Crassostrea virginica.* Stereogram of the 4 demibranchs with their included water tubes. The 5 proximal food collecting furrows lie immediately under the 5 gill axes at the bases of the demibranchs.

P.f.c.f., proximal food collecting furrow; W.t., water tube; O.l., outer lamella; I.l., inner lamella; O. of w.ch., opening of water chamber of plica into water tube; M.f.c.f., marginal food collecting furrow; Pl., plica; G.a., gill axes; I.j., interlamellar junction. (After Nelson, 1960, redrawn.)

exhalant siphon to be three times the speed in the inhalant siphon. Furthermore, the smallest angle between the two siphons in feeding animals was 90°, the inhalant siphon being situated posteriorly. Also rhythmical contractions of the adductor muscles, described in the oyster by Hopkins (1936a) and Nelson (1938) can help in carrying food-depleted water away from the inhalant current.

The development of tentacles around the apertures may vary with the habitat, presumably correlated with the amounts of suspended matter, perhaps mainly of larger-sized particles, in the water. Within the Veneracea, for instance, species typically found in clean sandy or gravelly bottoms offshore where the amounts of suspended matter is low, have poorly developed tentacles; species found in localities with moderate amounts of suspended matter have well-developed tentacles, and species normally occurring on beaches where large amounts of suspended matter may be brought into suspension by wave action, have the greatest development of tentacles (Ansell, 1961).

The gills, which both propel and filter the water in suspension feeding bivalves are typically composed of descending and ascending filaments (Fig. 1.31). The anterior and posterior faces of the filaments are equipped with lateral cilia (Fig. 1.32), which drive the water through the interfilamentary slits from the inhalant to the exhalant chambers. The latero-frontal cilia may more or less completely overbridge the interfilamentary slits. Atkins (1938) distinguishes between two types of latero-frontal cilia, eu- and micro-latero-frontal cilia. The large eu-latero-frontal cilia are found, for instance, in the Mytilidae among the filibranchs and in eulamellibranchs generally. The micro-latero-frontal cilia she described in forms that were previously stated to lack latero-frontal cilia, Arcidae, Anomiidae, Pectinidae, and other families. But Purchon (1958a) considers the micro-latero-frontal cilia to be identical with the so-called pro-latero-frontal cilia which are also present on the filaments of the forms that possess eu-latero-frontal cilia; Purchon, therefore, suggests that the Microciliobranchia of Atkins are forms that have lost the latero-frontal cilia during the evolution. On the frontal surface of the gill filaments are tracts of cilia beating either dorsally towards the gill axis or towards the free ventral margin of the demibranchs.

The shape of the gills, and the food currents on the frontal surface of the filaments and along the gill axis or ventral margins vary strongly, but Atkins (1937b) distinguishes between seven main types, which are further subdivided in a number of varieties.

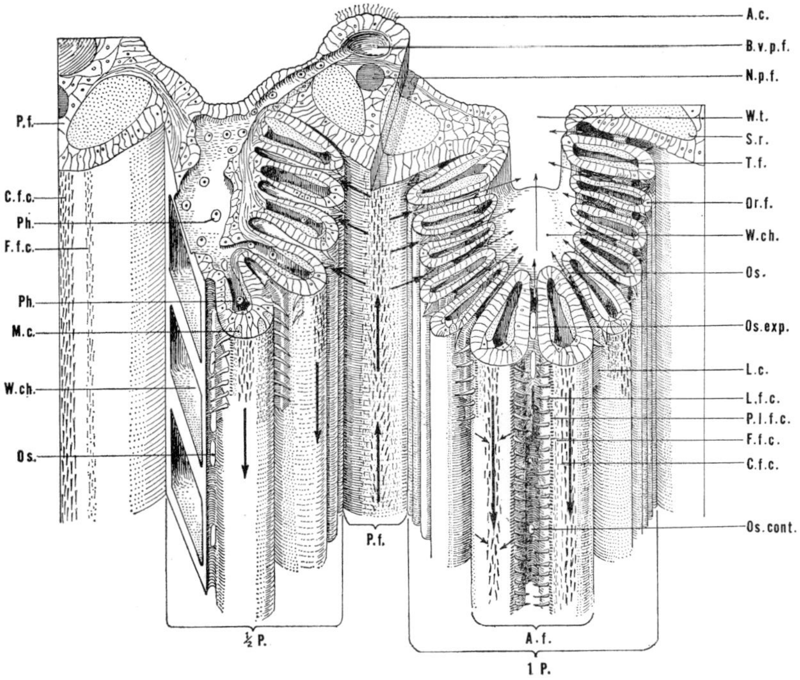

FIG. 1.32. *Crassostrea virginica*. Stereogram of a plica and a half with connecting principal filament. The skeletal rods of the principal filament become flattened plates lining the outer three-quarters approximately of the blood lacunae of the ordinary filaments. Since these blood channels end blindly they are referred to as lacunae rather than as blood vessels.

P.f., principal filament; C.f.c., Coarse frontal cilia; Ph., phagocyte; F.f.c., fine frontal cilia; M.c., mucus gland; W.ch., water chamber of plica; Os., ostium; A.c., abfrontal cilia; B.v.p.f., blood vessel of principal filament; N.p.f., Nerve of principal filament; W.t., water tube; S.r., skeletal rod; T.f., transitional filament; Or.f., ordinary filament; Os.exp., ostium expanded; L.c., lateral cilia., L.f.c., laterofrontal cilia; P.l.f.c., para laterofrontal cilia; Os.cont., ostium contracted; ½ p., ½ plica; A.f., apical filament; 1.p., one plica. Arrows indicate direction of transport along ciliary tracts and of water currents produced by the lateral cilia. (After Nelson, 1960, redrawn).

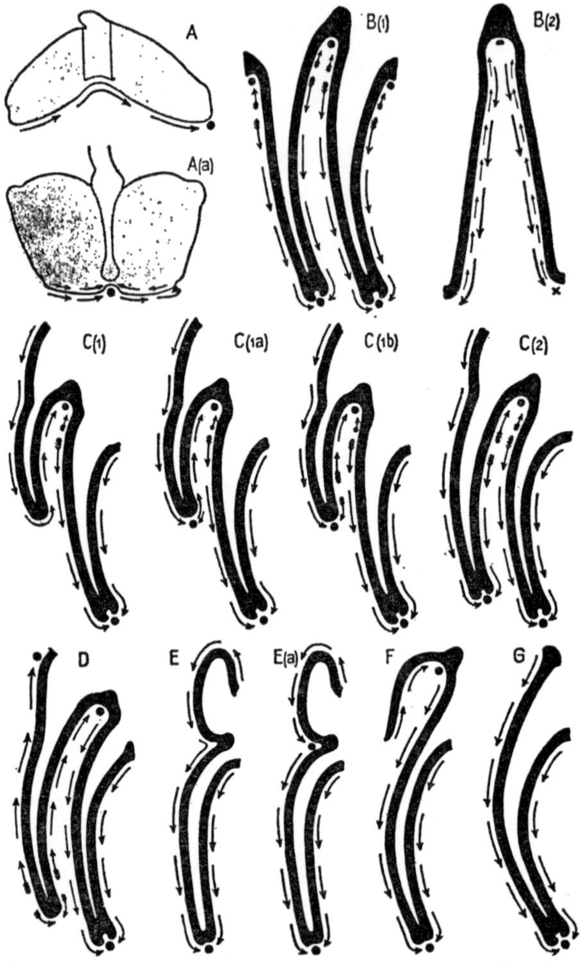

FIG. 1.33. Diagrammatic transverse sections of types of bivalve gills with their food currents. The inner demibranch is on the right in all sections.

A. Protobranchia: e.g. *Nucula*: A(a), *Nuculana minuta*.
B. Filibranchia and Pseudolamellibranchia: B (1), Mytilidae, Pinnidae; B (1a), Arcidae, Anomiidae; B (1b), most Pseudolamellibranchia: I, ordinary and II, principal filaments; B (2) *Heteranomia*.
C. Many Eulamellibranchia: C (1), many Eulamellibranchia; C (1a), e.g. *Pholadidea loscombiana*; C (1b), e.g. *Venus fasciata*; C (1c), e.g. *Barnea candida*; C (1d), *Cultellus pellucidus*; C (2), e.g. *Lutraria lutraria*; C (2a), *Solen* and *Ensis*; I, ordinary and II, principal filaments.
D. Unionidae.
E. Tellinidae, Semelidae, Anatinacea: E(a), *Tellina crassa, Scrobicularia plana*.
F. *Lasaea rubra*.
G. Lucinidae, Montacutidae, Teredinidae.
(After Atkins, 1937b.)

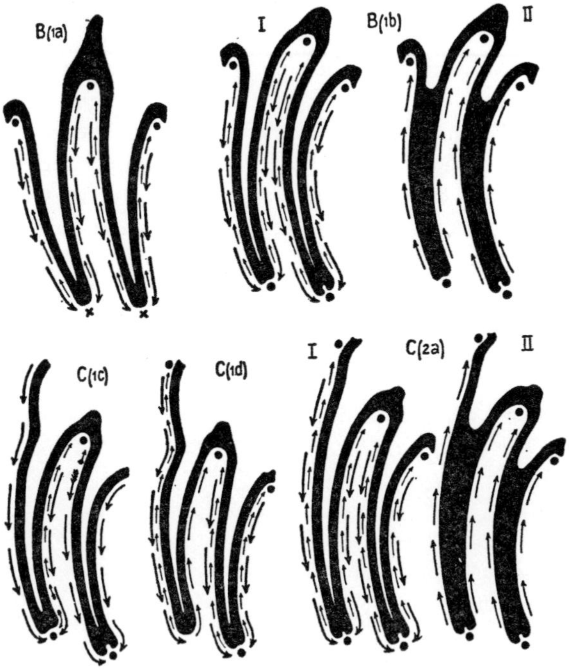

From Fig. 1.33 it appears that ciliary currents beating dorsally and ventrally may occur on the frontal surface of the same filament (Anomiidae, Arcidae, Pectinidae, Ostreidae, and others). The ventrally beating cilia are then coarser and larger than those dorsally beating. Bivalve gills are often plicate, in which cases the ciliation on the filaments in the bottom of the troughs between adjacent plicae, so-called principal filaments, may differ from that on the ordinary filaments: on the frontal surface of the principal filaments only dorsally beating cilia may be present, but both dorsally and ventrally beating cilia on the ordinary filaments. Typically, ciliary currents run along all gill axes and ventral margin of the demibranchs, but in some forms one or more may be absent. Generally, transport is anteriorly along the tracts, but in a few instances the tract along a gill margin transports posteriorly. Further details about the ciliary food currents and their variations are given below when dealing with the function of the gills. The two demibranchs may differ in shape; the outer is often smaller than the inner, and may be strongly reduced or absent.

Food is conveyed from the gills to the mouth by two pairs of mouth palps, which vary from small to large according to species (Fig. 1.34). The inner surfaces of the palps are covered by transversally arranged folds which carry complicated systems of ciliary tracts for transport and sorting of the material picked up from the gills. At the united bases of each pair of palps a

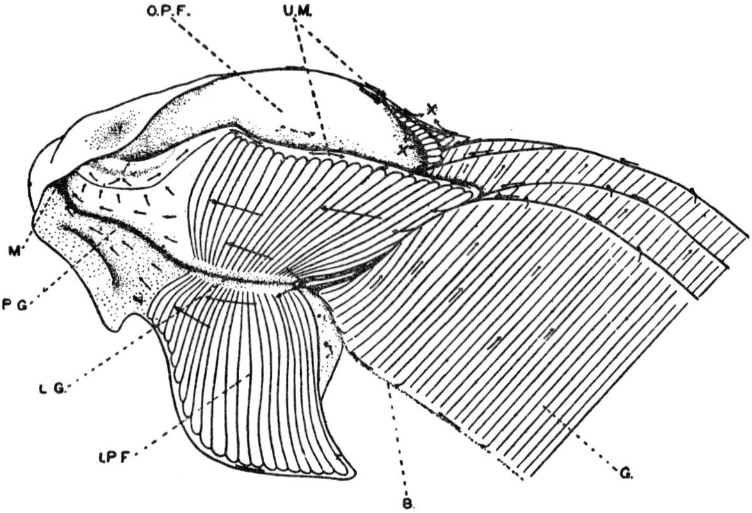

FIG. 1.34. *Ostrea edulis.* Junction of palps and gills; right palps opened out so as to expose inner ridged surfaces.
B., Base of gill demibranch; G., gill; I.P.F., inner palp face; L.G., lateral oral groove; M., mouth; O.P.F., outer palp face; P.G., proximal oral groove; U.M., upper margin of palps; X., point where material is rejected from palps. Arrows indicate the collection and rejection tracts of gills, palps and oral grooves. (After Yonge, 1928.)

ciliated oral groove is present, which connects gills and mouth. Material, especially that transported along the gill axes, can be conducted along the oral groove directly towards the mouth, thus by-passing the sorting mechanisms of the palps.

The interaction of the different elements of gills and palps in the feeding process in adult bivalves has mostly been observed on surviving preparations where valve and mantle of one side have been removed. In such preparations particles of different kinds have been supplied to the gills, and the way in which they are dealt with by gills and palps has been watched. The function of gills and palps has also been observed directly through the transparent shells of young metamorphosed stages, e.g. by Churchill and Lewis (1924), Menzel (1955), Nelson (1923), Tammes and Dral (1955), and Yonge (1926a).

Many bivalve gills possess well-developed sorting mechanisms (Atkins, 1936, 1937a). In gills with both dorsally and ventrally beating frontal cilia on the same filaments it has been observed that small and light particles such as diatoms or carmine when dropped on the gill surfaces are generally carried dorsally to the groove along the gill axis. This groove constitutes a comparatively safe and direct pathway to the oral groove and the mouth. Quite exceptionally the tract along the gill axis functions as a rejection tract; in *Tridacna,* according to Purchon (1955b), particles travelling forward in the tract do not seem to have any opportunity of reaching the mouth. Heavy and large particles, or masses of particles entangled in mucus, are carried ventrally. The ventrally beating tracts are active only when stimulated by the presence of heavy or large particles on the gills. The dorsally beating tracts function constantly. This arrangement makes it possible for the gills to carry rather large, but light objects dorsally. In *Arca,* Atkins observed that a piece of filter paper, 1mm^2, was transported dorsally, whereas tin foil of the same size stimulated the coarse ventrally beating cilia and was consequently carried ventrally. On arrival at the ventral margins of the demibranchs material is carried further on by ciliary tracts which are either superficial or more or less hidden in grooves. In Arcidae and some Anomiidae, where no grooves are present at the ventral margin, the ciliary tracts are directed posteriorly, and the transported material is thus rejected. In the large majority of bivalves, however, the ventral groove also transports towards the labial palps. But material is more easily dropped from the ventral groove on to the mantle than it is from the dorsal groove. Material falling on the mantle leaves the animal as so-called pseudofaeces.

In most bivalves with plicate gills, dorsal movement of particles takes place mainly in the troughs between the plicae, whereas ciliary tracts carrying ventrally are chiefly on the crests of the plicae. Dorsally beating cilia may even be absent on the one or two apical filaments. Plicate gills of this type are found in *Pecten, Ostrea, Solen, Ensis,* and others. Small, light particles are mainly carried dorsally in the plical troughs, while coarse particles or particle masses go on the crests towards the tracts or grooves along the ventral margin. Setna (1930) observed in the *Pecten* gill that when coarse particles drop into the plical troughs, characteristic flapping movements were released resulting in the throwing of such particles on to the crests. If too heavy loads of particles collect on the plicate gill, the whole demibranch contracts, whereby the troughs disappear and only the crests are exposed. All material is consequently carried ventrally to the free margin if not previously thrown off by vigorous movement of the whole gill. Special methods of cleansing the gill surfaces of too heavy loads of deposited material are presumably of common occurrence. Heavy loads of particles on the gills stimulate a strong secretion of mucus. In *Rocellaria,* Purchon (1954) observed the mucus with the entangled particles to form a sheet over the gill surface. The gill suddenly filled with water and then contracted, blowing off the mucus sheet from the gill surface. A similar method was also observed in *Tridacna* (Purchon, 1955b).

Some bivalves (*Pinna, Lyonsia, Trachia,* and others) with plicate gills possess only ventrally beating frontal cilia (Atkins, 1936, 1937a). Nevertheless, they can use the plicae for sorting. Fine particles are observed to be carried predominantly in the plical troughs, to be directed into the deep, protected canal inside the marginal groove (Fig. 1.35), whereas coarse and bulky material travels on the crests and further on to the superficial tract on the roof of the marginal food groove. The side and the roof of the marginal groove are formed by ventral extensions of the plicae. The groove can be actively opened or closed. Also in other forms, without plicate gills but with deep marginal grooves, opening and closing of the groove have been observed, e.g. in *Mya, Cardium, Zirphaea,* and others. Numerous bivalves have the marginal groove roofed over by guarding cilia (Atkins, 1937a). A typical arrangement of such cilia is seen in *Modiolaria marmorata* (Fig. 1.36). The ventral end of each filament is bulb-shaped and almost touching the neighbouring filaments. The frontal surfaces of the enlarged ends of the filaments are covered by stout cirrus-like cilia, except for narrow tracts leading through small gaps between the filaments into the marginal groove, well protected by the fan-shaped, guarding cilia. Only small particles or particle masses can pass this way. Bulky masses are caught by

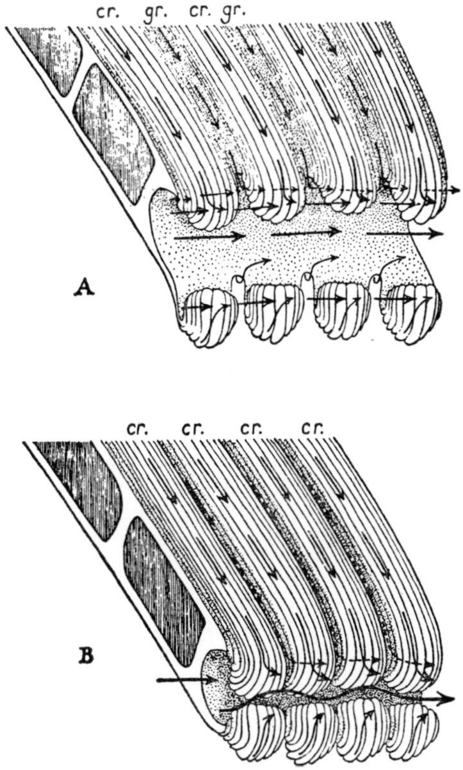

FIG. 1.35. *Pinna fragilis.* Diagrams of the ventral margin of a demibranch to show the utilization of plicae and marginal grooves in sorting.

A. Marginal groove open; the plicae expanded. B. Marginal groove closed; the plicae contracted. Broken arrows indicate the direction of deep hidden food streams. cr., Plical crest; gr., plical groove. (After Atkins, 1937a.)

the stout cilia and travel the more risky way outside the guarding cilia toward the labial palps, which may or may not be safely reached.

Whether collected material shall or shall not pass from the

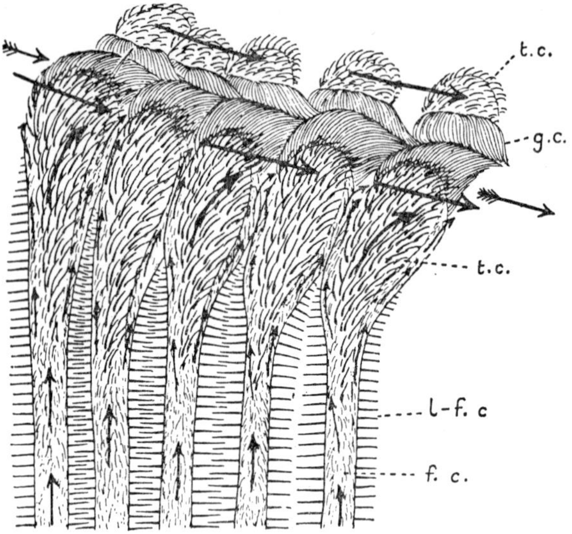

FIG. 1.36. *Modiolaria marmorata.* Marginal groove with guarding cilia. Small arrows indicate the path taken by fine particles, large arrows by coarse particles. Feathered arrows show the direction of the current within the groove.
f.c., Frontal cilia; g.c., guarding cilia; l-f.c., latero-frontal cilia; t.c., terminal frontal cilia. (After Atkins, 1937a.)

gills on to the palps is determined by the position of these organs. When material is to be rejected the palps spread apart and assume a transverse position, so that material from the gills cannot reach the inner palp surfaces (MacGinitie, 1941). Material is transferred to the palps only when these are stretched backwards, establishing contact with the margins of the demibranchs. Direct observations have shown that particles placed onto the inner surface of an isolated palp come under the influence of the elaborate system of ciliary tracts on the folds. The number of folds can vary strongly between species, and in the details the course and function of the ciliary tracts on the palps may also differ, but fundamentally the function of the palps is the same in all suspension feeding bivalves. The ciliary currents have been considered under three headings: rejection currents, acceptance currents and re-sorting currents (Purchon, 1955c). The number of ciliary tracts that can be recognized on a fold vary from six to eight between species (Fig. 1.37). The cilia on the crest of the folds always beat towards the mouth and constitute the main acceptance current; whereas the cilia in the bottom of the troughs between the folds beat towards the ventral margin of the palp, along which runs an important rejection tract. Re-sorting currents occur up the sides of the folds

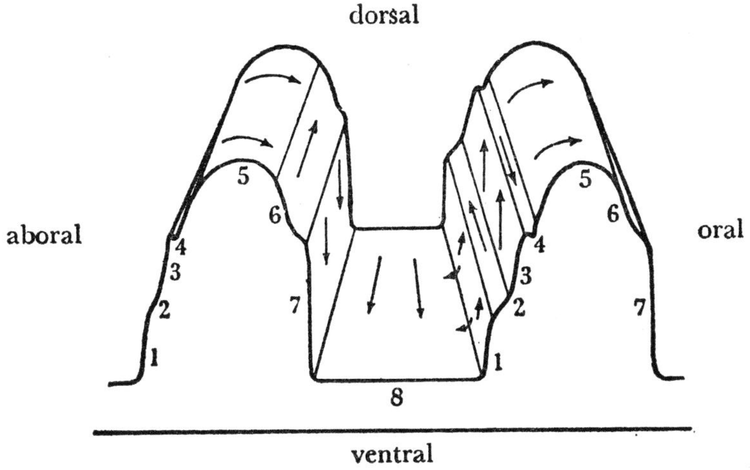

FIG. 1.37. Diagrammatic representation of two folds on the inner face of the palps of a member of the Ungulinidae or Thyasiridae showing the direction of beat of the cilia of the eight tracts. (After Allen, 1958a.)

and may transport material dorsally or ventrally. Details of the various ciliary tracts and their direction of beat are summarized in Table 1.5 (Allen, 1958a).

The mucus production and the muscular activity of the folds play a dominant role in the function of the palps. When the muscles of the folds are relaxed, the folds overlap and expose the anteriorly beating cilia of the crests, whereas the remaining ciliary currents are hidden; the whole surface will act as an acceptance tract for the material transported along the palp. This situation typically applies when small amounts of small and light particles are placed on the palp, under which condition sorting of material may thus be absent. On application of increasing amounts of material or of large or heavy particles, the folds may erect and

TABLE 1.5. A COMPARISON OF THE DIRECTION OF BEAT OF THE CILIA ON THE TRACTS OF THE PALP FOLDS IN VARIOUS BIVALVES (compare Fig. 1.37) (After Allen, 1958a)

dorsal

	1	2	3	4	5	6	7	8
Diplodonta and *Thyasira*	← →	↑	↗	↓	→	↑	→	↓
Ensis siliqua	→	↑	↗	↓	→	↑	→	↓
Mya arenaria	·	↑	↗	↓	→	↑	← ↓	↓
Ostrea edulis	←	↑	↗	↓	→	?	?	↓
Pholas dactylus	→	↑	→	↓	→	·	→	↓
Pholadidea loscombiana	→	↑	→	↓	→	·	→	↓
Barnea candida	aboral ↓ →	↑	→	↓	→	·	→	↓ oral
Barnea parva	→	↑	→	↓	→	·	→ ↓	↓
Zirphaea crispata	→	↑	·	↓	→	·	→ ↓	↓
Petricola pholadiformis	·	↑	→	↓	→	·	→	↓
Martesia striata	·	↑	→	↓	→	·	→	↓
Tridacna crocea	←	·	→	↓	→	↑	↓	↓
Rocellaria cuneformis	→	·	→	↓	→	·	→	↓

ventral

undulate gently, exposing the re-sorting and rejection tracts of the sides and bottom between the folds. Large and heavy particles are especially prone to reach the deep rejection tracts, whereas other particles may be subjected to the re-sorting currents; small and light particles may simultaneously be carried by the currents on the crests from fold to fold towards the mouth. Large amounts of material on the palps may stimulate a copious secretion of mucus, and all material may be rejected. If the palps become too heavily loaded they may also react strongly by contracting into spiral form or by making twisting movements, which will remove all material. The observations of the palp reactions towards various particulate matter indicate that the functioning of the palps is mainly determined by weight (density), size and quantity of particles, as emphasized by Wallengren (1905) and further substantiated by later studies on a large number of bivalves both from freshwater and the sea (Kellogg, 1915; Nelson, 1923; Churchill and Lewis, 1924; Yonge, 1928; Graham, 1931; Purchon, 1954, 1955a, c; Allen, 1958a; Ansell, 1961). However, Menzel (1955) also found sorting according to quality on the palps of *Crassostrea virginica*. He studied young, mostly 5 to 6mm long, stages attached to microscope slides and watched through the left attached valve. He found that little of the plankton, carmine particles, charcoal, etc., fed to and retained by the young oyster was carried directly to the mouth. Mostly the particles were going to the apex of the palps to be revolved by ciliary action into a ball. From this ball single particles were seen breaking away. Organic plankton particles tended to pass between the opposed palp surfaces to be transported to the mouth and swallowed, whereas useless particles, as carmine or charcoal, were usually added to the food ball again, eventually to be rejected as pseudofaeces. Similar qualitative sorting was not observed in young stages of *Ostrea equestris,* which accepted the materials indiscriminately.

Previously, reversal of ciliary beat was considered important in the functioning of the palps (Grave, 1916). According to Wallengren (1905) reversal of ciliary beat on the mouth palps of bivalves was first described by Purkinje and Vallentin in 1835 and later confirmed by Engelmann and Vignon, but Wallengren himself was unable to find such reversal in any species investigated even under varied experimental conditions. He suggests that the previous authors may have been misled by the many tracts on the folds with cilia partly beating in opposite directions. Reversal of ciliary beat has not been recorded by later investigators of the bivalve mouth palps.

The activity of the palps thus ensures that food reaches the mouth only in small amounts at a time. The mouth makes the final decision as to whether the food shall be ingested or rejected. If the mouth is closed the food eventually accumulates in amounts that cannot be accommodated in the narrow oral groove. It is consequently caught by rejection tracts that surround this groove.

Sorting of the particles is not necessarily finished when they have reached the mouth and have been ingested, but further sorting takes place in the more or less elaborate stomach, which secures that undigestible particles are preferentially transferred directly into the intestine, and digestible particles into the digestive diverticula. It appears that in the various bivalves the main sorting organ may be either the gills, the palps, or the stomach (Owen, 1955; Purchon, 1958b; Nelson, 1960).

From observations mainly made on preparations where the gill surfaces have been exposed by removal of valve and mantle, we thus obtain the following picture of the feeding process. Water is drawn through the interfilamentary slits by the action of the lateral cilia. Particulate matter in the water is retained on the gill, either by directly touching the frontal surfaces of the filaments where they are trapped in mucus, or by the straining action of the large latero-frontal cilia when such are present. The latero-frontal cilia beat towards the frontal surface of the filaments, thus throwing particles on the frontal ciliary tracts. Intermittent beating of the latero-frontal cilia has been observed both in *Venus* and *Mytilus*; the resting cilia are stated to form a sieve straining particles from the current passing between the filaments, whereas the active cilia throw the collected particles on to the frontal tracts of the adjacent filaments (Ansell, 1961). In *Mytilus,* the particles have further been described to adhere to the latero-frontal cilia, which thus seem to be sticky, and to be wiped off on to the frontal ciliary tracts (Wallengren, 1905; Tammes and Dral, 1955). Presumably, the often well-developed gill musculature is of importance in regulating the porosity of the gill filter by adjusting the width of the interfilamentary slits. Longitudinal muscle fibres along the gill axis may regulate the length of the gill and thus the distances between the filaments; and contraction of muscles within the filaments may shorten and broaden these, thus reducing the space between the filaments. Also the degree of blood-filling of the gills must influence their porosity (Elsey, 1935; Atkins, 1943; Purchon, 1954; Nelson, 1960). The interfilamentary slits should be narrow enough for effective straining of food particles from the water, and large enough for the unrestricted activity of the lateral cilia, thus maintaining a sufficient flow of water through the gills.

On touching the surfaces of the filaments the particles stimulate the numerous mucus glands to secrete. The larger and the more numerous the particles, the more copious is the secretion produced (Kellogg, 1915; Nelson, 1960). Mucus thus plays an important role in the feeding process. The particles that are caught on the gill surface may now be subjected to a sorting process performed by a variety of different mechanisms, as described above. A common feature of all sorting devices is that small particles or particle masses are preferentially sorted out and carried to the labial palps and the mouth along protected pathways, whereas large particles or particle masses are transported along superficially situated tracts, whence they can easily be deposited on the mantle and eventually rejected.

The correctness of this picture of feeding in bivalves has been strongly questioned by MacGinitie (1941), who used an improved technique for observing the feeding process in adult bivalves. Glass-covered windows, through which gills and palps could be watched during feeding, were made in the shells of the animals under experiment. MacGinitie found that disturbances or mutilations of the bivalves seriously interfered with the normal feeding process. The animals operated on, therefore, had to be given sufficient time for recovery, usually long enough to begin regeneration of the shell and mantle, before their feeding behaviour became normal again. During the observations the animals were not removed from their aquarium or disturbed in any other way. Observations of feeding were made on four different forms of bivalves, viz. the burrowing *Schizothaerus nuttallii*, the surface form *Pecten circularis*, the above-surface form *Ostrea lurida,* and the open-coast form *Mytilus californianus.* MacGinitie found that when feeding undisturbed, the bivalves covered the entire gill surfaces by a sheet of mucus. The mucus was secreted either at the upper edges of the gills or more or less uniformly over the entire gill surface. The mucus sheet was carried ventrally by the frontal cilia and transported in strings along the ventral margins of the demibranchs to the labial palps. The palps performed a selective function by partially removing undesirable particles. The rest of the material passed intact with the strings of mucus direct into the oesophagus. According to MacGinitie it is therefore not the gill proper but the sheet of mucus that is responsible for the straining of food particles from the water, just as is the case in, e.g. filter feeding prosobranchs.

When disturbed, or when undesirable material such as carmine or carborundum powder was added to the water, the bivalves were seen to cut off the mucus sheet at the dorsal edges of the demibranchs. However, even under such conditions particles might be caught by the gills and transported to the palps, where they were generally rejected. MacGinitie is of the opinion that the ingestion of particles that may take place in disturbed or even mutilated bivalves is abnormal. He therefore disregards all observations that have been made on feeding in bivalves that are disturbed, or even have been cut open.

If bivalves are feeding only when the gills are covered by a mucus sheet, the elaborate sorting devices of the gills cannot be considered as essential parts of the normal feeding mechanism, since these sorting devices presumably cannot function when a sheet of mucus is being secreted. From an admittedly teleological reasoning it is, however, unlikely that the complicated sorting mechanisms of the bivalve gill should not play a role in the feeding process. It seems, therefore, necessary to attempt to unite these contrasting views on feeding in suspension feeding bivalves by assuming that feeding by means of mucus sheets occurs in water that contains only small amounts of suspended material, whereas when much suspended material is present in the water no mucus sheets are secreted, so that the sorting mechanisms can function. The sorting devices of the gills should then represent adaptations ensuring that feeding can go on even in the presence of excessive amounts of material in the water, for instance, stirred-up bottom material. Under such conditions the sorting mechanisms carry the bulk of the material collected on the gills along superficial tracts, which may turn out to be rejection tracts. The bivalve is thus well protected against the deleterious effects of too large amounts of suspended material in the surrounding water. However, even when the gills are mainly engaged in cleansing themselves from excess material, the special food-collecting tracts may be able to select a fraction of the material that is relatively rich in small and light particles. This fraction must also be taken to be of a relatively high nutritional value, as food particles will probably be comparatively small and light particles. It should be remembered that digestion in bivalves is mainly intracellular, so that only small particles can be utilized as food (Yonge, 1931, 1937a).

In clear water, which contains only small amounts of plankton and other suspended matter, generally less than a few milligrams per litre (see p. 192), practically all material is probably ingested. Under such conditions sorting processes are apparently superfluous, and the coarse frontal cilia and superficial ciliary tracts of the ventral margins of the demibranchs can take over the function of carrying the food-loaded mucus sheet to the palps and mouth. The presence of the mucus sheet, of course, greatly increases the efficiency with which very small particles are retained by the gills. The paradoxical situation therefore arises that ciliary tracts that are mainly rejection tracts when much suspended material is present in the water are perhaps the only feeding tracts in clear water, whereas the main feeding tracts in turbid water are prevented from functioning in clear water.

These considerations, of course, presuppose that feeding by means of mucus sheets really exists and is of general occurrence. Even though there is no reason for doubting the correctness of MacGinitie's observations, it is to be regretted that he has published only a rather incomplete account of his extensive investigations on feeding in the four representative types of bivalves. Further experiments are greatly needed, also on such forms as the *Anomiidae* and *Arcidae,* where the marginal ciliary tracts beat posteriorly. The feeding mechanisms in filter feeding bivalves will probably not be finally understood until it has been observed, by means of MacGinitie's technique, how gills and palps function under conditions that imitate those prevailing in the natural surroundings. The problem of the "normal" functioning of the bivalve gill will be dealt with again under the discussion of the porosity of the gill.

As mentioned above, most investigators find that sorting of particles is performed according to size, shape and specific gravity of the particles, and not according to their value as food. However, it has been stated by some authors that qualitative selection can also take place (see Fox, 1936). In the oyster, Grave (1916) thus found that the composition of phytoplankton in the stomach content differed from that in the surrounding water. Similar findings were made by Buley (1936), who throughout a period of seven months regularly investigated the composition of net-plankton and stomach content of *Mytilus californianus* taken from the same water where the phytoplankton was sampled. He found, by taking the average for the whole period, that the percentages of dinoflagellates and diatoms in the water were 2·4 and 97·6, respectively, but 97·4 and 2·6 in the stomach content, which was interpreted as a pronounced preference in *Mytilus* for dinoflagellates. However, the dominating dinoflagellate in the stomach, *Prorocentrum micans,* largely passed through the plankton net which was used. The losses are stated to be usually "less than 90 per cent". Further, the dominating diatoms in the sea were very spiny *Chaetoceros* species which were hardly ever met with in the stomach content, probably owing to their large size and shape. The experiments, therefore, do not provide conclusive evidence for qualitative selection. Loosanoff (1949), however, offered such evidence. He observed that oysters, *Crassostrea virginica,* fed on a mixed culture of algae, contaminated with the purple bacterium *Cromatium perty,* would produce pseudofaeces which principally consisted of this bacterium, whereas the true faeces contained the other micro-organisms, but only small numbers of the purple bacterium. It was further observed that the oyster was able to exclude yeast cells from a mixture with other micro-organisms. In these examples selection probably cannot have been working on physical factors such as size and shape, but the selective mechanism is not clear, even when the above-mentioned observations (p. 81) of Menzel (1955) on mechanisms on the palps that are capable of sorting according to quality of particles suggest a mechanism which may be responsible. It is noteworthy that on continued feeding with the purple bacterium, some oysters eventually ingested it apparently without discrimination, whereas others continued to reject it even after several weeks' feeding.

The importance of qualitative sorting under natural conditions in the sea is not known.

1.101. ADAPTATIONS OF FEEDING MECHANISMS AND HABITS

As already mentioned, size, shape and fine structure of gills and palps vary widely within the suspension feeding bivalves. This variation seems to be related more to habitat and mode of life than to systematic position; gills and palps may differ greatly between closely related forms and agree in only distantly related forms. Within the Pholadidea, for instance, the gills of *Barnea parva, Pholas dactylus* and *Zirphaea crispata* belong to Atkins' Type C(2) (Fig. 1.33), the gills of *Barnea candida* to Type C(1c) and of *Pholadidea loscombiana* to Type C(1a); but *Barnea candida* and the rockboring veneracean *Petricola pholadiformis,* which may live side by side and thus occupy the same habitat, are remarkable for both having opposing ciliary currents on the filaments of the descending lamellae of the outer demibranch (Purchon, 1955c).

The amounts and particle size of suspended matter in the water are presumably important factors in determining the direction of evolution of structure and function of gills and palps. Strongly developed sorting mechanisms are typical of bivalves from sheltered waters of high turbidity, or from open coasts where many particles of large size may be present in the water (Atkins, 1937a; Purchon, 1955a, 1956; Nelson, 1960). In a study of Veneridae from different habitats, Ansell (1961) found large palps with numerous ridges and effective sorting in *Venerupis* spp. from the littoral zone with much stirred up suspended matter, but simple palps with relatively little sorting in sublittoral species living in water with only little suspended matter. Also in Lucinacea living in environments with little suspended matter the sorting mechanisms are reduced, and consequently larger particles accepted. Lucinacea are found in a variety of substrata from mid-tide levels to depths over 1,500m (Allen, 1958a).

Peculiar habits have been adopted by *Cryptomya californica* (MacGinitie and MacGinitie, 1949). Like the related *Mya* it is a deep-burrowing form, but in contrast to *Mya,* which is sedentary and reaches the surface of the bottom by long siphons, *Cryptomya* is a highly-active burrower with very short siphons; it is typically found deep in the bottom adjacent to the burrows of the shrimps *Callianassa* and *Upogebia* and the echiuroid *Urechis,* and with the siphons projecting into the burrows which are kept ventilated by their inhabitants. *Callianassa* is a deposit feeder and therefore no competitor for food to *Cryptomya,* but *Upogebia* and *Urechis* are suspension feeders, and *Urechis* is probably at least as efficient as *Cryptomya* in straining particulate matter from the through-current. As *Cryptomya* inserts its siphon in the burrow behind *Urechis,* no food particles large enough to be retained by *Cryptomya* are probably present in the water as long as *Urechis* is feeding. This may be the reason why *Cryptomya* is much more abundant around the burrows of *Callianassa.* Yonge (1951a) found that *Cryptomya* has large gills which may be correlated with paucity of food in the habitat, but no measurements of the size of the water current produced through the gills are available.

Differentiations of other structures than gills and palps may be of importance in the adaptation to life in water of high turbidity. In oysters, the species of the Genus *Ostrea* do not tolerate turbid waters, which are however invaded by species of the Genus *Crassostrea.* This difference in tolerance towards high amounts of suspended matter in the surrounding water has been attributed mainly to the separation in *Crassostrea* of the right mantle lobe from the visceral mass resulting in the formation of the promyal chamber. This chamber increases strongly the volume of water that can be taken into the dorsal part of the mantle cavity, water which presumably can be used for cleansing of this part of the mantle cavity by rapid contraction of the adductor muscle. In the flat ostrea-type of oysters, the narrow parts of the dorsal mantle cavity may be vulnerable to the encroachment of mud and sand, since little water is available for flushing the region (Elsey, 1935; Nelson, 1938, 1960; Yonge, 1960).

In some primarily suspension feeding bivalve groups, feeding on suspended matter has been supplemented by feeding on other types of food; suspension feeding may even be subordinate, and in some groups it has been completely given up. In the Tridacnidae living on coral reefs symbiotic unicellular algae, zooxanthellae, live in blood cells in the highly specialized mantle edge; they apparently constitute an important food source for the host (Yonge, 1936). The Teredinidae and Xylophagidae nourish on the wood in which they are boring (Potts, 1923), but they may supplement their diet with suspended food from the respiratory current. Proteins in the suspended matter especially may be of nutritive importance to the woodboring bivalves, since the substrate may be deficient in essential aminoacids (Lasker and Lane, 1953; Drisko and Hochman, 1957); and *Teredo* apparently digest cellulose directly and not indirectly by means of bacteria that might serve as a source of essential and other aminoacids (Greenfield and Lane, 1953). Deschamps (1957) analyzed the total content of organic nitrogen in *Teredo navalis* and in an amount of wood corresponding to that removed by the boring activity of the animal, and found that the wood representing the maximal amount eaten contained less than half the nitrogen in the animal itself. Significant amounts of organic nitrogen therefore seem to be obtained from the suspended matter in the respiratory current. Woodboring *Xylophaga* have also been found boring in the gutta-percha sheath of submarine telegraph cables and have, therefore, been assumed to be able to feed exclusively on suspended matter in the water. However, Knudsen (1961) points out that the specimens found in the cables might have lived on the plant fibres used for isolation beneath the gutta-percha sheaths.

The burrowing Tellinacea have abandoned suspension feeding and use the long and flexible inhalant siphon for sucking bottom material into the mantle cavity, where the material undergoes sorting similar to that of the suspension feeding bivalves. The adaptations to deposit feeding in Tellinacea have especially been studied by Yonge (1950).

The specialized Septibranchia are feeding on large food (Yonge, 1927). However, according to Purchon (1958a, 1959) they have not evolved from suspension feeding ancestors, but from primarily deposit feeding forms related to the Protobranchia.

1.102. VELIGERS AND EARLY METAMORPHIC STAGES

Most bivalves with a pelagic development pass a veliger stage which feeds on suspended particles in the water in a way similar to that of the veliger stage in prosobranchs; the cilia of the velum direct food particles towards ciliary tracts leading to the mouth (Yonge, 1926a; Cole, 1937). On metamorphosis, the method of feeding must, therefore, change fundamentally. The greater part of the velum disintegrates, but the upper mouth palps apparently develop from its apical region (Cole, 1938; Ansell, 1962). The palps grow very rapidly and take over the function of the velum as the food collecting organ; in the early stages of metamorphosis the food collecting mechanisms of the gills have not yet developed, but the gill rudiments produce a water current which impinges directly on the palps (Ansell, 1962). Also the cilia of the mantle may contribute to the creation of the food current, while the gill filaments are still rudimentary (Nelson, 1938).

1.103. EFFICIENCY OF PARTICLE RETENTION

As already mentioned the bivalves possess a number of mechanisms by which they may presumably vary the porosity of the gills. It is therefore not surprising that widely varying results have been obtained in investigations on the efficiency with which particles of various sizes are retained by various bivalves.

In the oyster *Crassostrea virginica*, Galtsoff (1928a) found that diatoms and dinoflagellates were effectively retained, whereas 70–90 per cent of *Bacterium coli* escaped the gill filter. According to Loosanoff and Engle (1947), however, even cells of the flagellate *Euglena* of 60 μ could easily pass through the gills, sometimes only 15 per cent were strained from the water, and maximum retention was 80 per cent. The percentage of the 5 μ cells of *Chlorella* sp. retained by the gills varied from 0 to 85; usually more than 50 per cent passed through the gills. Little correlation was found between the size of different micro-organisms and the percentage of cells removed by the gills. Moreover, the efficiency of retention was independent of the cell concentration even at concentrations so low that the rate of water transport was unaffected. Jørgensen and Goldberg (1953), however, in measurements of the rates at which graphite suspensions were cleared, found that particles of sizes down to a few microns were effectively retained, whereas 80 per cent or more of 1–2 μ particles passed through the gills. The difference in results probably arises from differences in technique. The measurement of water transport by means of the direct methods used by Galtsoff, Loosanoff and their colleagues implies a separation of the inhalant and exhalant currents, achieved by wrapping round the oyster a piece of dental rubber dam to form a cone completely enclosing the exhalant aperture. The efficiency of retention found by means of the clearance method probably approximates more closely the natural efficiency of the oyster gill than does the other methods used.

This conclusion is apparently not supported by the following observations of Nelson (1960). He removed one valve without cutting any blood vessels or nerves, and kept specimens thus treated of three species of oysters, *Ostrea edulis, Crassostrea virginica* and *C. angulata* in an apparently normal condition in the running water from the tanks of the Plymouth Marine Biological Laboratory for over two months. In the clear water, the ostia between the gill filaments were fully open and the gills relaxed. On dropping a fine suspension of carmine or plankton upon the surface of a demibranch he observed that all passed through the ostia. However, upon repeated applications of suspended matter, the ostia and interfilamentar spaces would restrict until finally none of the particles passed through. According to this description, the undisturbed oysters in clear water should not be prepared for feeding, the wide-open ostia letting through most particulate matter of suitable size as food to the oysters, but preparation for feeding should be induced by a sufficient increase in the concentration of particles in the water. However, another interpretation seems possible. It might be, as suggested p. 85, that in water containing relatively little suspended matter the porosity of the gill is not determined by the gill proper, but by a mucus sheet covering the gill surface, as described by MacGinitie (1941). In the undisturbed oyster in clear water the gill may, therefore, be relaxed with wide open ostia and still efficiently retain small particles. However, when the water contains high concentrations of particles, filtering by means of mucus sheet is abandoned, and efficient straining is, therefore, only possible upon restriction of the interfilamentar space and ostia.

In the mussel *Mytilus edulis,* Tammes and Dral (1955) observed widely varying retention of particles less than 30–40 μ. Thus, the retention of 7–8 μ blood corpuscles, and other particles of similar sizes, varied from 0 to 98 per cent; and from 0 to 50 per cent of 1·5–2·5 μ flagellates. However, in measurements of clearances of particles of various sizes, *Mytilus edulis* was found capable of efficiently retaining particles down to about 1 micron (Jørgensen, 1960; Jørgensen and Goldberg, 1953). Again the difference in results may be due to differences in technique: in Tammes and Dral's experiments, the mussels were furnished with a plastic tube in the exhalant aperture and embedded in plaster of Paris. It therefore appears that only undisturbed mussels efficiently retain particles down to about 1 μ; when exposed to conditions that are too artificial, the gills become more, but variably, porous.

Lasaea rubra cleared suspensions of a number of algae and other particles with equal rates independent of the size of particles, ranging from 1 to 50 μ. *Lasaea,* therefore, efficiently retain particles down to 1 μ. Also, the freshwater bivalves *Sphaerium, Musculium* and *Pisidium* presumably efficiently retain about 1 μ particles, since these bivalves were found to grow on a diet of pure bacteria at the concentrations prevailing under natural conditions (Rodina, 1948b).

In *Pecten irradians,* Chipman and Hopkins (1954) measured the clearances of ^{32}P labelled algae. The experiments suggested that the efficiency of retention might vary, but often high and constant clearances were found, with no difference in values for the flagellate *Chlamydomonas* sp., 7 μ in size, and the diatom *Nitzschia closterium*, 56 μ in length. *P. irradians* therefore seems to be able to efficiently retain particles down to at least 7 μ in diameter. In *Pecten latiauratus,* Jørgensen (1960) found a low, but varying retention of colloidal graphite particles. However, the scallops survived for only about a month in the laboratory, and they were probably very soon after arrival adversely affected by the laboratory conditions; further experiments are needed on scallops in order to determine the porosity of the gill of undisturbed animals.

So far, the presumably normally functioning gill of a number of bivalves was found to retain efficiently particles down to about one or a few microns. It is, perhaps, likely that under natural conditions suspension feeding bivalves, generally, will extract almost completely all particles from the water passing the gills down to the sizes of bacteria and other minute organisms. However, in *Mytilus edulis* large molecules such as lobster haemocyanin largely passed through gills that simultaneously efficiently retained 1 μ particles (Jørgensen, 1960). Also *Lasaea rubra* retained inefficiently the large molecules of *Arenicola* haemoglobin (Ballentine and Morton, 1956), but in these experiments particles just above the critical size were not present in the suspensions simultaneously with the haemoglobin, so that it is not definitely known whether or not the presence of the haemoglobin itself changed the porosity of the gills.

In experiments on the uptake of aminoacids at low concentrations by various marine invertebrates, Stephens and Schinske (1961) measured the gylcine concentrations in water drawn from the exhalant and inhalant siphon of *Spisula solidissima*. The ratio of the concentrations varied between 0·86 and 1·03, but on an average about 3 per cent of the glycin was found to have been removed from the through-current. Present evidence, therefore, suggests that colloidal or dissolved organic matter in the surrounding water may be only slightly accessible even to undisturbed, normally feeding bivalves.

It is not yet possible in any detail to correlate the experimentally found porosities of bivalve gills with the structures and mechanisms which are determining these porosities. Presumably, passage through the gills of large particles means that the interfilamentar spaces are kept correspondingly wide, whereas retention of such particles or even smaller ones indicates that the effective spaces have been reduced in width; but it is not finally agreed which are the straining structures when small particles are efficiently retained by the gills. However, the rather sharp size limit which could be found between retainable and non-retainable graphite particles in *Mytilus* and *Crassostrea* (Jørgensen, 1949; Jørgensen and Goldberg, 1953) indicates that it is the size of particles, and therefore the porosity of a filter, that is responsible for the retention, and not primarily adhesion to the latero-frontal cilia or adsorption to mucus on the gill, e.g. due to the electrical charge of the particles. The graphite particles used in the experiments were positively charged and might therefore have been adsorbed to the gill mucus, as are apparently polyvalent inorganic cations like strontium and calcium in *Mytilus* (Fretter, 1953; Rao and Goldberg, 1954). Yet, if adsorption takes place it can be of only minor importance for the retention of the graphite particles.

The critical size for retention of particles in *Crassostrea virginica* corresponds to the distance between the latero-frontal cilia, which according to Atkins (1936) is 1·5 to 3·7 μ. But in *Mytilus* the distance is also about 3 μ (J. Verwey, personal communication) and should, therefore, hardly allow anything like complete retention of 1 μ particles, as has actually been found. Verwey is of the opinion that the critical particle size for retention by the latero-frontal cilia in *Mytilus* is even as high as 6 μ, i.e. twice the distance between neighbouring latero-frontal cilia, since these beat intermittently.

As already mentioned some filter feeding bivalves have no large latero-frontal cilia, for instance Pectinidae, and Atkins (1938) questions that the micro-latero-frontal cilia of these forms do function as straining cilia. But also in *Pecten* constant low porosity of the gills can be established and maintained for some time, even though, as it seems, with greater difficulty than in forms with eu-latero-frontal cilia. It is, therefore, indicated that the gill filaments and their ciliary mechanisms cannot account for the straining efficiency of the gills when these are highly retentive, but a mucus sheet as suggested by MacGinitie (1941) might. Further determinations of retention of particles by undisturbed bivalves, with and without eu-latero-frontal cilia, are greatly needed, especially of particles ranging in size from a few microns down to less than one micron.

The bivalve veligers can only consume small particles, but since they are not filter feeders, probably no well defined minimum size exists below which food particles are not readily available. Measurements of width of the oesophagus indicate that only particles less than about 10 microns can pass into the stomach (Thorson, 1950). In the stomach of veligers of *Mercenaria mercenaria*, Carriker (1961) did not observe larger than 1–2 μ freshly consumed food particles, but the veligers of *Ostrea edulis* seemed to remove flagellates of 2–10 μ more efficiently from the surrounding water than those of 1–2 μ (Cole, 1937). The veligers of *Crassostrea virginica* were found to preferentially ingest the larger cells of *Chlamydomonas* for the small cells of *Porphyridium* (Loosanoff *et al.*, 1955).

1.104. RATE OF WATER TRANSPORT

Rates of water transport have been determined repeatedly in suspension feeding bivalves, especially in oysters and mussels. A reliable technique was first introduced by Nelson (1935, 1936)

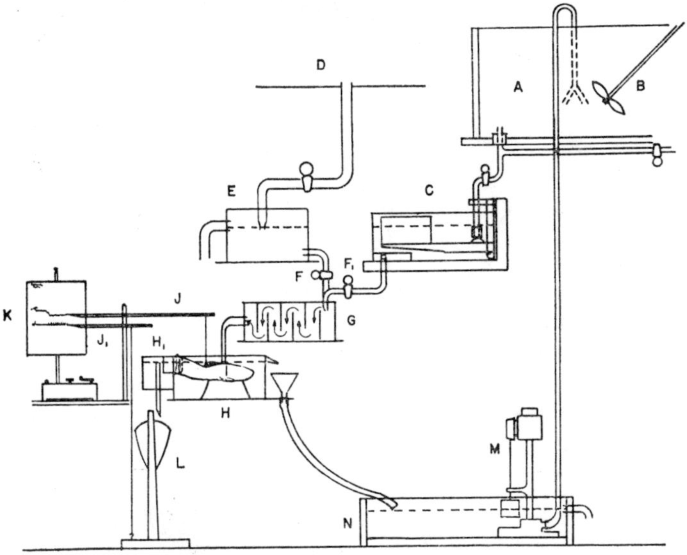

FIG. 1.38. Diagram of the apparatus employed to record the rate of water transport and shell movements of the oyster.
A, Aquarium; B, stirrer; C, constant level unit; D, sea water storage tank; E, overflow chamber; F and F_1, flow adjusting valves; G, mixing trough; H and H_1, two compartments of the oyster chamber; I, oyster; J., lever recording shell movements of the oyster; J_1, lever recording movements of the tripping vessel L; K, kymograph; L, tripping vessel; M, pump; N, basin. (After Loosanoff and Engle, 1947.)

in investigations on *Crassostrea virginica*. He used the so-called apron method (Moore, 1910) in combination with Galtsoff's (1926, 1928a) constant-level tank method. Apparatus of this type (Figs. 1.38, 1.39), in which the through-current is directly measured, have been used extensively in various modifications

(Galtsoff, 1946; Loosanoff and Engle, 1947; Collier and Ray, 1948).

Estimates of rates of water transport from the rates at which the animals clear the surrounding water of suspended matter were made already by Viallanes (1892) and Dodgson (1928). In bivalves, the principle was first used as a more exact method by Fox *et al.*

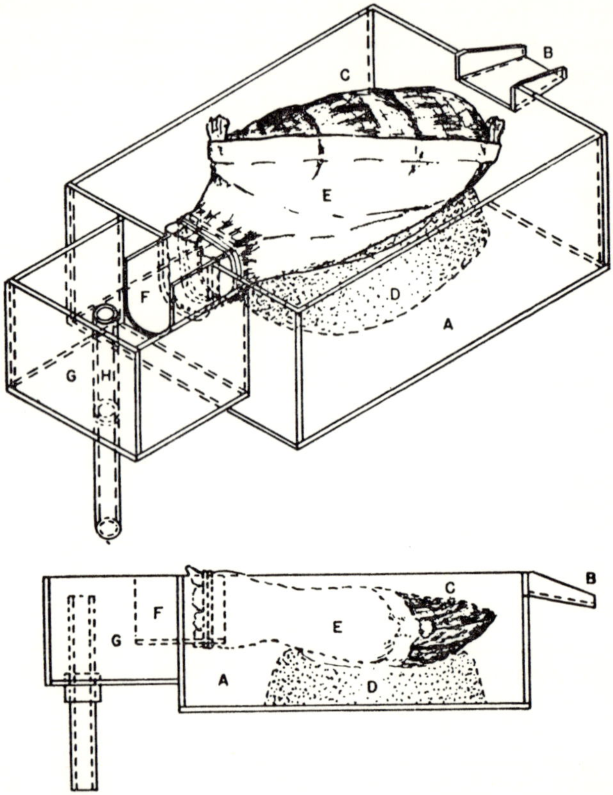

FIG. 1.39. A, Oyster chamber (Compare Fig. 1.38); B, overflow for excess water not transported by the oyster; C, oyster; D, concrete base on which the oyster is lying; E, rubber apron enclosing the exhalant aperture; F, trough to which apron is tightly attached and into which the oyster transports water; G, overflow chamber; H, stand pipe adjusted in height to a level slightly above that of the level, B, of the oyster chamber, to prevent a siphoning effect from the oyster chamber to the overflow chamber. (After Loosanoff and Engle, 1947).

(1937), who used finely ground calcium carbonate in suspension. Jørgensen (1943, 1949a) used suspensions of unicellular algae and colloidal graphite, thus reducing the dry matter in suspension from the very high value of about 3 g/l. using calcium carbonate to the more natural values of a few mg/l. Before the method can be used in determining the total through-current, it has to be ascertained that the suspended matter is retained completely by the gills. The concentration of the suspended algae and other particles can be determined photometrically (Jørgensen, 1943), or by measuring the radioactivity if labelled algal cultures have been used (Jørgensen and Goldberg, 1953; Chipman and Hopkins, 1954).

A different method has been developed by Hersh (1960). He kept the bivalves in narrow aquaria in water to which was added suspensions of aluminum dust. The particles were 1 μ thick and 10–100 μ large. The movements of the particles in the inhalant and exhalant currents were recorded by multiple photographic exposure, the particles in the currents being photographed by means of reflected light. Measurements of the movements of the particles on the photographic exposures and of the size of the siphon openings allowed calculations to be made of the rate of the through-current. The great advantage of the method is that measurements can be made without disturbing the animals and with very little suspended matter in the water; concentrations as low as 20 particles per ml could be used.

Results of recent work on rates of water transport in bivalves are summarized in Fig. 1.40. All rates have been expressed per gram wet weight of the soft tissues: when size, but not weight, of the animal was stated in the original the weight has, where possible, been estimated. All experiments on *Crassostrea virginica* have been made on adult individuals; generally the weight is not stated, but is here assumed to have been 20g. The weights of *M. californianus* used in the experiments of Fox *et al.* (1937) were estimated from the relationship between length and weight found by Fox and Coe (1943); weights of *M. edulis* used by Jørgensen (1949) and Willemsen (1952) were estimated from the relationship between length and weight found by Jørgensen (1960).

It appears from the figure that large mussels have been found to transport water at much lower rates than oysters and scallops of comparable weight, and than small mussels. However, bivalves are very sensitive towards changes in their environment; even slight chemical or mechanical stimulation may cause partial or complete closure of the valves, and thus corresponding reduction or stoppage of the water transport. Obviously, experimental determinations of the water transport are not practicable without some deviation from the normal environment and some disturbance of the animals. Flow rates measured in the laboratory therefore tend to be lower than under natural conditions. Not all bivalves are equally sensitive towards the experimental conditions, but in the case of *M. edulis* the present writer has been able to secure normal water transport only in small individuals (Jørgensen, 1960). The low figures for large mussels recorded in Fig. 1.40 can, therefore, be suspected not to represent normal values.

The variations found in the rate of water transport in the

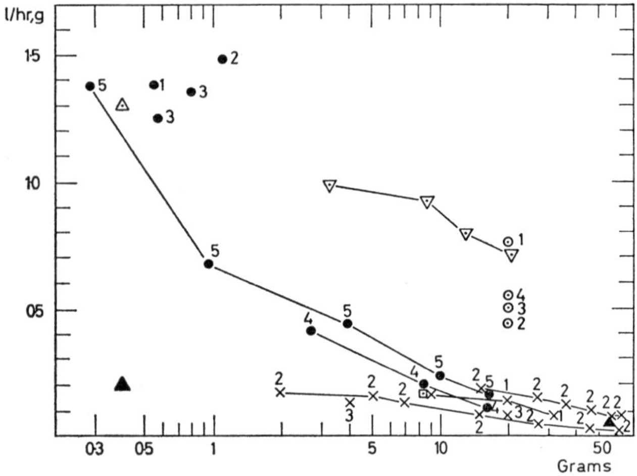

FIG. 1.40. Relationship between rate of water transport and wet weight of soft parts in various bivalves.

Crassostrea virginica

⊙[1] Loosanoff and Nomejko, 1946. Temp. 19–26°C. Water transport directly measured.

⊙[2] Loosanoff, 1958a. Temp. 16–28°C. Water transport directly measured.

⊙[3] Galtsoff *et al.*, 1947. Temp. 10–28°C. Water transport directly measured.

⊙[4] Jørgensen, 1952. Temp. 21–22°C. Clearance of colloidal graphite.

Ostrea edulis

△ Allen, 1962. Temp. 17–18°C. Clearance of *Phaeodactylum*. Wet weight estimated from dry weight, assuming 80 per cent body water.

oyster may to a great extent be due to the use of the assumed weight of 20g of all the experimental animals; within the accuracy of the determinations, the rates of water-transport were independent of the method used, whether direct (Galtsoff *et al.*, 1947; Loosanoff, 1958a; Loosanoff and Nomejko, 1946) or indirect (Jørgensen, 1952).

Amphidesma ventricosum transported water at low rates. During the experiments the animals were placed in a jar with sand; sea water rich in phytoplankton was added, and rate of removal of *Chaetoceros* colonies from the suspension was followed. The clearance presumably approximated the rate of water transport since the large *Chaetoceros* colonies are unlikely to pass through the gills. However, it is stated that only one out of several tests was successful, so more experiments are necessary to decide whether the calculated rate is normal (Rapson, 1954).

Mytilus edulis

●[1] Jørgensen, 1960. California. Temp. 14–15°C. Clearance of colloidal graphite.

●[2] Jørgensen, 1960. Maine. Temp. 13–14°C. Clearance of colloidal graphite.

●[3] Jørgensen, 1949. England. Temp. 16–19°C. Clearances of flagellates and diatoms. Weight estimated from Jørgensen (1960). Fig. 12.

●[4] Willemsen, 1952. Holland. Temp. 12–15°C. Clearance of silt. Weight estimated from Jørgensen (1960). Fig. 12.

●[5] Theede, 1963. Kieler Förde, Germany. Temp. 15°C. Clearance of graphite. Wet weight, estimated from dry weight.

Mytilus californianus

x[1] Fox, Sverdrup and Cunningham, 1937. Temp. 19–23°C. Clearance of calcareous mud or $CaCO_3$. Weight estimated from Fox and Coe (1943), Table I.

x[2] Rao, 1953. Temp. 10 and 20°C. Mussels from Los Angeles. Clearance of colloidal graphite.

x[3] Segal, Rao and James, 1953. Temp. 16°C. Mussels from the middle intertidal zone. Clearance of colloidal graphite.

Pecten irradians

▽ Chipman and Hopkins, 1954. Temp. 22–26°C. Clearance of flagellates and diatoms.

Amphidesma ventricosum

▲ Rapson, 1954. Clearance of diatoms.

Venus striatula

▼ Allen, see note to *O. edulis*.

Mya arenaria

◘ Allen, 1962, see note to *O. edulis*.

Ordinate: filtration rate in l./hr and g wet weight. Abscissa: wet weight of soft parts in g; logarithmic scale.

Also *Venus striatula* and *Mya arenaria* cleared suspensions of algae (*Phaeodactylum*) at low rates. The *Phaeodactylum* cells, which measure about 4 × 40 μ, are probably efficiently retained in the gills so that clearances to approximate rates of water transport. The initial algal concentrations, 30–60·10[6] cells/l., or about 1mg dry matter/l., were not unnaturally high, but during the experiments the animals were placed in containers without substratum. Also, Hersh (1960) kept his experimental animals in aquaria without substratum and found *Mya* and *Platyodon* to transport water at low rates. It is, therefore, not known whether the measured flow rates represent natural values or whether the experimental conditions were incompatible with normal activity. It is noteworthy that identical experiments performed with *Ostrea edulis,* which belongs to the epifauna, resulted in much higher clearances (Fig. 1.40). Indeed, all the high values of water transport are so far obtained in experiments on epifaunal or non-burrowing bivalves as oysters, mussels and *Pecten,* whereas burrowing forms have transported water at low rates. It would be of great interest to see whether this difference in feeding rates is typical of the habitats of the two groups, or is due to technical difficulties in securing normal activity in infaunal bivalves in the laboratory (see also p. 98).

Clearances have been measured in some other bivalves not included in Fig. 1.40. Thus, *Venerupis pullastra* cleared suspensions of iron oxide at low rates (D. B. Quayle, personal communication); and so did *Venus mercenaria* suspensions of a minute green alga and a small diatom (Rice and Smith, 1958). In neither instance was the efficiency of retention of the particles stated, so the rates of water transport are not known. The freshwater bivalves *Unio crassus* and *Anodonta piscinalis* cleared about 0·2 and 0·4l. water of suspended natural silt per hour. Weights of the animals are not stated, but *Unio* varied from 1 to 6cm in length and *Anodonta* from 7 to 9cm (Voskresiensky, 1959).

As in other ciliary suspension feeders, bivalves transport water at low pressures, but the maximum pressures measured in bivalves are relatively high, somewhat higher than in, for instance, sponges and ascidians, namely 7mm in *Crassostrea virginica* (Nelson, 1936), 5·5mm in *Mytilus edulis* (Tammes and Dral, 1955), and 5–17mm in *Teredo pedicellata* (Lane and Tierney, 1951).

The water transport through a bivalve is determined mainly by the activity of the lateral cilia, and by the position of the mantle edges and valves.

It has been assumed that the activity of the lateral and other cilia of the gill filaments of bivalves is controlled by specific substances. According to Bülbring *et al.* (1953) the substance may be acetylcholine. In the isolated gill of *Mytilus edulis* they found that the ciliary activity was stimulated by small amounts of eserine, whereas larger amounts inhibited the activity, which was also inhibited by atropin and d-tubocurarine. Furthermore, gill tissue was found to contain acetylcholine, choline acetylase and "true" cholinesterase; these observations all fitted the theory that ciliary activity may normally be controlled by acetylcholine liberated in the gill tissue.

A humoral control of ciliary activity was also indicated by the experiments of Aiello (1960) and Gosselin (1961); however, they find serotonin more likely to be the active principle. Gosselin (1961) found serotonin to be much more active than acetylcholine in stimulating the frequency of beating of the lateral cilia in *Mytilus edulis* and other bivalves. He also found that the gill tissue contains serotonin and the enzymes for its production and degradation. Gosselin suggests that serotonin acts as a local hormone produced by cells of the gill filaments, as did Bülbring *et al.* (1953) for acetylcholine, because the presence of nerves has not been demonstrated within the gills. Lucas (1931a, b) traced the distribution of the branchial nerves into the gills by means of serial sections, and he found that the nerves were limited entirely to the epithelial and connective tissues bordering the axis of the gills. He therefore concluded that the ciliated epithelium is not regulated by the nervous system, but that it acts as an autonomous system. However, after excision of the gills, the lateral cilia stop beating, but they begin to beat again after some hours. Aiello (1960) also found that the cilia temporarily stopped beating after section of the branchial nerves. He concludes that the branchial nerves have an activating role in ciliary movement, but the role of the nervous system in controlling the activity of the lateral cilia in bivalves seems not yet to be clear.

Little is known about the importance of variations in the activity of the lateral cilia on the rate of water transport through intact bivalves under natural conditions.

The importance of the movements of the mantle edges in determining the through-current was stressed by Hopkins (1933) and Nelson (1938), who showed that in oysters the edges can move in such a way as to stop the current, partially or completely, even when the valves are open. Nelson (1938) observed that when a heavy suspension of, for example, starch is fed to a water pumping *Crassostrea angulata* at any point along the inhalant area, the mantle border closes locally at the point of entrance of the suspension. In *Mytilus edulis,* Jørgensen (1960) tried to correlate the position of the mantle edges and the valves with the rate of the through-current. It was found that a small, but unmistakable reduction of the lumen of the inhalant aperture was followed by a reduction in the rate of water transport by some 30 per cent. A reaction confined to a slight change in the orientation of the tentacles presumably had a smaller effect; in undisturbed mussels, the tentacles along the mantle edges round the inhalant area are mostly turned outwards. In half-closed mussels, however, the rate of water transport was reduced to one-fifth of the rate of the normally open mussels.

1.105. LENGTH OF FEEDING TIME AND RHYTHMIC FEEDING

The water transporting activity of bivalves has often been stated to be rhythmic, or discontinuous, also in species or individuals living outside the tidal zone, and thus being permanently submerged. However, the most convincing evidence indicates that bivalves living under natural conditions are generally active most of the time. Thus, continuous registration of the movements of the shells of *Crassostrea virginica* has shown that this species keeps the valves open about 90 per cent of the time, when living under natural conditions in the sea. Loosanoff and Nomejko (1946) found that water was pumped at any time of the day, and freshly taken oysters contained food in the stomach irrespective of the tides or the varying light intensities of day and night.

It has been stated by Nelson and others (see Nelson, 1938) that the oyster is relatively inactive during ebb tide. According to Loosanoff and Nomejko it is possible that unfavourable conditions prevailed during ebb tide in the locality where Nelson made his investigations, and that this may explain the low intensity of feeding. Chestnut (1946) (cited from Korringa, 1952) observed that in Delaware Bay the oysters fed independently of the tides in the lower parts of the bay, but higher up the river, where the salinity was reduced during the ebb, food intake took place only during high tide. This agrees with Hopkins's (1936b) observations on the influence of salinity changes on feeding in the osyter (see below).

It should be noticed that the length of time during which the valves are kept open tends to be shorter in the laboratory than under natural conditions (Galtsoff, 1928a; Hopkins, 1931; Loosanoff and Nomejko, 1946; Galtsoff *et al.,* 1947). Even in the laboratory, however, it is stated that oysters seldom remain open for less than 10 hr/day. Hopkins (1931), for instance, reports that at optimal temperatures *Ostrea lurida* is open over 20 hr/day and *C. virginica* 10–14 hr.

Mytilus edulis living in natural surroundings feeds actively 97–99 per cent of the time, at suitable temperatures (Loosanoff, 1942). Under similar conditions *Venus mercenaria* remains open about 90 per cent of the time (Loosanoff, 1939).

Activity rhythms have been described in *Anodonta,* but the various authors are not in agreement. Koch and Hers (1943) found the ventilation of the gills to be regulated by the metabolic rate; at oxygen saturation in the surrounding water the valves were closed most of the time, and open only during short intervals. Barnes (1955) observed both slow and rapid rhythms in movements of the valves of animals kept in running water in laboratory sinks; still the valves were closed most of the time; in an example given about 90 per cent of the roughly 3-day period of registration. Koshtoyants and Salánki (1958), too, observed a slow rhythm under laboratory conditions, but the periods of activity lasted 2 to 4 days, and those of rest, with the valves closed, only 5 to 15 hrs. It is therefore possible that the freshwater bivalve *Anodonta* under natural conditions also transports water most of the time.

Rhythmic activities of complicated patterns have been described in oysters and quahogs kept in small volumes of water in which the valves were open for less than half of the time (Brown *et al.,* 1956). The results are therefore not directly applicable to the bivalves when living under natural conditions.

Rhythmic variation in activity without intervening periods of complete rest has been described by Rao (1954) in *Mytilus californianus* and *M. edulis,* which were found to vary the rate of water transport synchronously with the tidal cycles: high tides coincided with maxima in water transport and low tides with minima, as determined by clearances of colloidal graphite. The tidal rhythm persisted for several weeks in the laboratory. However, Jørgensen (1960) in experiments on *M. edulis* and using the same technique as Rao found no variation in water transport with the tidal cycle. A final explanation cannot be given on these contradictory results, obtained on *M. edulis* from the north-east coast of U.S.A. However, the experiments are only partially comparable. In Rao's experiments, the mussels were most likely transporting water at reduced rates, at least at rates considerably lower than in the experiments of Jørgensen in which the mussels probably filtered at about the rates of undisturbed animals in natural environments. Rao did not, therefore, investigate the relationship between the phases of the tidal cycle and the natural, unrestricted, water transport. The rhythm in water transport found by him may be secondary to a rhythm in the sensitivity towards those factors in the laboratory environment that were responsible for the low water transport; as the laboratory environment was constant throughout the experiments, it follows that such a rhythm in sensitivity must result in a synchronous rhythm in the measured clearances. Absence of a tidal rhythm in water transport was also found by Theede (1963) in *M. edulis* from the Baltic Sea and the North Sea.

1.106. ENVIRONMENTAL FACTORS INFLUENCING FEEDING ACTIVITY

A number of environmental factors are known that influence the activity of the gills and mantle or shells, and thus the flow rate, for instance, temperature, latitude and composition of the surrounding water.

Feeding takes place only within certain temperatures. In *C. virginica* Galtsoff (1928b) showed that the optimum temperature for the mechanical activity of the gills was about 30°C. Hopkins (1933, 1935) also found that maximum work was performed by the gills at this temperature. However, in intact oysters the optimum temperatures for water transport were around 20°C. At higher temperatures valves and mantle borders started to close, thus diminishing the diameter of the inhalant and exhalant apertures, so that the increase in activity of the lateral cilia did not result in a corresponding increase in flow rate. In fact, Loosanoff (1958a) observed no marked fluctuations in rate of pumping in *C. virginica* between 16 and 28°C. From the data presented in Figs. 61 and 62, and Table 52 in the paper of Galtsoff *et al.* (1947) it appears that the rates of water transport varied independently of the temperature within the even wider range of 10 to 28°C. As the rate of oxygen consumption increases with the temperature, rates of water transport in relation to oxygen consumption must fall proportionally. On average, the oxygen consumption in various species of oysters was found to be tripled when the temperature rose from 10 to 28°C, and doubled when it rose from 16 to 28°C (Gaarder and Eliassen, 1954; Pedersen, 1947; *Ostrea edulis*: Nozawa, 1929; *O. circumpicta*: Mitchell, 1914; see also Galtsoff and Whipple, 1931; *Crassostrea virginica*). Consequently, the feeding conditions should be the poorer the higher the temperature within the natural temperature range in which *C. virginica* lives; a result that invites further investigations.

Optimum temperatures for rates of water transport need not be the same throughout the whole range of distribution of the species. In *Mytilus californianus,* Rao (1953) measured the rate of water transport as a function of the latitude, by comparing the activity of specimens from three localities along the Californian west coast, namely, from Los Angeles (34°00′N), Fort Ross (38°31′N) and Friday Harbor (48°27′N). He found that the flow rate was greater at any temperature in mussels from higher latitudes than in those of the same weight from lower latitudes. Mussels from Friday Harbor showed the same pumping rate at 6·5°C as those from Fort Ross at 10°C, or from Los Angeles at 12°C. This means that the animals under natural conditions maintain about the same pumping rates in all localities, irrespective of the normal differences which exist in temperatures.

Normal temperatures are taken as 6·5, 10 and 16°C for Friday Harbor, Fort Ross and Los Angeles, respectively.

The differences in optimum temperatures for *M. californianus* from different localities need not be racial, at least not exclusively, because similar differences have been found in the flow rates within populations from different intertidal levels (Segal, Rao and James, 1953). Animals from the low intertidal level pumped, at the same temperature, faster than mussels from a higher level with a higher mean temperature, but slower than animals from about 10m depth with a lower mean temperature. In this case the different populations mix during reproduction, since the larvae are pelagic; the relation between activity and temperature that is characteristic for populations from different tidal heights is therefore established during the lifetime of the single individuals.*

In *M. edulis,* no difference was found in the rates at which graphite suspensions were cleared by mussels from the mid-tidal zone and from the lowest level of the tidal zone (Jørgensen, 1960).

The extent to which the rate of water transport is affected by the amounts and quality of suspended matter has often been discussed in the literature on feeding in oysters and other bivalves. It is widely agreed that the rate of water transport is largely independent of the concentration of particles when only small amounts of suspended matter are present, as in clear water, and also independent of the quality of the particles, i.e. whether they have food value or not. For example, in *Crassostrea virginica,* Lund (1957) found that 10mg/l. of suspended matter, Fuller's earth, did not measurably influence the activity, and a suspension containing about 5–10mg/l. of a mixture of Fuller's earth and chloroplasts caused hardly any production of pseudofaeces; practically all retained material was therefore ingested at these concentrations. However, at higher concentrations a reduction in the rates of water transport can be observed. Loosanoff and Tommers (1948) found that in *C. virginica,* 100mg/l. of silt, calcium carbonate, kaolin, or Fuller's earth would depress the rates of water transport by about 50 per cent, and 1,000mg/l. by more than 80 per cent. Similar results were obtained with suspensions of

* As pointed out by A. J. Southward, differences in the activity of animals from different tide levels need not entirely be a matter of temperature, but other factors in the habitat should be investigated too.

various algae or yeast (Loosanoff and Engle, 1947). It was further seen that ingestion of food particles was impaired or stopped by high concentrations of micro-organisms in the water, even when large amounts of pseudofaeces were produced. However, in nature *C. virginica* is known to be capable of feeding even at high concentrations of silt in the water. Nelson (1921) thus found oysters to continue feeding in water containing 400mg silt per litre (cit. Nelson, 1960), and Mackin (1962) found oysters to feed without any apparent disadvantage in waters containing up to 700mg/l. of silt. However, in these experiments the effect of the high loads of silt on the rates of water transport are unknown. *C. virginica* and other *Crassostrea* species are presumably exceptionally tolerant to high concentrations of suspended matter in the water, as the species are especially abundant in inshore coastal waters, and very often in regions in which the silt content of the water may run high. In contrast, the oysters of the genus *Ostrea,* are largely confined to clear waters [Yonge, 1960; Nelson, 1960 (see also p. 88)]. It might, therefore, be expected that in the species of *Ostrea,* water transport is more affected by high concentrations of suspended matter than in the species of *Crassostrea*; but in *Ostrea gigas,* Chiba and Oshima (1957) found no effect on clearances of suspended bentonite when the initial concentration varied from 50 to 1,000mg/l. Similar variations in initial concentrations of bentonite were also without effect on rates of water transport in *Mytilus edulis, Mercenaria meretrix* and *Venerupis semidecussata.* In the last species, even very low initial concentrations of bentonite did not result in higher clearances. However, in all the experiments the rates of water transport, as measured from the clearances of bentonite, seem to have been low. Thus, 34mm large *Mytilus* cleared 0·15–0·33l./hr, 55mm *Mytilus* 0·84–1·25l./hr, and 75–90mm *Ostrea* 0·24–0·81l./hr, or only about one-tenth the optimal values recorded for mussels and oysters of similar sizes in Fig. 1.40. The experiments, therefore, do not seem to be conclusive as regards the effect of increased concentrations of suspended matter on rate of water transport in the undisturbed bivalves.

We must be prepared to find that different species of bivalves are not equally sensitive to excessive amounts of suspended matter, but that the amounts tolerated will depend upon the habitats occupied. Perhaps, even the same species may show differences in tolerance towards the presence of suspended matter, depending upon the condition in their habitat, but comparative investigations on undisturbed bivalves are lacking.

The effect of salinity on the feeding rate in *Crassostrea virginica* was investigated by Hopkins (1936b). Normal pumping rates were registered at salinities ranging from 25 to $39^{0}/_{00}$. Adaptation to $20^{0}/_{00}$ was possible, but the activity was never as great as at higher salinities. Effective pumping was not observed below $13^{0}/_{00}$.

Loosanoff and Tommers (1947) studied the effect of changes in *p*H of the medium on *C. virginica*. Lowering of the *p*H from 7·75 to 6·75–7·0 was followed by a transient increase in pumping rate, which was later followed by a decrease to below the normal values. At a *p*H of 6·5 and lower the activity was always reduced, at *p*H 4·25 to only 10 per cent of that of the controls.

Sperm of *C. virginica* carries a substance, named diantlin, that stimulates the rate of water transport (Nelson, 1941a, b; Nelson and Allison, 1940). The substance acts primarily by relaxing the intraplical musculature, thus causing a separation of the filaments within each plica (Nelson, 1941a). It has also been observed to speed up the frontal cilia of the marginal grooves of the demibranchs, from which it might be inferred that perhaps also the water transporting lateral cilia are stimulated (Nelson and Allison, 1940). Sperm of *O. cristata* has also a strong action on the water pumping in *C. virginica,* whereas sperm of, for example, *Mya, Nereis,* or man is without effect.

It has been reported that natural sea water from the Gulf of Mexico contains a factor which is necessary for normal pumping in the oyster (Collier, 1959; Collier *et al.,* 1953). This factor could be measured photometrically with the *N*–ethyl–carbazole reagent which can be used for estimating minute amounts of carbohydrates (see p. 205). The concentration of the factor, expressed in arabinose equivalents, ranged from 2 up to 100mg/l. No oyster was observed to remain open at concentrations below 4·8mg/l. At higher temperatures the threshold was higher, rising to 12 mg/l. at 28°C. Above the threshold values a close correlation was found between flow rate and concentration of the carbohydrate factor. During the passage of water through the oyster from 5 to 15 per cent of the substance was removed.

The amounts of the carbohydrate measured by Collier *et al.* are very large, and larger than the total amounts of organic matter generally found to be present in sea water, viz. about 4–5mg/l.; even though much higher values of dissolved organic matter have been obtained in estuarine water, for instance, from Texas where the work of Collier has been performed (Wilson, 1961) (see Table 2; 19.A). Furthermore, if the oyster retains about 10 per cent of the carbohydrate factor, it might easily ingest 1mg or more of it for every litre of water filtered, or several times its total food requirements (see also p. 267).

The carbohydrate factor cannot be of general importance in the regulation of the water transport in *C. virginica* throughout its range of distribution; Long Island Sound is well known for its oyster beds, but no carbohydrate factor could be demonstrated in its waters throughout the greater part of the year (Wangersky, 1959); when demonstrable amounts were present, the concentration never rose to the values necessary to induce activity in the oysters studied by Collier and his colleagues.

1.11. ECHINODERMATA

A ciliary epithelium with mucus producing cells is of widespread occurrence within the echinoderms, and seems to be lacking only in the Holothurioidea. In the Crinoidea, which are typically suspension feeders, the ciliary tracts constitute important parts of the feeding mechanisms, whereas in the Asteroidea, Ophiuroidea and Echinoidea the ciliary systems serve to clean the surfaces of the animals of deposited matter. However, many examples are known where cleansing tracts have evolved into food collecting tracts, mostly of deposited food; in a few instances suspension feeding has been adapted as a supplementary food gathering mechanism.

The Crinoidea mostly inhabit deep waters, but some of the stalkless crinoids penetrate into shallow waters. One of these, *Antedon bifida,* survives laboratory conditions, and its mode of feeding has been observed (Gislén, 1924; Nichols, 1960). The undisturbed *Antedon* attaches itself to the substratum by its cirri and stretches its ten arms out horizontally. The pinnules, which are placed at distances of 1·0–1·4mm from each other along the arms, are spread out at almost right angles. Arms and pinnules bear tube feet in groups of three on each side of the food groove; each group consists of a short, a medium-sized and a large tube foot (Fig. 1.41). The large tube feet are stretched at

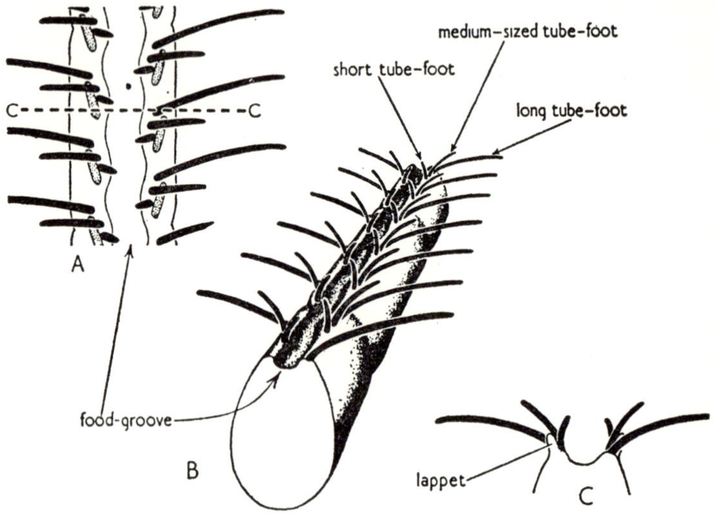

FIG. 1.41. *Antedon bifida.* Diagrams to show the arrangement of the tube-feet on a pinnule.
A, plan view of part of a pinnule to show the relationship between the tube-feet (silhouetted) and the lappets (stippled). B, perspective view, looking distally, of part of a pinnule, showing the angle at which the tube-feet are held during feeding. The lappets have been omitted for clarity. C, transverse section across the line CC in A, showing the angle at which the tube-feet are held in relation to the lappets during feeding. (After Nichols, 1960.)

right angles to the pinnules, so that the interpinnular space is bridged by the feet that serve the catching of food particles. The food-collecting area, therefore, corresponds closely to the area covered by the extended arms. *Antedon* is not a passive suspension feeder; food particles must be present in the water in order to stimulate the feeding activity. Addition of plankton to the water initiates rapid beating of the tube feet towards the food groove, apparently due to stimulation of sensory cells on the tube feet by touching food particles. The particles are entangled in mucus secreted by the tube feet and carried to the food groove by the shorter tube feet, which also make rapid bending movements towards the food groove. After a rapid bending the tube feet return more slowly to the "waiting position". In the food groove further mucus is added and the ciliary current carries the particle-laden mucus towards the mouth, which is kept wide open during feeding. In the non-feeding *Antedon* the mouth is a narrow slit, and the margins of the food groove are pressed together closing the groove.

Magnus (1963) has studied the feeding behaviour of another stalkless crinoid, *Heterometra savignyi,* by direct observations during diving of a population of animals living in shallow coastal waters of the Red Sea. The crinoids were found to spend the daytime inactive with coiled arms, but at sunset the arms uncoil and are oriented across the direction of the current with the aboral side turned against the current. If the direction of current changes during the night the orientation of the arms is readjusted accordingly. At sunrise the animals again stop feeding, roll up the arms, and hide in crevices if such are available.

It seems to be unknown how small particles may stimulate the tube feet to activity, and thus be caught and ingested; however, the fact that zooplankton organisms and crushed crab liver and other material stimulate feeding activity suggests that *Antedon,* and possibly other crinoids, may mainly feed on larger food particles than the majority of the suspension feeders. Analyses of content of intestinal tracts of crinoids from various parts of the sea also show that zooplankton organisms and large particles constitute an important part of the ingested materials (Gislén, 1924).

In the Echinoidea, suspension feeding is described by MacGinitie and MacGinitie (1949) in the sand dollar *Dendraster excentricus* living in the tidal zone. It stands on edge crosswise to the direction of the tidal flow, with about one-third of the test buried in the sand. Particles in the streaming water are caught on ciliary tracts of the surface of the animal, entrapped in mucus and carried to the mouth. Other sand dollars feed by similar mechanisms, but are apparently exclusively deposit feeders. In *Melitta,* mainly particles less than 20 μ were selected and carried along the acceptance tracts, a large part were only about 1 μ (Goodbody, 1960).

No suspension feeding Asteroidea seem to be known, but within the Ophiuroidea *Ophiothrix fragilis* and *Ophiopholis aculeata* may be predominantly suspension feeders (Wintzell, 1918). The long flexible arms can be raised from the substratum and exposed to the water currents flowing along the bottom. Particles may be caught by strong ciliary currents along the ventral side of the arms between the rows of arm spines. The mucus-entangled particles accumulate at the bases of the arms from where they are transferred to the mouth by means of tube feet (Gislén, 1924). The importance of suspension feeding in *Ophiotrix* and *Ophiopholis* has been substantiated by feeding experiments (Roushdy and Hansen, 1960). In an experiment where the animals were offered carbon-14 labelled cultures of *Sceletonema costatum,* the 4 μ large cells were efficiently removed from the water and ingested. *Ophiocomina nigra,* which also possesses strong ciliary currents along the arms, did not ingest phytoplankton.

Many Holothurioidea feed by moving the tentacles surrounding the mouth over the substratum; particles adhering to the mucus-secreting surface of the tentacles are swept off by pushing the tentacles down into the gullet. However, holothurians living attached to rocks, as *Cucumaria curata* and *Psolus chitonoides,* may extend their tentacles freely into the surrounding water and thus feed on suspended matter caught on the sticky surface (MacGinitie and MacGinitie, 1949). Also the pelagic holothurian *Pelagothuria* is a suspension feeder (Gislén, 1924).

1.12. TUNICATA

The tunicates are all marine forms; they are almost all exclusively suspension feeders, but the ascidians *Octacnemus, Hexacrobylus,* and some other deep-sea forms take larger prey, as indicated by the aberrant structure of the feeding organs and by the presence of e.g. copepods in the alimentary canal. *Octacnemus* seems capable of swimming, in contrast to other ascidians (Madsen, 1947; Millar, 1959).

1.121. ASCIDIACEA

The Ascidiacea include all the sessile tunicates. When feeding, they pass a current of water through the pharynx or branchial basket, which communicates with the atrial cavity through numerous ostia in the pharynx wall. The water current enters through the inhalant or branchial aperture and leaves through the exhalant aperture or atriopore. The pharynx serves for feeding as well as respiration. The structures specialized for effective feeding are similar in all ascidians (Fig. 1.42).

Ventrally the endostyle with numerous mucus glands extends almost throughout the whole length of the pharynx. Opposite the endostyle the dorsal lamina with its ciliary tracts conveys collected food particles into the oesophagus. Anteriorly the peripharyngeal grooves run on both sides of the pharynx from the ventral to the dorsal midlines. The ciliary tracts of the two grooves are continuous with the dorsal lamina. In the pharynx the ostia are arranged in transverse bands, with blood vessels enclosed in the separating bars. Longitudinal vessels are also present. In quite small ascidians the inside of the branchial sac is of rather simple structure, but in larger species complex structures generally project into the lumen. At the junctions between the transverse and longitudinal blood vessels proliferations develop, first in the form of bifurcate blind tubules that bend anteriorly and posteriorly, and fuse with the adjacent tubules. In this way secondary longitudinal bars with blood vessels arise. From these secondary bars, blood-filled papillae may project. The whole pharynx may be folded longitudinally in some species.

The inside of the pharynx is ciliated. The cilia of the papillae and the transverse bars transport towards the dorsal lamina. The cilia covering the lateral edges of the ostia are responsible for the water current through the animal.

Direct observations on feeding have been made on transparent species, or on specimens with part of the test and pharynx removed; probably the feeding mechanism is similar in all ascidians. The description of feeding most often met with in the literature can be summarized as follows (Fol, 1876; Roule, 1884; Orton, 1913; Hecht, 1918a; Plough and Jones, 1939; van Weel, 1940; Berrill, 1950). Mucus produced by the glands of the endostyle is lashed on to the inner surface of the pharynx by means of ciliary tracts that run throughout the length of the endostyle. Particles in the feeding current are entrapped in the mucus, which is carried around the pharynx towards the dorsal lamina. Apart from the endostyle, mucus may also be produced in small amounts by gland cells in the pharynx. According to this description the straining of particles from the water passing the pharynx wall is mainly performed by the ostia. Special functions in the feeding process have also been ascribed to the bars and papillae. Thus, Orton (1913) mentions that feeding in *Ascidia mentula* is assisted by the waving action of the longitudinal bars in a transverse direction, so that the food masses are pushed on as well as lashed by cilia towards the dorsal lamina. Hecht (1918a) observed in *A. atra* how particles retained in ostia were carried by cilia to the tips of the papillae, and there incorporated in strings of mucus. Waves of contraction brought two rows of papillae together, and by the action of the cilia the food cord passed from one row to the next.

This picture of feeding in ascidians has been modified by the work of MacGinitie (1939b), Werner and Werner (1954) and Werner (1959), who showed that in undisturbed transparent specimens of *Ciona intestinalis, Ascidia californica* and the colonial *Diplosoma pizoni* and *Clavelina lepadiformis,* the straining of particles was not performed by the ostia, but by continuous sheets of mucus produced by the endostyle, one on each side of the pharynx. These unbroken sheets of mucus are carried by the cilia of the bars, or papillae of the bars where such are present, and peri-pharyngeal grooves towards the dorsal lamina and there formed into a mucus string (Fig. 1.42). When disturbed, the animals cut off the secretion of the mucus sheet with the result that only larger particles are filtered by the ostia and entangled in the rather small amounts of mucus which may still be produced by the endostyle and branchial sac, especially in animals that are cut open for inspection. It thus appears that the feeding process observed by previous workers is not normal even when it results in the ingestion of particle-laden mucus.

MacGinitie's observations were confirmed by Jørgensen and Goldberg (1953), who found that colloidal graphite, added to undisturbed *Ciona intestinalis* or *Molgula manhattensis,* is completely retained by the branchial sac, whereas almost all graphite

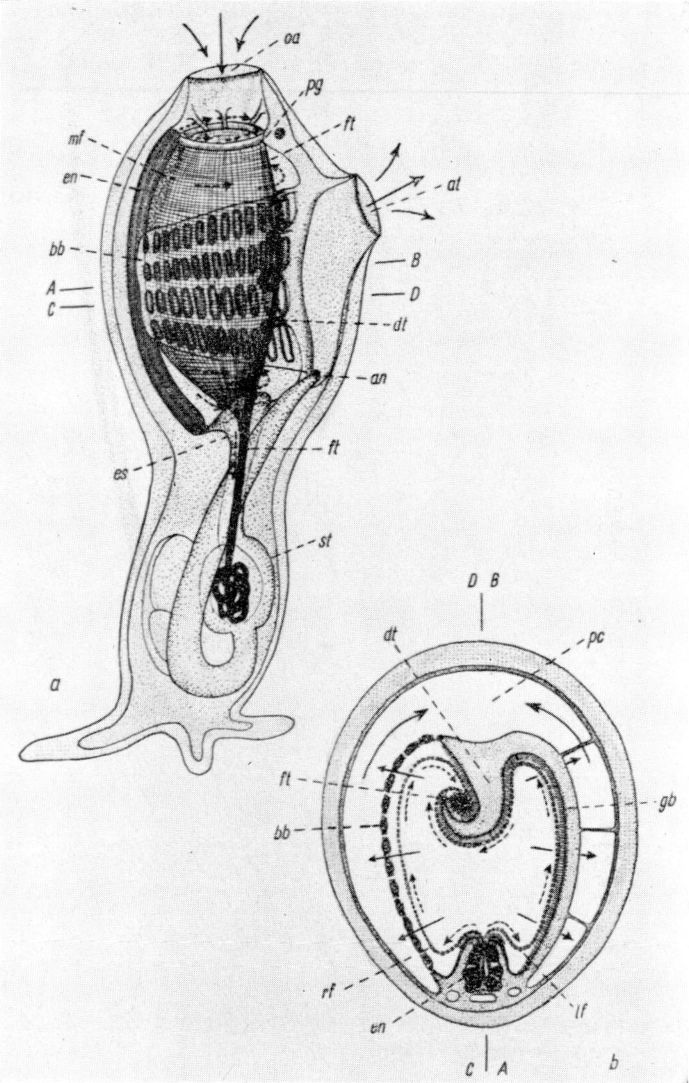

FIG. 1.42. *Clavelina lepadiformis.*
a, Lateral view of a young animal. b, Diagrammatic transverse section along the lines A to B and C to D. In the right part of the figure the section is at the level of one of the ciliary tracts which transport the mucus sheet (A to B); in the left part of the figure at the level of a row of ostia (C to D), through which the water is passing. an, Anus; at, atriopore; bb, branchial basket with ostia; dt, tongue of the dorsal lamina; en, endostyle; es, oesophagus; ft, food string; gb, gill bar; lf, left half of the mucus filter; mf, mucus filter; oa, oral aperture; pc, peribranchial chamber; pg, peripharyngeal groove; rf, right half of the filter; st, stomach. Large arrows indicate direction of water currents; broken arrows direction of transport of mucus filter. (After Werner, 1959.)

particles would pass through the ostia when the animals were disturbed. That a continuous mucus sheet was produced in the former case could also be demonstrated by addition of excess graphite. A contraction of the animals then occurred, and large flakes of graphite-blackened mucus were expelled through the inhalant siphon. Thus the sheets of mucus that are produced by the endostyle are stable structures, which do not deteriorate even when ejected by the violent contractions of the animals.

It has been taken for granted by most investigators that the feeding current is produced exclusively by the cilia lining the ostia. However, Hoyle (1953) has stated that the spontaneous rhythmic squirting observed in *Phallusia mammillata* and other monascidians is an integral part of their feeding mechanism. During the spontaneous squirting in *Phallusia* the branchial and atrial sacs expel synchronously about two-thirds of their contents at each squirt. Hoyle measured simultaneously the ciliary through-current and the rate of renewal of the water in the branchial sac by the squirting. He arrived at a ciliary current of about 1ml/min, against 1·5ml/min as renewal rate by means of squirting. It was further found that whereas the ciliary current was independent of the amount of the food present in the surrounding water, the rhythm of squirting increased during starvation to about twice the normal rate. Addition of food restored the normal frequency in about 2 hr. As, under experimental conditions, the squirting brought larger amounts of water into the pharynx than did the ciliary current, it was quite naturally assumed that squirting is of great importance in the normal feeding process. However, it will be shown below that the measured through-current is probably only a small fraction of the rate under normal conditions.

It is not stated how the particles taken into the pharynx as a result of the squirting become available as food to the animal. The mere intake of water presumably results in little deposition or adsorption of the suspended particulate matter on to the feeding mucus lining the wall of the pharynx, especially if the water has entered with the ciliary current produced by the action of the ostial cilia. The water expelled from the pharynx through the mouth during squirting therefore probably contains nearly the same amounts of suspended matter as that taken in, in which case the spontaneous squirting would be of negligible importance in feeding. Perhaps its function is to assist in bringing fresh water within reach of the inhalant current, and thus to avoid recirculation of water that has already passed through the filters of the same or neighbouring ascidians. This explanation fits with the observation that squirting is more frequent when only little food is present in the water; under such conditions it would seem especially important that the region of intake is kept free of exhaled water. In *Ciona intestinalis* and *Molgula manhattensis* it has been observed that the spontaneous squirting does not interfere with normal feeding. The mucus sheets are not dislodged from their position on the surface of the branchial sac. Water is, therefore, not exchanged to any degree between the atrial and pharyngeal cavities during squirting (Jørgensen, 1955).

The ascidians are mainly non-selective filter feeders, even though MacGinitie (1939b) observed that larger particles may be pushed out of the feeding mucus by means of special large cilia. The rejected particles are subsequently expelled by squirting. Further, the ring of tentacles that projects into the lumen of the inhalant siphon prevents the largest particles from entering. Slight mechanical stimulation of the inside of the inhalant siphon causes closure of the atriopore, thus stopping the through-current. Strong stimulation is followed by contraction of the whole animal (Hecht, 1918b; Bacq, 1935). This last-mentioned reflex is also elicited by over-loading of the filters, and it thus prevents over-filling of the intestinal tract. The strong contraction is preceded by closure of the exhalant aperture. Both the water of the atrial and pharyngeal cavities is therefore expelled through the inhalant aperture, carrying with it the particle-loaded mucus sheets. The reversal of the through-current under this type of contraction thus prevents the ostia from becoming clogged by the feeding mucus (van Weel, 1940; Jørgensen and Goldberg, 1953).

1.1211. EFFICIENCY OF PARTICLE RETENTION

The mucus filter of ascidians (*Ciona* and *Molgula*) completely keeps back graphite particles about 1 μ in diameter (Jørgensen, 1949b, 1952a). In contrast, a low degree of retention of blood proteins was observed. Thus, only a small percentage of haemoglobin was removed from the water during its passage through the mucus sheet. Similar results were obtained with haemocyanin from the gastropod *Haliotis,* whereas in the only experiment performed with crab haemocyanin 25 per cent was retained (Jørgensen and Goldberg, 1953).

Also vanadium, which is accumulated in blood cells in certain ascidians, is inefficiently retained by the mucus sheets; only a few per cent were found to be extracted from the water passing the pharyngeal wall of *Ciona* (Goldberg *et al.*, 1951).

In the absence of the mucus sheet, the smallest particles that can be efficiently retained depend upon width of the ostia and the length of the lateral cilia. The cilia are about 20 μ in length in a number of ascidians. Figures of ostia with cilia indicate that the width of the ostia can be about three times the length of the cilia, in which case probably only the larger phytoplankton organisms would not be able easily to pass the pharynx wall proper (Werner and Werner, 1954).

1.1212. RATE OF WATER TRANSPORT

The flow of water through ascidians has been measured by Hecht (1916), Jørgensen (1949b, 1952a), Goldberg *et al.* (1951) and Hoyle (1953). Hecht introduced a glass tube into the oral siphon of *Ascidia atra*. From measurements of the rate at which carmine particles were sucked through the tube the water transport was calculated to be about 80ml/hr/g wet weight, but the true value may be only half of this amount (Hoyle, 1953). The weight of the animals varied from 10 to 155g. The same rate of water transport was found in *Ciona intestinalis* by Goldberg *et al.,* who measured the water flow directly by means of the constant-level tank method.

Jørgensen (1949b) calculated from the rates at which water was cleared of graphite particles that the flow rates were 110 ml/hr/mg nitrogen in *Ciona intestinalis* weighing about 2·6g, and 150ml/hr/mg nitrogen in *Molgula* sp. weighing about 0·5g (16–19°C). Computed on a weight basis *Ciona* cleared 230ml/hr/g and *Molgula* 540ml/hr/g. The filtration rate was independent of the concentration of graphite particles in the water.

These results on flow rates in ascidians are in contrast to the finding of Hoyle, who measured a ciliary current of only 60ml/hr in *Phallusia mammillata*. As mentioned above the spontaneous rhythmic squirting in different ascidians which has been recorded by Hoyle, Hecht (1918b), Yamaguchi (1931), and others, was found to bring larger amounts of water into the pharynx than did the ciliary current. However, the measurements of the ciliary current are open to criticism. Rates for the through-current and for the renewal of pharynx water by squirting were determined by a method similar to that used by Wells and Dales (1951) for measuring ventilation of polychaete tubes. The apparatus registers only minimum values for the through-current, because the water is transported against higher pressures than under normal conditions. In *Ascidia atra,* Hecht (1916) measured the pressure that could be produced by the ciliary mechanism, by inserting a tube into the atrial siphon. The pressure varied from 1 to 2mm sea water. In Hoyle's experiments the pressure set up by the ciliary activity and against which water was transported was about 2mm. Presumably, therefore, the ciliary current measured by Hoyle in *Phallusia* is only a small fraction of the true through-current. On the other hand, the recordings of the volumes of water delivered during squirting were probably close to the true values, because in this case water was transported by strong muscular contractions, which produce pressures much higher than 2mm.

Furthermore, it can be shown that in, for example, *Ciona intestinalis* renewal of water in the pharynx by spontaneous squirting is of no importance in feeding. In *Phallusia* squirting resulted in a ventilation of the pharynx of about 90ml/hr. The ventilation is most certainly not greater in *Ciona,* which produced a ciliary through-current of about 3,000ml/hr (Jørgensen, 1952a), or at least 30 times the volume pumped in and out of the branchial sac by means of the rhythmic contractions.

1.1213. LENGTH OF FEEDING TIME

It seems generally implied that undisturbed ascidians are feeding continuously, but intermittent feeding has also been indicated. According to Carlisle (cited by Bone, 1958a) several species of ascidians occasionally show both inhalant and exhalant aperture open with no water current passing through. I have not observed such behaviour in healthy undisturbed *Ciona* or *Molgula*; when tested, expanded animals with open apertures have always been found to be producing mucus filters and maintaining a constant through-current. But systematic investigations of a possible intermittence in feeding in the Ascidiacea are lacking.

1.122. THALIACEA

The pelagic, mostly oceanic, Thaliacea include the orders Pyrosomida, Doliolida and Salpida (Berrill, 1950). The feeding organs of the Pyrosomida are similar to those of the Ascidiacea, but their mode of functioning seems not to have been studied.

In Doliolida feeding has been described by Fol (1872, 1876) and Fedele (1923) and reviewed by Neumann (1935). The ostia-bearing pharyngeal wall is reduced and is present only posteriorly to the endostyle, so that the wall cannot support and transport the feeding mucus. The function of supporting the feeding mucus is taken over entirely by the peripharyngeal grooves. The mucus is produced within the tube-formed endostyle and carried forward to the peripharyngeal grooves from which it projects posteriorly as long threads into the pharyngeal cavity.* The convergent mucus threads often form a funnel-shaped net, the tip of which is continuously drawn into the oesophagus. Mucus is produced by the endostyle in varying amounts; production seems to be stimulated by food in the water. Differences in the intensity of production apparently determine whether a sheet-like structure or only scattered mucus threads are available for filtration of food particles suspended in the water. The through-current is produced by the cilia of the ostia and, perhaps mainly, by the rhythmic contractions of the eight bands of muscles, which completely encircle the body. At the same time as forcing a current of water through the interior, the muscular contractions cause the individual to move forward (Berrill, 1950).

The feeding process in Salpida is similar to that of Doliolida. However, the water current is produced exclusively by rhythmical contractions of the circular muscles of the body wall. Atrial and buccal circular muscles function as valves which permit passage of the water only in one direction (Fedele, 1933; Carlisle,

* Also in the young stages of ascidians, until the pharynx walls are sufficiently developed, the feeding mucus is carried forward to the peripharyngeal grooves (Werner and Werner, 1954). A forward transport of mucus has even been described in adult ascidians and amphioxus, especially in publications that appeared prior to the investigations of Orton.

1950). A pharyngeal wall with ostia is absent. As in Doliolida, feeding may be slow or fast, depending upon the amounts of particulate matter in the water. A high concentration of particles independent of their value as food stimulates the rate of water transport and the production of feeding mucus. During fast feeding the mucus produced by the endostyle forms a complete funnel-shaped net extending backwards from the peripharyngeal grooves. The net is continuously carried backwards, especially by the cilia of the oesophagus, which twist the mucus net into a string (Fig. 1.43).

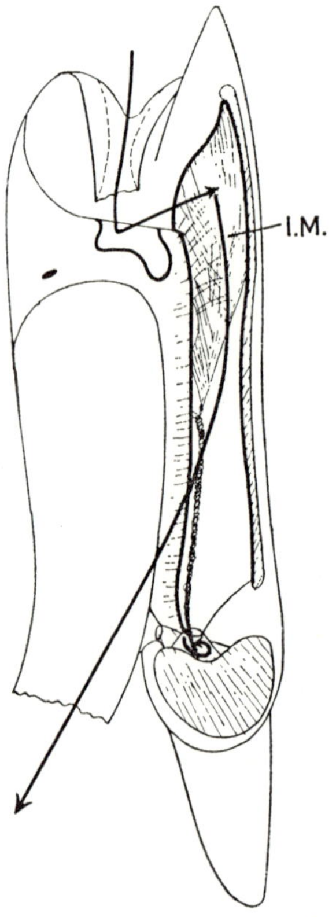

FIG. 1.43. *Salpa africana-maxima,* seen from the right side: diagram of rapid feeding. The arrows indicate the direction of the through-current which passes the mucus filter (I.M.). The endostyle is shown on the right of the mucus filter. (After Carlisle, 1950.)

1.123. LARVACEA

The feeding mechanism of appendicularians, except Kowalevskaiidae, is of the familiar tunicate type. According to Fol (1872) mucus produced by the endostyle is carried to the peripharyngeal tracts and formed into a funnel-shaped feeding net that extends posteriorly towards the oesophagus. The apex of the net is twisted into a string which continuously enters through the oesophagus. A feeding current through the pharynx is maintained by the activity of the ciliated ostium or spiracle, which connects the posterior part of the pharynx with the exterior.

In Kowalevskaiidae the endostyle and the peripharyngeal tracts have disappeared, but the feeding current passes special ciliated structures on the pharynx wall, in which structures

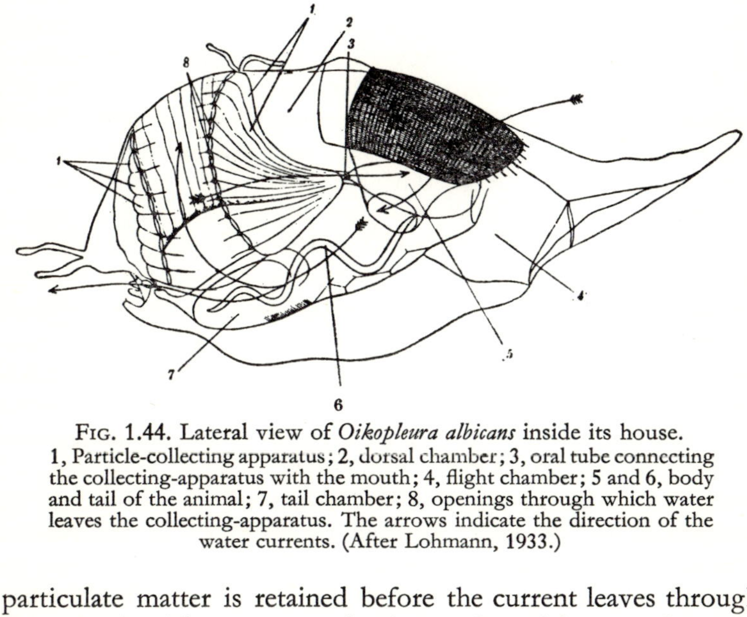

FIG. 1.44. Lateral view of *Oikopleura albicans* inside its house. 1, Particle-collecting apparatus; 2, dorsal chamber; 3, oral tube connecting the collecting-apparatus with the mouth; 4, flight chamber; 5 and 6, body and tail of the animal; 7, tail chamber; 8, openings through which water leaves the collecting-apparatus. The arrows indicate the direction of the water currents. (After Lohmann, 1933.)

particulate matter is retained before the current leaves through the spiracle. The exact mechanisms of particle retention and ingestion are not known (Brien, 1948).

It was believed by Fol that the appendicularians, like other tunicates, inspire the surrounding water directly. However, through the well-known investigations of Lohmann it was shown that before the water enters through the mouth a concentration of food particles can take place in a special structure secreted by the oikoplast epithelium of the body. Only in *Oikopleura* has the function of the structure, "the house", been investigated in detail (Lohmann, 1899a, b, 1909, 1933; Körner, 1952).

The houses of *O. albicans* is shown in Fig. 1.44. The arrows indicate the direction of the through-current which is produced by undulating movements of the tail of the animal. Water enters the house through a funnel, the wide external opening of which is covered by a filter. This filter, which is absent in some Oikopleuridae, allows only particles of less than about 35 μ in diameter to pass, varying with the species and size of the house (Lohmann, 1903). The whole or part of the through-current passes a particle-collecting apparatus before leaving through the narrow exit of the house via the dorsal chamber. The expulsion of water through the exit is responsible for the forward movement of the house.

The collecting apparatus has been described as consisting of a ventral and a dorsal layer of parallel running tubes; the two layers are separated by a membrane. Fine fibres are stated to cross the lumen of the tubes of the dorsal layer throughout their length; Lohmann (1899a) termed this part of the collecting apparatus the trap (Reuse) (Figs. 1.45 and 1.46). The tubes of both layers and from both sides open medially into a common channel, one end of which is closed and the other connected with the inhalant aperture or mouth of the animal. The tubes of the ventral layer open laterally into the tail chamber, and the tubes of the dorsal layer into the dorsal chamber. The hydrostatic pressure in the tail chamber produced by the movements of the tail forces food containing water into the ventral tubes, whereas food depleted water leaves by the dorsal tubes. Eventually the food particles suspended in the water become concentrated in the median channel, but how is not entirely clear. Most particles accumulate in the channel *below* the "filter" of the dorsal tubes; particles that do enter this "filter", the trap, are not accessible as food to the *Oikopleura* (Lohmann, 1909, 1933). Therefore, the concentration of the food particles cannot take place by filtration by means of the trap as widely assumed (see for instance Jørgensen, 1955; Jordan, 1913; Nicol, 1960; Yonge, 1928). Körner (1952) speaks of the accumulation of particulate food both by the filtering action of the trap and by sedimentation of particles in the median channel: when the mouth is closed "tritt im Hauptkanal (median

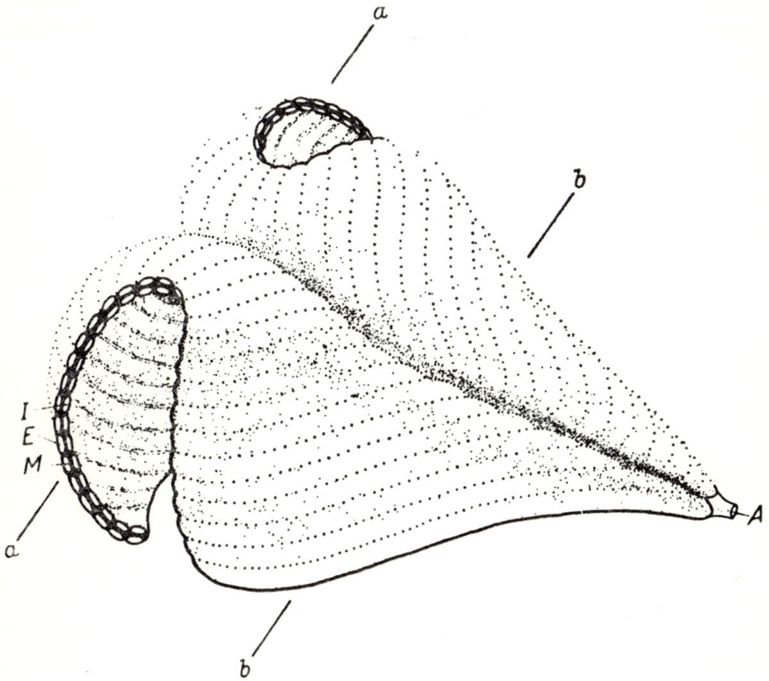

FIG. 1.45. *Oikopleura dioica*. Diagram of particle-collecting apparatus. I, inhalant apertures; E, exhalant apertures; M, membrane which separates inhalant and exhalant parts of the collecting-apparatus; A, tube to which the mouth can be attached. (After Körner, 1952.)

channel) fast ein Stillstand des einströmenden Wassers ein, und ein grosser Teil des Nannoplanktons sedimentiert schon hier in dieser kurzen Zeit auf den Boden des Hauptkanals". It is, however, not easily understood how conditions for effective sedimentation of suspended particles can be established in the median channel. If all water entering the ventral tubes passes the channel before leaving by the dorsal tubes, the rate of water flow across the channel cannot be very much slower than in the tubes, since the cross sectional area available to the water current is apparently not much increased in the median channel. Furthermore, the water coming from the two sides meet in the ventral channel (see Fig. 1.46). The resulting water movements presumably do not favour sedimentation.

However, if the membrane separating the ventral and dorsal tubes were permeable to the water, but not to the suspended particles, the collecting apparatus would concentrate the particles

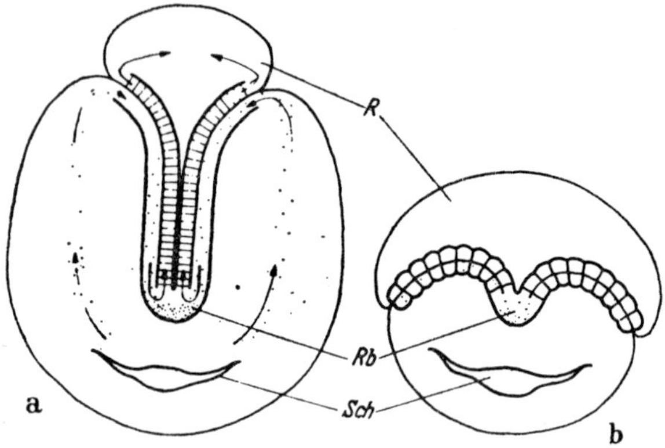

Fig. 1.46. *Oikopleura dioica.* Diagrammatic transverse sections through the collecting-apparatus.

a, Section across the line a–a in Fig. 1.45; b, across the b–b. R, Dorsal chamber; Rb, part of the collecting-apparatus from which the concentrated suspension of particulate matter can be sucked into the pharynx of the animal; Sch, transverse section through the tail in the tail chamber. The arrows indicate the direction of the water currents. (After Körner, 1952.)

as they travelled up the ventral tubes. The energy to the concentrating process would be supplied by the hydrostatic pressure forcing water into the ventral tubes. If the process were operating, water flow in the median channel would be slow, or might even stop if the animal discontinued feeding on the content of the channel. It is therefore of interest to know the fine structure of the membrane separating the ventral and dorsal tubes. Mr. E. J. Fjerdingstand is making an electron microscopic study of the structure of the collecting apparatus in the house of *Oikopleura* sp. and has found the membrane to be a net-like structure, built of thick longitudinal and fine transversal fibrils. The distances between the thick fibrils are about 0·8 μ, and between the thin fibrils about 0·1 μ; i.e. the membrane is freely permeable to water

FIG. 1.47. *Oikopleura* sp. Electron microscopic section of the net-like membrane which separates the incurrent and excurrent parts of the collecting-apparatus, fixed in 1 per cent potassium permanganate. The meshes are about 0·8 × 0·1 μ large. (E. J. Fjerdingstad, unpublished.)

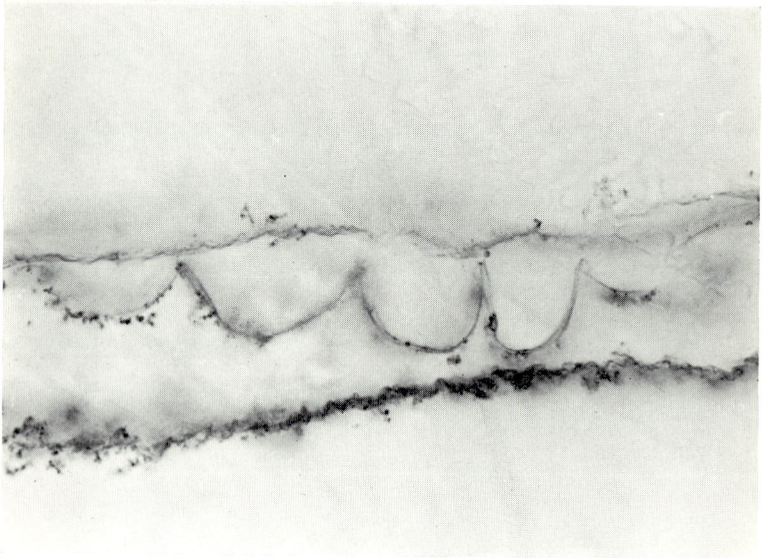

FIG. 1.48. *Oikopleura* sp. Transverse section through the incurrent and excurrent parts of the collecting-apparatus, fixed in 1 per cent potassium permanganate. It is seen that the incurrent and excurrent parts are not built of separate tubes, but are narrow continuous spaces, only separated by the folding, highly porous membrane. Excurrent space above, incurrent below the folding membrane. Distance between the dorsal and ventral walls about 30–40 μ. (E. J. Fjerdingstad, unpublished.)

(Fig. 1.47). It further appeared that the ventral, incurrent, portion of the collecting apparatus is not built of tubes, but is one narrow slit. The dorsal, excurrent, portion is separated from the ventral by the folding median membrane, which subdivides the dorsal portion into a system of parallel tubes, as described by Lohmann and Körner. However, the tubes are presumably not completely separate, and furthermore no fibrils have been found to cross the lumen (Fig. 1.48).

It thus seems that the median membrane in the collecting apparatus acts as a filter which prevents particles larger than about 0·1 μ in diameter from passing through, and that these particles are subsequently sucked into the median channel and further into the pharynx with the current arising from the activity of the cilia of the spiracle. The extent to which the food particles become concentrated in the feeding current is not known. Lohmann estimates the concentration of particulate matter to have increased 20–30 times in the median channel, but apparently on irrelevant evidence (Lohmann, 1899a, p. 358). The animal can presumably regulate the intake into the pharynx by raising or sinking the lower lip of the mouth, which projects into the lumen of the median channel. The intake is stated to be intermittent, but no details are given about length of feeding time or about the possible co-ordination between the activities of tail and pharynx in feeding (Lohmann, 1899a, 1933; Körner, 1952).

The particles in the suspension sucked in from the median channel are normally caught in the feeding mucus, but if the water contains non-acceptable materials, reversal of the direction of the current has been observed, so that the pharynx content is expelled through the mouth. The reversal of the current is due to reversal of the beat of the cilia of the ostium (Lohmann, 1909); in contrast to most tunicates, *Oikopleura* lacks muscles in the body wall to contract the pharynx.

The presence of obnoxious material, for instance carmine particles, in the water may also lead to stoppage of the production of feeding mucus, so that the particles will leave through the ostum. The endostyle is thus as important in the feeding process of most appendicularians as it is in other tunicates, even when the organ is reduced in size. The reduction is apparently related to the enrichment of the feeding current that occurs in the collecting apparatus.

The role played by food uptake outside the house is uncertain. In the laboratory, food has been found to be taken up by animals without the house (Körner, 1952); and Lohmann (1899a) assumed that *Oikopleura* may also in nature feed outside the house. He suggested that house building may be induced by low levels of food particles in the water, whereas at high levels feeding may take place by directly transporting the sea water through the pharynx. But in a later publication (Lohmann, 1933) he is doubtful whether normal feeding is possible outside the house. Support for feeding directly on the surrounding water was offered by the finding in the intestinal tracts of appendicularians from plankton samples of diatoms too large to pass the filter at the inhalant aperture of the house (Körner, 1952).

The species of Fritillariidae and Kowalevskaiidae also build houses, but of types that differ greatly from one another and from that of Oikopleuridae (Fol, 1872). Presumably the houses are of importance in the collection of food, but their mode of functioning is only poorly understood.

1.13. ACRANIA

The feeding organs of the Acrania are of the tunicate type, but the function of the organs seems to differ in important points from their function in the tunicates. Feeding has especially been studied in amphioxus (*Branchiostoma lanceolatum*); when feeding, it lies buried in the bottom sand or gravel with the ventral side turned upwards and with the mouth opening free of the bottom, exposed to the water above. A current of water is maintained through the animal by the activity of the lateral cilia of the branchial bars. Water enters through the mouth and leaves the atrial cavity through the atriopore. The mouth is surrounded by movable and sticky cirri, which bend inwards during feeding, thus preventing the coarser particles in the water from entering with the current. The buccal cavity contains the wheel organ, which consists of ciliated grooves. Particles in the incoming water may be caught in the wheel organ, carried back towards the peripharyngeal bands and further on to the dorsal groove, which extends to the oesophagus (Fig. 1.49). The endostyle in the ventral part of the pharynx is similar in structure to the ascidian endostyle (Barrington, 1958). The opening between the buccal cavity and the pharynx can be adjusted in size by the velar muscles. On passing the opening, the water is brought in close contact with the velar

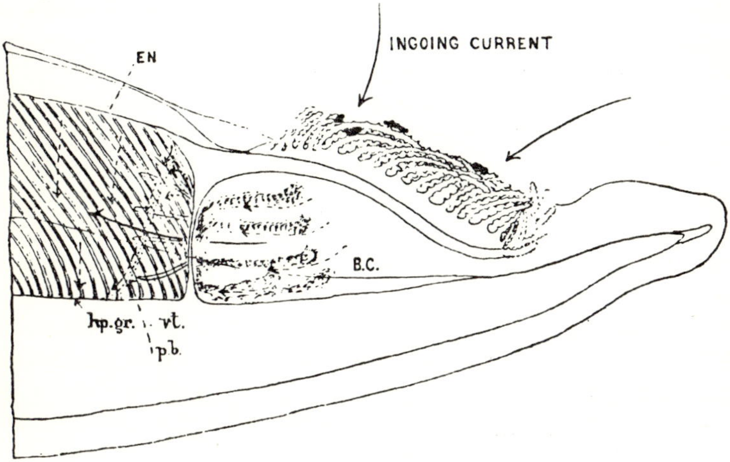

FIG. 1.49. Lateral view of the anterior end of Amphioxus. B.C., Buccal cavity; EN., endostyle; hp.gr., dorsal groove; p.b., peripharyngeal band; v.t., velar tentacles. The broken arrows indicate the direction of the transport of food particles along the ciliated grooves of the wheel organ, along the peripharyngeal band, and across the pharynx. Large arrows indicate direction of the water current. (After Orton, 1913.)

tentacles which normally hang across the current. Both cirri and velar tentacles are well equipped with tactile receptors, and probably also with chemoreceptors (Bone, 1961). In the pharynx, the water enters between the branchial bars into the atrial cavity, which completely surrounds the pharynx. The floor of the atrial wall contains cross-striated muscles which normally contract as a whole, expelling water from the atrial cavity. The exhalant opening, the atriopore, is surrounded by musculature by means of which the size of the opening is controlled.

The functioning of the pharynx in feeding was not correctly understood until the investigations of Orton (1913). Previously it was stated that food particles in the through-current were caught by the endostyle and carried anteriorly to the peripharyngeal bands. Orton observed, however, that the ciliary tracts of the endostyle were beating transversely, not longitudinally. Their function is not to catch particles in the through-current, but to carry the mucus produced by the endostyle on to the ciliary tracts of the internal surface of the pharynx. These tracts then transport the mucus with its entrapped particles to the dorsal groove. Orton figures the mucus with embedded particles as separate, often cylindrical masses (Fig. 1.50). He states, however,

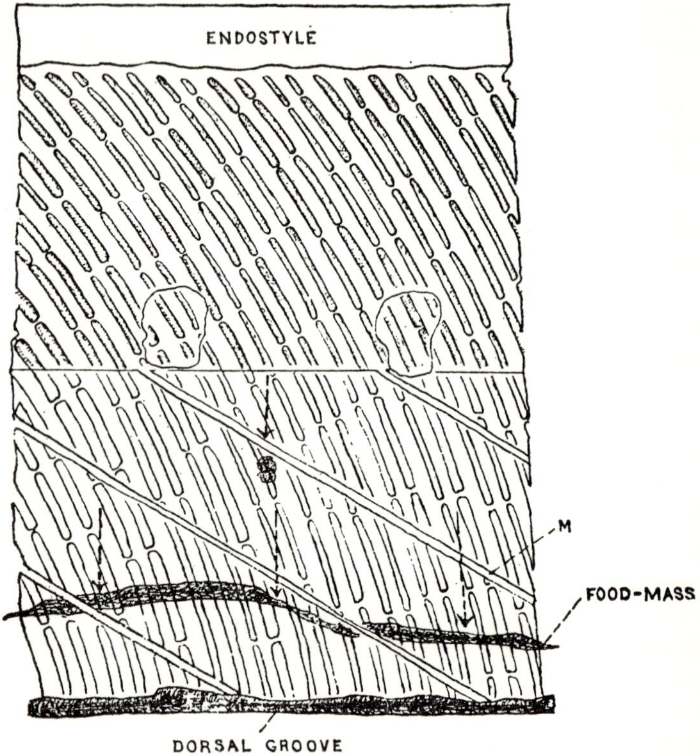

FIG. 1.50. View through the body wall and pharynx of Amphioxus. M., Boundary of adjacent myotomes. Arrows indicate the direction of transport of mucus masses with imbedded food particles across the pharynx wall. (After Orton, 1913.)

that "the particles become caught in thin sheets of mucus". Two such small sheets are also depicted in the middle of the figure. These sheets are apparently converted into thread-formed masses during the further transport dorsally. It is close at hand to suggest that in the undisturbed normally feeding amphioxus the endostyle does not deliver the mucus in the form of small separate mucus sheets, but that a continuous sheet of mucus may be produced and transported as such across the branchial wall to the dorsal groove, as in normally feeding ascidians. The mucus production of the endostyle is probably under nervous control (Bone, 1961), but only little seems to be known about the effect of various experimental conditions on the function of the endostyle and especially on the form in which mucus is delivered and carried across the pharynx wall. Feeding in amphioxus was more recently investigated by van Weel (1937), who however makes no reference to the work of Orton. According to van Weel the catching and transport of particles in the pharynx take place only in the region of the dorsal groove and the endostyle. Particles are assumed to be directed against dorsal groove and endostyle by means of the lateral cilia of the gill bars.

Amphioxus possesses well developed mechanisms for protection against clogging of the mouth and pharynx when much suspended matter is present in the water entering the inhalant aperture. Clogging of the buccal cirri with particulate matter results in closure of the atriopore and a rapid contraction of the muscles in the atrial floor starting at the atriopore, and forcing water out of the mouth (Dennell, 1950; Bone, 1961). A similar reflex can be elicited by mechanical stimulation of the velar tentacles, for instance, caused by large suspended particles. The reversed water current may dislodge and remove material deposited on the mouth cirri and other places. According to Bone (1961) this "cough" reaction is evoked specifically by stimulation of the anterior sensory system, resulting in a cleansing of the buccal region of the animal. But clogging of the pharynx walls by retained particulate matter may also elicit a forcible ejection through the mouth of water contained within the atrial and pharyngeal cavities, thus removing the material interfering with the unrestricted passage of water through the gill slits (Dennell, 1950). Whether or not a cleansing reflex results from clogging of the pharynx seems to depend upon the type of the clogging material; thus the expulsion of water was not seen in graphite suspensions (Bone, personal communication). In this amphioxus differs from *Ciona* (see above).

Feeding in amphioxus seems to be less automatic and stereotype than in most other suspension feeders, the tunicates included. According to Bone (1958a, 1961, and personal communication) amphioxus can distinguish between suspensions of food particles as algae and of inorganic particles as colloidal graphite. Even in dense suspension of the inert graphite particles the water was transported at maximum rates, as when no or few particles are present in suspension. Graphite particles embedded in mucus were observed continuously to pass through the animal. In suspensions of algae, however, the food did not pass directly through, but accumulated in the digestive canal. When a sufficient amount had been collected, water transport and thus feeding was discontinued for at least 8–9 hr. Feeding was first resumed when the food previously collected had been digested. Furthermore, the rate of water transport through the pharynx was found to depend upon the concentration of suspended food particles, decreasing with increasing concentration. Amphioxus can regulate the rate of the water transport by adjusting the size of the inhalant and exhalant apertures and the rate of beating of the lateral cilia, which are under nervous control.

In algal suspensions, the through-current slowed at regular intervals, usually about every 10–16 min, and the animal contracted the muscles of the atrial floor, forcing water out of the mouth. Feeding was then almost immediately resumed at its original rate. These rhythmic contractions recall those described in feeding ascidians, but their functional significance is not finally revealed.

No data seem to be available on absolute amounts of water filtered, or on the efficiency with which particles of various sizes are retained in the filters of amphioxus or other Acrania.

1.131. FEEDING IN ACRANIATE LARVAE

Some species of Acrania have pelagic larvae, in other species the larvae mainly live on the bottom, but may migrate towards the surface during the dark hours (Wickstead and Bone, 1959). Feeding has been studied in the early amphioxus larva by Bone (1958b, c); it starts after the first gill slit has developed ventrally. The body surface of the larval stages is covered by cilia, which produce currents of water along the body, from head to tail. At the same time the young larva is moving slowly forward in the water. In the older larva the action of the cilia, which persists until metamorphosis, does not suffice to move it. The lateral cilia of the gill slit, or slits in later stages, produce a current of water which enters through the mouth and leaves through the gill slit. Particles in the current may pass out again by the gill slit; but the usual course is that they stick to the peripharyngeal band, from where the mucus-embedded particles move backward over the gill slit and pass into the intestine.

When unpalatable substances as carmine suspension were added to the water, the gill slit was found to contract and close so that the current of water through the pharynx stopped. The gill slit thus serves the function of the atriopore in the metamorphosed amphioxus. Temporary closure of the gill slits and thus cessation of food uptake may be a regular feature in normal feeding. Bone observed that once amphioxus larvae in a suspension of algae had filled the hind gut they would stop feeding, apparently until the gut content was digested. However, even at algal concentrations much higher than in nature over 11 hr. were taken to fill the hind gut. The larvae would therefore seem to have been feeding at low rates. Investigations of gut content of acraniate larvae from the plankton or immediately above the bottom indicated that feeding is also intermittent under natural conditions. The gut of the migrating larvae of *Branchiostoma belcheri* mainly contained bottom dwelling algae, indicating that the larvae feed only when on the bottom. The gut of the pelagic larvae of *B. malayana* were generally almost empty (Wickstead and Bone, 1959).

1.14. DISCUSSION AND CONCLUSIONS

1.141. TYPES OF SUSPENSION FEEDING

Most suspension feeders feed by passing the surrounding water through structures acting as filters that retain suspended particles with efficiencies depending upon size and shape of the particles. In other suspension feeders, the water with its content of suspended particles is not truly filtered, but is carried along surfaces capable of retaining particles that obtain contact with the surfaces. Presumably, secretion of mucus is of importance in retaining the suspended particles, and ciliary activity in renewing the water current along the particle collecting surfaces, as well as in carrying the collected particles towards the mouth; but much remains to be learnt about the mechanisms of particle retention in the various groups of suspension feeders belonging to this feeding type. The members are often referred to as whirlers, "Strudler" in German, a term which may be used for lack of any better.*

Suspension feeders belonging to some large phyla or classes in which suspension feeding predominates are filter feeders, e.g. Porifera, Tunicata, and probably Bivalvia, but also other classes contain filter feeders, e.g. Gastropoda (many Prosobranchia), Echiuroidea (*Urechis*), Polychaeta (*Chaetopterus*), Crustacea (Phyllopoda, Cladocera, Copepoda, Cirripedia, Euphausiacea). Non-filtering suspension feeding is characteristic of several small phyla as Brachiopoda, Entoprocta, Polyzoa and Phoronida (van Gansen, 1960), but elsewhere mostly encountered within groups where suspension feeding is of more erratic occurrence, e.g. Coelenterata, Rotifera, Opistobranchia, Polychaeta, Echinodermata, and others. The suspension feeding pelagic larval stages of the bottom invertebrates are also whirlers.

1.142. EFFICIENCY OF PARTICLE RETENTION

Presumably, the efficiency with which the whirlers retain suspended particles touching the particle collecting surfaces does not decrease abruptly at a certain minimum size of the particles, but

* As mentioned, the various authors often do not distinguish between filtering and non-filtering suspension feeders, but use the terms "suspension feeding" and "filter feeding" as synonyms, according to whether emphasis is on state of food or on mode of food uptake, e.g. Jordan and Hirsch (1927), Jørgensen (1955), van Gansen (1960). Also, the term "whirlers", has been used as almost synonymous with "suspension feeders", to include typical filter feeders as sponges, tunicates, copepods and cladocerans, e.g. by Jordan (1913), in one of the first attempts to subdivide the animal kingdom according to feeding types, and by Remane (1952) who distinguishes between "innere Strudler" as bivalves and amphioxus and "aussere Strudler" as bryozoans and some polychaetes. The partition of the suspension feeders into "filterers" and "whirlers" adopted here agrees on the whole with that of Werner (1959); but his subdivisions are based on other criteria: in whirlers the production of the water currents and the food uptake are performed by the same organ, in filterers by separate organs. Werner consequently ranges the Cirripedia amongst the whirlers.

no systematic studies seem to have been made on the relationship between size, and other properties of particles, and clearance in whirlers. In the filter feeders, however, the porosity of the filters mostly sets a well defined minimum size for particles that are efficiently retained from the water passing through the filters.

Also an upper size limit exists, above which particles are not accessible by filter feeding. This limit may be determined by screen-like structures preventing larger particles from entering the feeding organs.

In several groups of filter feeders, the porosity of the filters can vary according to the experimental conditions, e.g. in bivalves and tunicates. Previous findings of highly, but variably porous filters in such forms turned out not to be characteristic of undisturbed feeding, during which the filters were found to be more retentive. In normally feeding tunicates and filter feeding prosobranchs the low porosity of the feeding organs depends upon the formation of continuously secreted, "endless", mucus filters covering the pharyngeal walls and gills, respectively. Such mucus sheets are perhaps also produced in Acrania. Mucus filters have further been described in *Chaetopterus variopedatus* and *Urechis caupo*. The minimum size of particles retained by the mucus filters is not precisely known, and may vary between filters. *Crepidula* even produces two filters, one coarse and one fine, working in series. Generally, however, mucus filters seem efficiently to retain particles down to less than 1 μ. It is a matter of discussion whether a continuous mucus filter can also be formed in bivalves, or whether the particle retention is always performed by the gill proper. The porosity of the "filter" of even undisturbed bivalves seems to be somewhat higher than that of the mucus filter of, e.g. tunicates, the smallest particles efficiently retained being about one, or even a few, microns large. In the Porifera (Spongillidae), the whole system of incurrent canals acts as a filter, the ultimate filtration of the through-current being performed by the collars of the choanocytes. The collars are built of parallel fibrils placed about 0·2 μ apart. Particles larger than this size must therefore be efficiently retained. Filtration by means of setae as met with in Crustacea permits efficient retention of about one micron large particles in Phyllopoda, Cladocera, and the younger stages of Copepoda, whereas in the older stages of the larger copepods species, and in *Euphausia* particles smaller than 5 to 10 μ may easily pass the filters. Also the filters in the different suspension feeding insects are often rather coarse.

Thus, most filter feeders seem efficiently to retain particles of sizes down to the sizes of the smallest living organisms present in the water, as nanno-phytoplankters and bacteria, whereas no filter feeder is known with certainty to efficiently retain colloids such as large proteins. The retention of haemocyanin observed by the mucus filters of *Chaetopterus, Urechis* and *Ciona* needs reinvestigation, because the aggregational state of the haemocyanin molecules and thus the size of the particles may not have been controlled (Jørgensen, 1960). Colloids that may naturally be present in the surrounding water therefore seem not to be easily accessible by filtration to the filter feeders.

1.143. CHARACTERISTICS OF SUSPENSION FEEDING

Suspension feeders are typically non-selective feeders, which clear the surrounding water of particles at rates that are independent of the concentration of the particles below certain levels and of their value as food, and which feed continuously when undisturbed. This statement needs comments and qualifications.

CONTINUOUS OR DISCONTINUOUS FEEDING

In most suspension feeders, feeding does not seem to be regulated by the need for food (appetite). In undisturbed suspension feeding Porifera, Coelenterata (*Aurelia*), Serpulimorpha, Copepoda, Prosobranchia, Bivalvia, and Ascidiacea, feeding activities as water transport and collection of particulate matter are mostly stated to go on more or less continuously, independently of particle concentration in the water and thus of amount of material collected. In several instances where feeding was found by some investigators to be discontinued for longer or shorter periods of time, e.g. bivalves and copepods, it seems likely that the intermittent feeding was due to adverse laboratory conditions. In some suspension feeders, *Urechis, Chaetopterus* and, probably, the Larvacea, the feeding mechanisms themselves are such that collection of food is regularly interrupted, in *Urechis* and *Chaetopterus* on ingestion of the feeding nets, in Larvacea during the discarding and rebuilding of the house. But little is known about the length of time spent on feeding under natural conditions in these forms, and about the factors regulating the periods of feeding and rest. It is thus not known whether need for food, e.g. absence of food in the intestinal tract is of any importance in inducing feeding. However, food uptake regulated according to need has been described in Acrania, which possess feeding structures basically similar to those of the tunicates, and which should thus be capable of continuous feeding; in amphioxus, water transport and food uptake was found to be discontinued when enough food has been collected and ingested, and resumed after digestion of the food.

RATES OF WATER TRANSPORT INDEPENDENT OF CONCENTRATION AND QUALITY OF SUSPENDED MATTER

The suspension feeders are generally found to transport water at rates that are independent of concentration and quality of particles suspended in the water below certain levels of particle concentrations (Porifera, Rotatoria, Copepoda, Bivalvia and Ascidiacea). Mostly, water is transported at unchanged rates even in the absence of any particles in the water; but in some species, which are only facultative suspension feeders, the feeding current may stop when no or too little food is present in suspension in the surrounding water, and the alternative mode of feeding is resorted to. Such behaviour has been described in the gastropod *Bithynia* and in copepods. Sufficiently high particle concentrations will necessarily result in reduced rates of water transport due to clogging of the filters. Mechanisms of protection against overloading of the feeding organs are numerous. Some groups, e.g. sponges, bivalves and ascidians, may close up and partially or completely stop the through-current at high particle concentrations.

Exceptions to the rule that suspension feeders transport water at rates that are widely independent of the concentration of particles have been reported from e.g. Cladocera, Thaliacea, and amphioxus. The reports on Cladocera have already been discussed (p. 34). Thaliacea, i.e. Doliolida and Salpida, are peculiar in being stimulated to increased activity by the presence of particles in the water, irrespective of their food value: both addition of graphite and food particles increased the water transport. In amphioxus, on the other hand, the filtration rate decreased with increasing concentration of food particles, whereas addition of graphite particles did not reduce the rate. Amphioxus thus appeared to be able to distinguish between particles with and without food value. This ability was also demonstrated in *Phallusia,* which reduced the frequency of squirting when algae, but not when graphite, were added to the water. Filtration rates in ascidians as *Ciona* and, probably, *Molgula* were, however, found to be independent of the quality, as well as concentration, of suspended particles. *Ciona* cleared suspensions of graphite and diatoms at the same rates, and *Molgula* cleared graphite suspensions at about the rates found in *Ciona* (Jørgensen, 1952; Jørgensen and Goldberg, 1953). Considering the basic similarity in structure of the feeding organs in Ascidiacea, Thaliacea and Acrania, the great differences described in their mode of functioning were unexpected and suggest divergent evolution in feeding habits in these groups. However, few investigators have been working on more than one group or type, so comparative studies are lacking.

NON-SELECTIVE FEEDING

In suspension feeders in which rates of water transport are independent of whether particles are present or not in the water, the rates of collection of particles are necessarily independent of the quality of particles, as long as these are inert. Still, ingestion of the particles might be selective according to food value of the particles. The presence or absence of food selectivity has mostly been studied by comparing the composition of the particulate matter in the intestinal tract of freshly caught suspension feeders with that of the suspended matter found in samples of the surrounding water. For various reasons this method can be of restricted value. Particulate matter, e.g. phytoplankton is often heterogeneously, patchy, distributed so that the sample of water examined may differ in composition from the water mass from which the content of the intestinal tract is obtained. Therefore, even large differences in composition does not necessarily prove qualitative sorting (Beklemishev, 1954). Often the particulate matter ingested by suspension feeders is of similar composition as that present in the surrounding water, indicating that qualitative sorting is at most of minor importance in nature. However, experiments on food uptake by suspension feeders kept in mixed suspensions of particles, selected according to quality, but of similar physical properties, are needed. Most laboratory experiments aimed at investigating the importance of food value or other qualitative properties of suspended particles in influencing the feeding behaviour have been made with suspensions containing only one type of particle at a time, or with mixtures of particles differing in size or density, as well as in quality.

The conclusion that food selection by qualitative sorting is absent or of little importance in most suspension feeders is also supported by the fact that mechanisms for sorting of particles according to quality have not been described in any suspension feeder except young stages of *Crassostrea* (p. 81); known sorting mechanisms apparently function according to such factors as size, shape and density of the particles. However, sorting according to such physical factors may result in the separation and ingestion of fractions enriched with respect to particles of food value (see p. 85).

Whereas true sorting according to quality in the feeding organs seems to be the exception (see *Aurelia aurita*, p. 6, and Loosanoff, p. 86), selection in terms of acceptance and ingestion or rejection in bulk of the collected material is the rule. Whether collected material is being ingested or not appears mainly to be determined by the degree of the filling of the intestinal tract and of the volume of material presented, whereas in most suspension feeders the value as food, or other quality of the material, is of minor importance, as long as the single constituents of the collected material are inert. Suspension feeders therefore appear to be typically non-selective.

1.144. RATE OF WATER TRANSPORT IN RELATION TO FOOD REQUIREMENTS (OXYGEN CONSUMPTION)

Even in suspension feeders in which transport of water is not regulated by the concentration and quality of the particles in the surrounding water to meet the actual need of the organisms, the rate of water transport must nevertheless be adapted to the concentration of food particles in the water. The rate must be such that enough food is being collected to cover the requirements for maintenance, growth and reproduction, at least during part of the year. In environments where the concentration of food varies strongly throughout the year, a water transport that does not vary with the food concentrations must result in equally strongly varying amounts of food being collected. It is reasonable to assume that in such environments the rates of water transport are adapted to secure optimal feeding at medium concentrations of food particles in the water. Concentrations of food particles above a certain level may result in superfluous feeding (Beklemishev, 1962). This level is determined by the amounts of food that can be ingested, digested and absorbed. Below a certain level, too little food is collected to secure maintenance. This level will depend upon the maintenance requirements of the animal and its ability to reduce these requirements without concomitant reduction in the rate of water transport.*

Rates of water transport in suspension feeders have mostly been expressed in relation to size or weight of the animals; selected values for the various groups of suspension feeders have been cited above. However, in suspension feeders transporting water at rates that are not regulated by the actual food concentration in the water, it is probably of even greater interest to know the rates of water transport in relation to the energy requirements of the animals. From this relation it is possible to estimate the concentration of food that must be present in the water to cover the requirements of the animal, and to compare this with the concentrations actually present of the various fractions of organic matter, and thus to appreciate the possible potential significance of these fractions as sources of food. In Table 1.6 are collated data on the amounts of water transported in relation to the energy metabolism, expressed in terms of oxygen consumption, in a number of suspension feeders. With the exception of the polychaetes, only filter feeders are represented. The table only includes

* The rate of water transport must to a great extent depend upon the structure and size of the feeding organs, and thus be more or less genetically fixed. It is not known to what extent ontogenetic adaptation of rate of water transport to the food concentration in the surrounding water is possible in the various suspension feeders, but probably the rates largely represent phylogenetical adaptations, in which case food conditions similar to those of the present day must have existed for long periods of time.

data that may be representative of undisturbed feeding; still, many of the data are not very reliable, especially those obtained by comparing rates of water transport and of oxygen uptake derived from unrelated experiments performed by different investigators. Measurements of both parameters on the same individuals of the various types of suspension feeders from various localities and biotops are highly needed. The following comments should be made to the data of the table.

PORIFERA

Rates of water transport and oxygen consumption were measured on different genera of siliceous sponges, the figure expressing filtration rate in terms of oxygen consumption should therefore be considered suggestive only.

URECHIS

Considering the great discrepancy in rates of water transport determined by Hall and Jørgensen, and the uncertainty concerning the normal length of feeding time, and the factors influencing it, the value of 20l. of water transported per ml oxygen uptake is only suggestive.

POLYCHAETA

Only the figures for *Schizobranchia, Myxicola* and *Sabella* are calculated from measurements of clearance and oxygen uptake performed on the same species, and only in the case of *Schizobranchia* have the measurements been made in the same series of experiments. However, the indirectly obtained figures do not differ greatly from those more directly calculated. The figures may need some correction, because the animals were found to be feeding discontinuously. However, the discontinuous feeding may, at least partially, be due to the laboratory conditions, since undisturbed sabellids and serpulids are stated to be feeding most of the time (see p. 20). It appears that the figures are in the lower end of the range of the suspension feeders represented in the table, varying from about 0·5 in *Schizobranchia* to 10 in *Salmiacina* with no significant correlation to body size. The concentrations of graphite used in determining the clearances are stated to be rather high; it therefore remains to be seen whether clearance may

Table 1.6. Relationships Between Rates of Water Transport and Oxygen Uptake in Various Suspension Feeders

	Species	Locality	Litres of water transported for each ml of oxygen consumed
	Porifera		
(1)	*Grantia compressa*	Rocky coast, Plymouth, England	13
(2)	*Halichondria panicea*	Rocky coast, Plymouth, England	20
	Echiuroidae		
(3)	*Urechis caupo*	Mud flat, Corona del Mar, California, U.S.A.	20
	Polychaeta		
(4)	*Myxicola infundibulum*	Off Plymouth, England	3 (1)
(5)	*Schizobranchia insignis*	Off Friday Harbor, Washington, U.S.A.	0·5 (1)
(6)	*Sabella pavonina*	Off Plymouth, England	8 (3)
(7)	*Pomatoceros triqueter*	Off Plymouth, England	(9)
(8)	*Hydroides norvegica*	Off Plymouth, England	(6)
(9)	*Spirorbis borealis*	Off Plymouth, England	(4)
(10)	*Salmacina dysteri*	Off Plymouth, England	(10)
	Cladocera		
(11)	*Penilia avirostris*	Black Sea, coastal waters	270
(12)	*Daphnia longispina*	Tributary to Volga, U.S.S.R.	4
(13)	*Simocephalus vetulus*	Lake Naroch, U.S.S.R.	14

	Copepoda		
(14)	*Calanus finmarchicus*	English and Scottish coastal waters	8
(15)	*Pseudocalanus minutus*	English and Scottish coastal waters	6
(16)	*Centropages hamatus*	English and Scottish coastal waters	8
(17)	*Temora longicornis*	English and Scottish coastal waters	9
(18)	*Acartia clausi*	English and Scottish coastal waters	max. 8
	Acartia clausi	Long Island Sound, Connecticut, U.S.A.	1
(19)	*Acartia tonsa*	Long Island Sound, Connecticut, U.S.A.	0·5
(20)	*Metridia lucens*	English and Scottish coastal waters	max. 0·2
(21)	*Oithona similis*	English and Scottish coastal waters	max. 0·04
	Gastropoda		
(22)	*Bithynia tentaculata*	River Rhein and Lahn, Germany	2·7
	Bivalvia		
(23)	*Ostrea edulis* (veliger)	Reared in the laboratory	40,15
(24)	*Crassostrea virginica*	Off Woods Hole, Massachusetts, U.S.A.	16
(25)	*Pecten latiauratus*	Estuarine waters, Mission Bay, California, U.S.A.	17
(26)	*Pecten irradians*	Off Beaufort, North Carolina, U.S.A.	10
(27)	*Lasaea rubra*	Tidal zone, Plymouth, England	2
(28a)	*Mytilus edulis,* 2–3 cm	Rocky Coast, England, U.S.A. (Maine and California)	12
(28b)	*Mytilus edulis,* 4·5 cm	Kieler Förde and coast of North Sea, Germany	3
	Tunicata		
(29)	*Ciona intestinalis*	Estuarine waters, Mission Bay, California, Woods Hole, Mass., U.S.A.	13

Footnotes: See overleaf

(1) *Grantia*. Clearance of graphite by Jørgensen (1949b). Oxygen consumption by Hyman (1925).

(2) *Halichondria*. Clearance of graphite by Jørgensen (1949b). Oxygen consumption in various species of siliceous sponges from v. Buddenbrock (1939, p. 624) and Pütter (1914). See also Jørgensen (1955, text to Table 3).

(3) *Urechis*. Mean oxygen consumption 0·01ml/hr/g (Hall, 1931). The figure of 20l./ml. O_2 uptake is arrived at by assuming that the animal is maintaining the feeding current for about 20 per cent of the time. The figure would be only about 0·7 if Hall's (1931) values for the rate of water transport represent normal flow rates (see p. 11).

(4–10) *Polychaeta*. Clearances of graphite by Dales (1957, 1961). (4), (6) oxygen consumptions by Wells (1952), cited from Dales (1957); and (5) by Dales (1961). Values in brackets are calculated from rates of oxygen consumptions as estimated from the relations between oxygen uptake and body weight in various worms, given by Zeuthen (1947, Fig. 34).

(11) *Penilia*. Clearance of flagellates by Pavlova (1959). Oxygen consumptions, personnal communication, Dr. E. V. Pavlova.

(12) *Daphnia*. Clearance of bacteria by Manuilova (1958). Oxygen consumption from Richman (1958).

(13) *Simocephalus*. Clearance of *Chlorella* by Sushchenya (1958a, b). Oxygen consumption from Richman (1958).

(14) *Calanus*. Clearance of *Chlamydomonas* by Gauld (1951). Oxygen consumption by Raymont and Gauld (1951) and Marshall *et al.* (1935).

(15) *Pseudocalanus*. Clearance of flagellates, > 10 μ, and oxygen consumption by Marshall and Orr (1961).

(16) *Centropages*. Clearance of *Chlamydomonas* by Gauld (1951). Oxygen consumption by Raymont and Gauld (1951).

(17) *Temora*. Clearance of *Skeletonema* chains and oxygen consumption by Berner (1962).

(18) *Acartia clausi*. English waters: Marshall and Orr (1961). Long Island Sound: Conover (1956).

(19) *Acartia tonsa*. Conover (1956).

(20) *Metridia* and (21) *Oithona*. Marshall and Orr (1961).

(22) *Bithynia*. Clearance of charcoal particles by Schäfer (1953). Oxygen consumption by Berg and Ockelmann (1959).

(23) *Ostrea veliger*. Clearance of flagellates by Bruce, Knight and Parke (1940), Jørgensen (1943), and Walne (1956). Oxygen consumption in *Mytilus* veliger by Zeuthen (1947). As *Mytilus* and *Ostrea* veligers of the sizes used in the experiments are similar in shape and organization, it was considered possible to apply the values for oxygen uptake in *Mytilus* in the calculations.

(24) *Crassostrea*. Clearance of graphite and oxygen consumption measured in the same individuals (Jørgensen, 1952).

(25) *Pecten latiauratus*. Clearance of *Asterionella* colonies and oxygen consumption measured in the same individuals (Jørgensen, 1960).

(26) *Pecten irradians*. Clearances of flagellates and diatoms by Chipman and Hopkins (1954). Oxygen consumption by v. Dam (1954).

(27) *Lasaea*. Clearances of algae and oxygen consumption by Ballentine and Morton (1956) and Morton *et al.* (1957).

(28a) *Mytilus*. Clearances of graphite and oxygen consumption partly in the same individuals (Jørgensen, 1960).

(28b) *Mytilus*. Clearances of graphite and oxygen consumption in the same individuals (Theede, 1963).

(29) *Ciona*. Clearances of graphite or diatoms and oxygen consumption in the same individuals (Jørgensen, 1952, 1955).

increase with lower concentrations of graphite, or with other kinds of particles.

CLADOCERA

As mentioned above, many investigators have found the rates with which various Cladocera are clearing suspensions to vary with the concentration of the suspended particles. However, it seems that the finding of decreasing clearances with increasing concentrations may not express true adaptation of filtering rate to food concentration, but may be due to the use of excessive concentrations of suspended matter, often in combination with too small volumes of water (p. 36). The table only includes values derived from experiments in which the initial concentration of suspended algae or bacteria have been *ca.* 1mg wet weight, or less, per litre of water, and in which the animals have been kept in large volumes of water. The exceptionally high figure of 270 for *Penilia* is calculated from the means of a great number of values on clearances and oxygen uptakes.

COPEPODA

The values are all calculated from clearances of middle-sized algae, which are presumably efficiently retained by the filters of the copepods, but are not taken by direct seizing; the clearances should therefore express rates of water transport through the feeding organs. It is seen that the amounts of water filtered per ml oxygen consumed in copepods fall in the lower end of the range of the suspension feeders represented in the table. In a number of species, *Calanus, Pseudocalanus, Centropages* and *Temora,* figures varying from about 6 to about 9 are encountered. The variation was independent of the size of the species, and no significant difference appeared between stages IV, V and adult females. Very low figures were found for *Metridia* and *Oithona,* maximally 0·2 and 0·04, indicating that in these two species filter feeding may be of minor importance.

In copepods, the normal filtering rates are of special interest, since if the average rates in the sea are much below 10 or even 1 litre of water per ml oxygen uptake, too little food would be secured by filter feeding alone at most natural phytoplankton concentrations at least in species and stages that only inefficiently retain algae smaller than 5 to 10 μ (see p. 33). It would therefore seem that in such cases either the copepods must be able to utilize other suspended organic fractions than phytoplankton, or raptorial feeding must serve as a necessary supplementary mode of food uptake. If, however, filtration rates amount to about 10 or more litres of water per ml oxygen uptake, ample amounts of food are presumably being obtained from filter feeding alone during large parts of the year (see p. 251). Raptorial feeding may supplement filter feeding during periods of low phytoplankton concentration, or when the phytoplankton is dominated by diatoms or dinoflagellates, too large to be caught by filtration. According to Petipa (1959), *Acartia clausi* feed by filtering only when the concentration of small food particles is high in the water. When large phytoplankton algae or animal food up to 600 μ in size are present, the food particles are seized directly.

BITHYNIA

The concentration of charcoal used in determining filtration rate was very high; the figure of 2.7 is therefore only tentative.

BIVALVIA

Adult oysters and scallops, and oyster veligers and young mussels are seen to transport water at the relatively high rates of 8 to 40 litres for each ml of oxygen consumed, with no clear difference between species and age groups. In *Lasaea,* the values were only about a quarter of the values calculated for *Mytilus* from the same region, but it is not known to what extent laboratory conditions may be responsible for this difference. Relatively low filtration rates were also obtained by Theede (1963) in medium-sized mussels.

CIONA

The experiments on *Ciona* from Mission Bay were performed at Scripps Institution of Oceanography, La Jolla, California, on specimens that were acclimatized to the circulating sea water in the laboratory. Sometimes the turbidity of this water was high enough for clearances to be measured of the suspended matter which caused this turbidity; light extinctions were measured on a Beckman spectrophotometer using a 10cm long absorption cell. The clearances turned out to be constant with time, that is, not to vary with the decrease in turbidity, but owing to the low initial concentrations (extinctions) the clearances could only be rather inaccurately determined. However, they were consistently higher, on an average about 50 per cent, than the clearances of graphite and algae measured in the same individuals. The barely observable disturbance of the animals that was caused even by careful addition of suspensions of particles to the jar containing the animals thus seems to have depressed the rate of water transport measurably, and the undisturbed *Ciona* therefore probably filtered about 20 litres per ml of oxygen consumed.

In summary, it appears that Polychaeta, Copepoda and Gastropoda (Bithynia) clear the surrounding water at lower rates than most suspension feeders from the other groups represented in Table 1.6; but, it is not finally settled whether these relatively low clearances are typical for the species or groups, or whether they are due to adverse experimental conditions. Most filter feeding Porifera, Cladocera, Bivalvia and Ascidiacea (*Ciona*) have been found to filter about 10 to 20 litres of water for each ml of oxygen consumed, apparently independently of age, geographical distribution, or habitat: no clear difference can be observed between species from freshwater or sea water, or between epifaunal or pelagic species. Within the restricted range of habitats represented in the table, the variation in rates of water transport that can be causally related to ecological variations, such as differences in level of food concentration in the surrounding water, is therefore small compared with the variation due to other factors, such as varying experimental technique and varying ability of the animals to exhibit normal behaviour under laboratory conditions.

1.145. MUCUS AND SUSPENSION FEEDING

The important role which mucus plays in the feeding mechanisms of many, if not most, suspension feeders has been repeatedly stressed above. Here, some supplementing comments should be made.

Generally, mucus and cilia are intimately connected in the trapping and transport of suspended food particles. However, usually the mechanisms do not seem to have been established primarily to collect food, but to clear the surfaces of particulate matter which might enter with the respiratory current, or in other ways obtain contact with and settle on the body surface. Examples of feeding mechanisms that may have evolved from cleansing mechanisms in ancestral forms may be found in, for instance, bivalves, gastropods, echinoderms.

Cleansing mechanisms based on mucus secretion in combination with ciliary currents are still widely distributed, not only amongst aquatic metazoans, suspension feeders and non-suspension feeders, but also amongst terrestrial forms. It may, therefore, be of interest to look into the specific properties of mucus in order to appreciate its usefulness and wide application in particle-trapping mechanisms.

The chemical and physical properties do not seem to have been studied in mucus produced in the feeding organs of suspension feeders. However, even when mucus derived from various sources may vary in chemical composition and physical characteristics, the principal features seem to be similar. Thus, mucus has the properties of a highly viscous, elastic gel, irrespective of its source of origin, whether it is the nasal cavity or the gastro-intestinal tract of mammals, or the hypobranchial gland of a marine gastropod (Keiser-Nielsen, 1953; Kwart and Shachoua, 1957; Proetz, 1943). Especially important is the ability of mucus to form fine elastic fibres. From freshly extracted mucus of the hypobranchial gland in *Busycon,* about 25cm long fibres with a diameter of 1–2 μ could be drawn. Mucus owes its properties to the content of mucoproteins and mucopolysaccharides which possess extraordinary capacities in binding water and solutes; often less than 1 per cent of these mucins are present in the mucus gel.

Mucus has been ascribed an internal reticular structure formed by the mucoprotein and mucopolysaccharide fibres (Keiser-Nielsen, 1953; Kent, 1962). An important property of the fibres of this structure is their ability to adsorb particles. When particles, except such with lipoid surfaces, touch mucus, this will spread over the surface of the particles. Furthermore, if particles are mixed into a sample of mucus they are found not to move freely around in the mucus, but to be fixed to the internal structure. Upon agglutination of the mucins, e.g. by acidification of the mucus, the particles are included in the mucins indicating a strong adsorption to the fibres and reticulum formed by the mucins

(Keiser-Nielsen, 1953). This strong particle-binding property of mucus is, of course, of special importance to its function in the cleansing and suspension feeding mechanisms.

Of further interest to us is the ability of mucus to form sheets covering the mucus-producing surfaces. The sheets have especially been studied in the nasal cavity of higher vertebrates (Bang, 1961; Lucas and Douglas, 1934; Proetz, 1943; Wilska, 1953). Proetz described the mucus sheet as being "exceedingly thin, slippery, adhesive and tenacious". The thickness of the mucus sheet in the trachea of normal rats was measured by Dalhamn (1956) on freeze-dried sections, and found to be 5 μ, whereas the increased mucus flow caused by exposure to SO_2 resulted in a mucus sheet about 25 μ thick. The tenacity and elasticity of the mucus sheet normally covering the nasal cavity and transported by the ciliary currents to the pharynx or the vestibule was demonstrated by Proetz (1943) who found it "possible to restrain the progress of the blanket (mucus sheet) at one edge of the field allowing the cilia to tag at it and stretch it, until it finally snapped like a rubber band".

The well-defined mucus sheets of the respiratory tracts of higher vertebrates are transported by the metachronously beating cilia, which are stated to lie beneath the mucus sheet and to touch the sheet by their tip during the forward stroke. The highly viscous mucus sheet therefore seems to float on a serous fluid of low viscosity which surrounds the cilia and in which the greater part of the ciliary cycle is carried out (Lucas and Douglas, 1934).

Only little is known about the fine structure of the mucus sheets responsible for particle retention in such suspension feeders as tunicates, gastropods, and perhaps bivalves, and we do not know to what extent properties ascribed to mucus sheets from the nasal cavity of mammals apply to the mucus filters in suspension feeders. However, important differences are likely to appear. In one instance, the coarse mantle filter in *Crepidula,* Werner (1959) has described a secondary structure, the filter being built up as a net of mucus fibres (p. 56). Also the fine mucus sheets covering the gill of *Crepidula,* ascidians, etc., may contain true secondary structures and be net-like with meshes or holes, rather than being homogeneous sheets formed by the coalescing secretions from the mucus glands and with porosities determined by the mucoprotein and mucopolysaccharide reticulum. Werner (1959) suggests two ways in which such mucus nets may possibly be formed. Either glands may secrete mucus in the form of fine fibres which subsequently harden in the water and partially adhere to one another to form thin sheets of irregular reticular structure; or more regular nets may perhaps be formed by specialized structures, comparable to those producing the coarse mantle filter in *Crepidula*.

PART 2

POTENTIAL SOURCES OF FOOD FOR SUSPENSION FEEDERS

The potential food sources for suspension feeding organisms are conveniently classified in the following groups, phytoplankton, suspended particulate dead organic matter (organic detritus), dissolved or colloidal organic matter, and heterotrophic organisms (bacteria, fungi, yeasts, heterotrophic flagellates, etc.). To these groups could be added zooplankton, which itself mostly consists of suspension feeding organisms, but which may serve as food to larger suspension feeders, especially among the vertebrates. Furthermore, we are interested in knowing the amounts of particulate inorganic matter suspended in the water, since this matter may be retained in the food-collecting organs of the suspension feeders, together with the organic fractions, and may thus interfere with feeding.

In order to evaluate the role of the various organic fractions as food for the suspension feeders, we must know not only the concentration of these fractions, but also the rate at which the fractions are produced, because on these rates depend the populations of suspension feeders that may be sustained by the various fractions.

2.1. CONCENTRATIONS OF THE VARIOUS ORGANIC FRACTIONS IN THE SEA

The methods used for estimating the concentrations of the various organic fractions will be described in some detail in order to evaluate the reliability of the estimates (see also Strickland, 1960).

2.11. PHYTOPLANKTON

2.111. METHODS OF MEASURING STANDING STOCK OF PHYTOPLANKTON

Two types of methods have been used, viz. (1) counting of phytoplankton organisms combined with measurements of their dimensions, and (2) measurements of chemical constituents, especially pigments, in phytoplankton samples combined with measurements or estimates of the concentration of these constituents in the various organisms contained in the sample.

2.1111. *Cell counts and cell size determination*

Large difficulties are encountered in the direct measurements of the phytoplankton standing stock. The phytoplankton organisms do not possess specific sizes or gravities that would make possible their separation by filtration or centrifugation from the rest of the suspended particles in the water, so they must be counted and measured individually under the microscope. The microscopical examination is seldom feasible without concentration of the organisms in a sample of water, since the density of the organisms in the natural population is generally too low for direct enumeration under the microscope.

Filtration is the most commonly used principle for concentration of the phytoplankton. Quantitative sampling of plankton by filtration was introduced by Hensen (1887) who found nets of bolting silk most suitable for this purpose. The meshes of the finest bolting silk are about 50–75 μ in diameter. Hensen determined experimentally the volumes of water filtered by his nets, so that he was able to measure the concentration in the sea of those plankton organisms that were unable to pass the meshes of the nets.

The plankton net still constitutes a valuable tool in the quantitative measurements of larger phytoplankton organisms, the microplankton, ranging in size from about 50 μ to the largest phytoplankton cells or colonies which may be up to several hundred microns in diameter. The term "microplankton" was in fact introduced to distinguish the part of the phytoplankton that is retained by the bolting silk, in contrast to the nannoplankton which consists of phytoplankton organisms small enough to pass through such nets (Lohmann, 1903). Sometimes the part of the phytoplankton that consists of the smallest algae ranging in size from about 1 to 5–10 μ in diameter is termed the ultraplankton.

The great advantage of the plankton net is the rapidity with which it can filter the large samples of water needed to obtain reliable average values of the larger plankton organisms due to their low concentration and patchy distribution. However, the inadequacy of the net method in quantitative work intended to comprise the total phytoplankton population was soon made clear, especially by the work of Kofoid (1897) and Lohmann (1903, 1908) who found that large numbers of the phytoplankton organisms were small enough to pass through the meshes of even the finest silk nets. The particular importance of these small forms as food for the suspension feeders appeared from their abundance in the intestinal tracts of planktonic filter feeders and especially in the food-collecting apparatus of appendicularians. In fact, it was the discovery of the dominant role of the nannoplankton in the appendicularian house that led Lohmann to appreciate the importance of this fraction of the phytoplankton.

To measure quantitatively the total amount of phytoplankton, Lohmann combined the net method with filtration through hard filter paper and with centrifugation of water samples, a method at that time recently introduced in plankton research. The filter paper was found to retain many times more phytoplankton organisms than the net, yet heavy losses occurred of the smallest organisms, especially the naked flagellates, which might either pass through the filter paper or might be caught and destroyed in the meshes and pores of the filter. Concentration by centrifugation, however, made possible the separate counting of these organisms. Generally, Lohmann found good agreement between results obtained by the method of centrifugation and of filtering through paper.

Since the pioneer work of Lohmann, the reliability of centrifugation in quantitative phytoplankton investigations has often been discussed. Ballentine (1953) found that naked flagellates were efficiently recovered by centrifugation in the living state; and in a natural plankton population, Creitz and Richards (1955) obtained yields by centrifugation of 69 to 86 per cent of the phytoplankton mass that was retained by millipore filters. Dinoflagellates were less readily thrown down by centrifugation than diatoms and other plankton organisms. However, other investigators obtained far less satisfactory results with the centrifuge. Allen (1919) compared the number of organisms found by the centrifugation method and by a dilution-culturing method, which yielded values that were more than 30 times higher than those obtained by the centrifugation method. Steemann Nielsen (1933) compared the centrifugation method with the sedimentation method of Utermöhl (see below) using formalin fixed samples of marine phytoplankton. The Utermöhl technique yielded 5 to 10 times higher counts of a small diatom and a coccolithophorid, and also significantly higher values of larger diatoms, than the centrifugation method.

Obviously, methods implying concentration of organisms by centrifugation are bound on occasions to give too low results, since phytoplankton organisms may have specific gravities equal to or even below that of the surrounding water. However, the varying specific gravity of the plankton organisms cannot explain the low recovery in the experiments where a sedimentation method was compared with a centrifugation method. Steemann Nielsen (1933) suggests that the poor results often obtained by centrifuging plankton samples may be due to currents developing when the centrifuge is about to stop.

The efficiency of the continuous Foerst Electrical Centrifuge, which has been used as a standard technique in quantitative studies of plankton, was investigated by Hartman (1958) who found that one run through the centrifuge removed only about half of the total particulate organic matter that might be removed in five runs. Further runs, however, did not increase the recovery.

A great methodological improvement in the quantitative phytoplankton research was introduced by Utermöhl (1931). According to his method, the preserved plankton sample is transferred to a glass with a thin bottom. When the plankton is settled, it can be directly counted through the bottom of the tube by means of a specially built microscope with the objectives pointing upwards. The method is used in various modifications that allows the measurement of even the ultraplankton, and it is still regarded as the best standard method for obtaining quantitative records of such phytoplankton organisms that can be satisfactorily preserved (Braarud, 1958). However, the great disadvantage of the Utermöhl and all other methods based on the recording of preserved plankton samples is that no satisfactory preservative has yet been found for the important naked forms of the nanno- and ultraplankton.

Improved methods of estimating these naked forms were made possible by the introduction of the membrane filters. According to the method of Goldberg *et al.* (1952), the sample of the sea water is filtered through a cellulose acetate filter with pores of a suitable diameter, e.g. 0·5 μ. After filtration the membrane is dehydrated and cleared with cedar oil, so that the membrane with the retained organisms, which may be fixed and stained, can be mounted in balsam on a microscope slide. Again the method provides minimum counts only, because also in the membrane filters the fragile, naked ultraplankton organisms may disintegrate during the filtration process; up to 45 per cent losses are recorded (Lasker and Holmes, 1957).

Enumeration of cells has been extensively used as a standard method of quantitative estimation of phytoplankton. However, the mere number of algal cells is not a satisfactory measure of the standing stock because of the great geographic and seasonal

TABLE 2.1

	Cell volumes in μ^3
Diatoms	20–20,000,000
Dinoflagellates	500– 100,000
Flagellates	14– 500
Coccolithophores	14– 500

variation in the proportion of large and small species of algae which compose the phytoplankton. The diatoms especially exhibit extreme variation in size between species, but great differences are also found within the other major groups of the phytoplankton as appears from Table 2.1 (Harvey, 1950).

Even within the same species, the size of the individuals can vary considerably; thus in the diatom *Ditylum brightwelli,* the volume of the largest cells are some 30 times the volume of the smallest cells (Gross, 1940).

To obtain a satisfactory measure of the standing stock, it is therefore necessary to know the volume and the content of organic matter of the individual cells. This was realized already by Lohmann (1908) who introduced the methods of volume measurements of plankton organisms still in use. Volumes are easily calculated of phytoplankton cells of a simple geometrical shape as the sphere, cone or cylinder, or composed of parts of such simple shapes. More complicated shapes Lohmann reproduced in Plasticine from the contours of the objects drawn by means of the "drawing apparatus". This method was used for instance for peridineans and most zooplankton organisms. The volume of the original was calculated from the displacement volume of the model.

In most of the groups of the phytoplankton algae, cell volume and plasma volume do not differ enough to warrant a correction. But in the diatoms, the larger part of the cell is generally occupied by the siliceous shells and, even more, by vacuoles filled with cell sap. Moreover, the relative volume of the plasma varies with the species, and within the species with the size of the individuals. Thus, in small cells of *Ditylum brightwelli,* the plasma volume was 1/3–1/4 of the cell volume, but 1/20 in the largest cells (Gross, 1940). In the diatoms, Lohmann therefore directly estimated the plasma volume which was arrived at as the sum of the volumes of (1) the plasma membrane covering the inside of the shells, and estimated to be 1 μ thick in small and 2 μ thick in large diatoms; and (2) the strings of plasma crossing the vacuole and the central mass of plasma containing the nucleus; to this was added (3) an estimate of organic matter supposed to be present in the cell sap. The estimate was arbitrarily put to 10 per cent of the volume. The last value is probably very low. In the cell sap of the giant cells of the brackish water green alga *Chara ceratophylla,* dry organic matter amounted to at most 0·3 per cent (Collander, 1930). Obviously, the "plasma volumes" of diatoms calculated by Lohmann are only rough estimates.

Lohmann has tabulated the volumes of a series of important plankton organisms from the Baltic Sea. Unfortunately, the diatoms are only listed by their calculated "plasma volumes", whereas the dimensions of the cells are not stated. This introduces a great uncertainty in the estimates of later investigators who have used Lohmann's "plasma volumes" as standards in calculations of the standing stock of phytoplankton, since it cannot be determined how much the size of the diatoms, whose plasma volume is to be estimated, differ from the size of the diatoms measured by Lohmann.

Lohmann's method of measuring the phytoplankton standing stock in terms of plasma volumes has more recently been used by e.g. Gillbricht (1952, 1959) and Krey (1953). In other works, standing stock is expressed as total cell volume (Bursche *et al.*, 1959; Cushing, 1957; Overback, 1959; Riley, 1957). Cell volume can be easily converted into wet weight, also termed biomass, of the phytoplankton, since the density of the phytoplankton algae is approximately unity even in diatoms (Gross and Zeuthen, 1948). Often biomass is not calculated from the cell volume, but obtained as the wet weight after draining of the plankton sample. This wet weight will vary considerably with the technique used, depending upon how much water remains intercellularily. According to Strickland (1960) the weight is perhaps never less than twice the wet weight calculated from cell volumes. Strickland suggests calling the weight calculated from algal volumes "the algal weight" and to reserve the terms "wet weight" for the weight obtained directly after filtering or centrifuging the phytoplankton sample.

Difficulties arise in converting algal weights, and even more the directly obtained wet weights, into equivalents of dry organic matter, both because so few investigations are available on organic content of the various phytoplankton organisms and because the organic content of the standing stock of phytoplankton must vary strongly with the relative abundance of diatoms. Presumably, organic content of plasma volume is more constant, but it will probably vary with the physiological state of the algae. A number of data on dry weight and organic matter of natural phytoplankton populations and algal cultures are compiled in Table 2.2.

It appears from the table that the determinations by Grim (1939) of dry weight of phytoplankton vary from 20 to 117 per cent of the cell volume! The figures are, however, too high

TABLE 2.2. PER CENT PHYTOPLANKTON DRY MATTER ESTIMATED FROM MEASUREMENTS OF THE WET WEIGHT, THE CELL VOLUME, OR THE DISPLACEMENT VOLUME

Type of phytoplankton	Type of measurement	% dry weight	% dry organic matter	References
PLANKTON SAMPLES				
Mixed netplankton	Cell volume	39	28	Riley, 1941a*
Mostly diatoms	Cell volume	65	35	
Mostly peridineans	Cell volume	25–30	26	
Mostly flagellates	Cell volume	25	23	
Mostly cyanophyceans	Cell volume	20–25	21	Grim, 1939†
Mixed phytoplankton	Cell volume	47–117	—	
Mostly diatoms and ceratians	Wet weight	18	—	
Mostly *Melosira* (diatom)	Wet weight	20	—	
Mostly ceratians	Displacement volume	10·6	—	Brandt and Raben, 1920
ALGAL CULTURES:				
Chlorella pyrenoidosa	Wet weight	25	20	Myers and Johnston, 1949
Chlorella pyrenoidosa	Wet weight	11–42	—	Spoehr and Milner, 1949
Chlorella pyrenoidosa	Wet weight	25	—	Burlew, 1953
Chlorella pyrenoidosa	Cell volume	28	—	
Skeletonema costatum	Cell volume	—	26	Cushing and Nicholson, 1958
Nitzschia palea	Wet weight	43	*ca.* 38	von Denffer, 1948

* Figures corrected for particulate matter outside the algal cells.

† Figures not corrected for particulate matter outside the algal cells.

because the weight measurements, but not the volume measurements included unknown and probably varying amounts of detritus and zooplankton. Dry weight is only about 10 to 20 per cent of wet weight or displacement volume, which percentages represent average contents of dry matter of the phytoplankton, probably mixed with some detritus and zooplankton. Riley (1941a) corrected for the organic matter present outside the living algal cells by means of the regression equation of cell volume on dry weight or organic matter. Provided that dry weight and organic matter were constant fractions of the phytoplankton, they could be calculated from these equations. The volume determinations are stated to be rough, so the figures are probably not very accurate (see Table 2.2). The phytoplankton consisted in the greater part of large diatoms, therefore 28 seems a too high value for the percentage of organic matter.

A number of analyses of dry weight and ash is available on the freshwater green alga *Chlorella pyrenoidosa,* which is extensively used in mass culture. The data are obtained under conditions that are remote from those prevailing in natural waters and it is difficult to judge to what extent they are representative of natural phytoplankton populations. The cultures are enriched with organic and inorganic nutrients as well as carbon dioxide. Dry weights range from 11 to 42 per cent of the wet weight or cell volume. "The composition and concentration of the mineral nutrients appear to have an effect upon the per cent dry weight. Under a given set of conditions, the per cent dry weight tends to increase with the yield of cells and the age of the culture" (Spoehr and Milner, 1949). The strong increase in dry weight is due to accumulation of fat, and it is especially pronounced in old cultures growing in media deficient of nitrogen (Aach, 1952). Similar results were also obtained with the diatom *Nitzschia palea* the dry weight of which in old cultures with accumulated fat reached 43 per cent of the wet weight calculated from the cell volume (von Denffer, 1948).

In the present treatise, dry organic matter of the phytoplankton standing stock has been estimated, somewhat arbitrarily, by assuming the organic matter to constitute 20 per cent of the plasma weight and 10 per cent of algal weight or wet weight. Similar conversion factors have been used recently by e.g. Banse, Gillbricht and Krey. No account is taken of the fact that organic matter per unit of algal weight or wet weight varies with the composition of the plankton or experimental technique. Other conversion factors are suggested by Strickland (1960) and by the Plankton Symposium (1958).

Obviously, available methods are not very exact for determining the standing stock of phytoplankton from measurements of volume or weight. Even careful counting and measurements of the cells result in estimates of organic matter that may probably be wrong by a factor of 2–3. Moreover, the technique is very laborious and time consuming. No wonder, therefore, that it has been relatively little used, and that attempts have been made to measure chemically the standing stock.

2.1112. *Plant pigments and phytoplankton standing stock*

Standing stock of phytoplankton has been estimated from determinations of various compounds in samples of the plankton, for instance organic carbon, nitrogen or phosphorus (see Strickland, 1960). However, most of these compounds are not specific for phytoplankton, but may be contained in detritus, or zooplankton, included in the sample. In the search for compounds specific of phytoplankton organic matter, interest has naturally been focused on the characteristic plant pigments, especially chlorophyll which has been extensively used in estimations of phytoplankton standing stock.

The autotrophic plankton algae, as autotrophic plants generally, produce specific pigments active during photosynthesis. The most important are the chlorophylls and the carotenes (Table 2.3). Chlorophyll a and β-carotene are universally and abundantly present, whereas chlorophyll b is present as a second chlorophyll in green algae and in some flagellates, and chlorophyll c in diatoms and peridineans. The blue-green algae contain only chlorophyll a; they are further characterized by containing phycocyanin, a protein with a tetrapyrrolic group. The dominance of chlorophyll a over chlorophyll c in diatoms and peridineans indicated in Table 2.3 is probably not generally valid, as in plankton samples chlorophyll c is often present in no lower concentration than chlorophyll a (Currie, 1958; Humphrey, 1960; McAllister *et al.*, 1960).

TABLE 2.3. DISTRIBUTION OF PIGMENTS IN PHYTOPLANKTON ALGAE (After Strain, 1951)

Pigments	Diatoms (Bacillariophyceae)	Peridineans (Dinophyceae)	Flagellates (Euglenophyceae)	Green algae (Chlorophyceae)	Blue-green algae (Cyanophyceae)
CHLOROPHYLLS:					
Chlorophyll a	XXX	XXX	XXX	XXX	XXX
Chlorophyll b	—	—	X	XX	—
Chlorophyll c	X	X	—	—	—
CAROTENES:					
β-carotene	XXX	XXX	XXX	XXX	XXX
PHYCOBILINS:					
Phycocyanin	—	—	—	—	XXX

XXX indicates the principal pigment of the group.
XX indicates a pigment comprising less than half of the total pigments.
X indicates a pigment comprising a small fraction of the total pigments.
— indicates absence of the pigment.

Chlorophyll a is unique in being, probably, the only one of the pigments that transfer light energy directly to the photochemical reaction: light absorbed by the other pigments apparently passes through chlorophyll a. These pigments therefore enable the algae to utilize for photosynthesis light of wave lengths that are not absorbed by chlorophyll a (Strain, 1951).

Measurements of plant pigments. In most methods of pigment measurements, the pigments are extracted from the sample of phytoplankton by means of acetone or other organic solvent. The pigments in the extract are measured either summarily as total pigments or individually, and then generally only the chlorophylls are considered.

(1) *Measurements of total plant pigments* were introduced by Harvey (1934) who matched the colour of the extracts in acetone with an artificial yellowish-green standard consisting of a mixture of nickel sulphate and potassium chromate in dilute acid. However, the tint of the colour of the extracts vary with the composition of the plankton samples, and may especially be influenced by yellow pigments present in detritus and zooplankton. The tint of the extracts may differ so much from the standard that accurate matching is impossible (Harvey, 1950). Nevertheless, the method has been widely used.

(2) *Measurements of the individual pigments* utilizing the differences in light absorption of the various pigments were introduced by Kreps and Verjbinskaya (1930), who determined the chlorophyll content of netplankton spectrophotometrically after extraction with alcohol. The separate absorption maxima of the chlorophylls and the other pigments has made possible the measurements of several of the individual pigments in the unfractionated extract. Richards with Thompson (1952) devised a method for the separate determination of chlorophylls a, b and c, and of two fractions of carotenoids, the non-astacin fraction derived from the phytoplankton and the astacin fraction from the zooplankton. Chlorophyll a and b can be expressed in absolute measures as they are chemically identified and available in chemically pure form, so that their specific absorption coefficients can be determined. Chlorophyll c was not finally chemically identified until recently (Jeffrey, 1962), but it has been expressed in Specified Pigment

Units (SPU) defined so that one such unit is close to 1g of chlorophyll. The method of Richards and Thompson is being used more and more frequently, but the larger number of measurements of chlorophyll in phytoplankton has still been made with less exact methods.

In most methods in use, chlorophyll is measured from its absorption in red light, provided either by means of red filters that are often permeable to light of a broad range of wave lengths or by spectral light of wave length at about 660 mμ, the wave length of maximal absorption for solutions in acetone of chlorophyll a. In extracts of phytoplankton, chlorophyll a is probably almost always responsible for the larger part of the light absorption; the contribution of chlorophyll b and c, though, will be the greater the broader the spectrum of the red light applied.

The absorption of the extracts is compared with the absorption of standard solutions of chlorophyll. The standards may consist of chemically pure chlorophyll a, but more often less well-defined commercial preparations of chlorophyll have been used, mostly mixtures of chlorophyll a and b. The final value of the chlorophyll of the unknown sample will of course depend upon the kind of chlorophyll preparation used as a standard. Evidently, therefore, the results obtained by means of the various modifications of the chlorophyll method are not strictly comparable.* Yet, the variation in results due to differences in method is probably smaller than that due to real differences in chlorophyll content of the phytoplankton organisms (see below).

Relation between chlorophyll and units of plant pigment. The ratio between chlorophyll and units of plant pigment in plankton samples vary widely (Table 2.4). Some variation might have been anticipated because the different pigments can vary independently of each other, not only from species to species but also with the physiological state of the individuals composing the

* In the literature, chlorophyll measured in acetone extracts of plankton samples at 660 mμ and with chlorophyll a as a standard is often termed chlorophyll a even though chlorophyll b or c most probably have been present in the extract and therefore contributed to the absorption of light.

phytoplankton. However, a variation from 0·14 to 0·88 μg chlorophyll per unit of plant pigment (UPP), as found by the various investigators, is probably not exclusively due to normal variation in the relation between chlorophyll and the other

TABLE 2.4. RATIOS OBTAINED BETWEEN PHYTOPLANKTON CHLOROPHYLL AND HARVEY UNITS OF PLANT PIGMENT

μg chlorophyll per unit of plant pigment	References
0·095–0·24	Atkins and Parke, 1951
0·14	Margalef *et al.*, 1959
0·23	Riley and Conover, 1956
0·26	Riley, 1941a
0·55	Riley, 1938b
0·51	Riley, 1939
0·77	Riley, 1939
0·88	Riley, 1938a

pigments in the living phytoplankton. It is noteworthy that if only measurements made since 1941 are considered, the variation is reduced to from 0·14 to 0·26 μg chlorophyll per UPP.

2.1113. *Concentration of chlorophyll in phytoplankton algae*

The fact that most pigment analyses aim at measuring the chlorophyll a content of the phytoplankton indicates that chlorophyll a has generally been considered the most satisfactory measure of the standing stock of phytoplankton or of its photosynthetic potential. However, recent work shows that the production of phytoplankton may perhaps be closer correlated with total pigments than with chlorophyll a alone (Currie, 1958; McAllister *et al.*, 1960; Strickland, 1960). Total pigments may, therefore, turn out to constitute a more constant fraction of total phytoplankton organic matter than does chlorophyll a. But until more analyses of total pigments are available, attempts at estimating standing stock of phytoplankton must be based on the available data on chlorophyll.

The amounts of chlorophyll that may be present in plankton

algae have been determined both in cultures of algae and in natural phytoplankton populations. Data from algal cultures show that the chlorophyll concentrations vary widely. Generally, the concentrations are high during the growth phase of the culture, but decline rapidly during the stagnation phase. Extreme variation in concentration has been encountered in *Chlorella pyrenoidosa* with chlorophyll concentrations ranging from 0·003 to 6 per cent of the dry weight (Spoehr and Milner, 1949). The extremely low concentrations were, however, only observed under conditions not realized in nature, such as aeration with CO_2-enriched air in senescent cultures. Low concentrations of chlorophyll in algae were also produced in media deficient in nutrients, as e.g. iron, nitrogen and phosphorus. It is not known, though, to what extent nutrient deficiency may influence chlorophyll concentrations in natural phytoplankton populations.

Under more normal conditions of growth, the chlorophyll concentration in *Chlorella* ranged from *ca.* 2 to 5 per cent, the higher values prevailing in cultures grown in weak light, and the lower values in cultures grown in strong light (Rabinowitch, 1945). The diatom *Nitzschia closterium* contained on an average 2·4 per cent of chlorophyll (Pace, 1941). Also in cultures of *Nitzschia*, the concentration of pigment varied with the intensity of light (Harvey, 1953).

In a number of species belonging to various groups Parsons (1961) obtained concentrations of chlorophyll a ranging from 0·5 to 2·4 per cent of the dry organic matter, in cultures under the exponential growth phase (Table 2.5).

Chlorophyll concentrations in a number of algae have been calculated from data provided by Atkins and Parke (1951) on chlorophyll contents, and number and size of the cells in the cultures. The volumes were calculated from the formula 4/3 $\pi r_1 r_2 r_3$, inserting for the radii r_1,r_2 and r_3 both the sets of minimum and maximum values in order to obtain the average volumes. Dry organic matter was assumed to be 20 per cent of the algal weight (plasma weight), which was obtained by assuming a specific gravity of 1. The percentages of chlorophyll estimated in this way ranged from 2·9 in *Gymnodinium* to 20 in *Chlamydomonas*. However, the values are probably not very exact and may rather indicate how unreliable can be the methods of determining phytoplankton standing stock from measurements of the size of the cells constituting the population.

In natural phytoplankton populations, attempts have been made at estimating chlorophyll concentrations from regression

TABLE 2.5. CHLOROPHYLL A CONTENT OF PHYTOPLANKTON ALGAE GROWN IN CULTURES (After Parsons, 1961)*

Species	Chlorophyll a in per cent of dry organic matter
Chlorophyceae	
Tetraselmis maculata	1·5
Dunaliella salina	2·4
Chrysophyceae	
Monochrysis lutheri	0·59
Syracosphaera carterae	1·0
Bacillariophyceae	
Skeletonema costatum	1·6
Coscinodiscus sp.	0·53
Phaeodactylum tricornutum	1·6
Chaetoceros sp.	0·97
Dinophyceae	
Amphidinium carteri	1·0
Exuviella sp.	0·78
Myxophyceae	
Agmenellum quadruplicatum	1·0

* Vigorous cultures in the exponential phase of growth; illumination *ca.* 0·04 cal/cm^2/min.

equations of chlorophyll on cell volume of the phytoplankton (Riley, 1941a; Gillbricht, 1952; Banse, 1956) or on total particulate organic carbon (Steele and Baird, 1961, 1962a). Significant relationships were found between chlorophyll and particulate organic matter in periods of the year of high productivity. The percentages of chlorophyll obtained by this evidently inaccurate method range from 0·67 in netplankton from the North Atlantic to 3·6 for diatoms in the Baltic Sea.

More recently, a promising method of determining the pigment concentration in natural phytoplankton populations has been introduced by McAllister *et al.* (1960). Oceanic water filtered through net of 70–100 μ pore size was transferred to a large polyethylene vat on the deck and covered by polyethylene plates so that the illumination in the vat corresponded to a depth of 20m below the surface. The phytoplankton developed rapidly, and the increase in organic matter and pigments was followed. The phytoplankton, mostly coccolithophores, which thus grew under approximately normal conditions, contained about 1 per cent of chlorophyll a. A similar experiment was later made with coastal water transferred to a spherical plastic bag containing 125m^3 of water; the bag was kept with its centre 5·5m beneath the sea surface (McAllister *et al.,* 1961). During the first 12 days of growth, diatoms completely dominated and chlorophyll a amounted to 2·6–3·6 per cent of phytoplankton organic matter. On the 15th day nitrogen was depleted and growth had stopped; dinoflagellates now contributed significantly besides the diatoms, and the percentage of chlorophyll a amounted to 2·3. During the following period of stagnation of growth the chlorophyll a concentration decreased to about 1 per cent. Illumination varied during the period of growth, but was mostly low.

Obviously, the available data on chlorophyll contents of the phytoplankton organisms are too incomplete to indicate exact factors to be used in converting measurements of chlorophyll into phytoplankton standing stock. To estimate accurately standing stock from analyses of chlorophyll, a number of conversion factors would of course be needed because the factor would depend upon the species composition of the phytoplankton and on the conditions of growth. A serious error introduced by using only one standard factor in calculating standing stocks in natural waters from chlorophyll measurements arises from the effect of different light intensities on chlorophyll concentrations. The experiments on *Chlorella* and *Nitzschia* showed a twofold difference in chlorophyll concentrations between cultures grown at low and high light intensities. Observations on natural phytoplankton populations indicate that chlorophyll concentrations can adapt rapidly to changes in illumination. In East Sound, Washington, Yentsch and Scagel (1958) thus found that the chlorophyll a content per cell was high during the night and diminished two- to threefold during daylight hours, but only in the upper layers of the euphotic zone.* Under different environmental conditions, the pattern of variation in chlorophyll content of the cells may also be different. In a dense, static population of diatoms in a small tidal creek, Ryther *et al.* (1958) found that chlorophyll a was reduced during the night to about half the day value.

The error due to the effect of different light intensities on chlorophyll concentration is probably most serious when chlorophyll data from tropical and temperate waters are to be compared. Riley *et al.* (1949) assume a threefold difference in chlorophyll concentration of phytoplankton from tropical and temperate waters, but exact data on the effect of latitude on chlorophyll concentrations are still lacking. The fact that in tropical waters also a considerable part of the phytoplankton standing stock is adjusted to photosynthesis at the weak light of the deeper layer of the euphotic zone must tend to reduce the effect of differences in light intensity due to difference in altitude.

A further error arising from the use of a standard conversion factor is due to the fact that measurements of chlorophyll made by means of the different methods are not directly comparable, as indicated above. "Chlorophyll a", of course, constitutes a smaller percentage of phytoplankton organic matter than does "unspecified chlorophyll". However, the error is small because chlorophyll a most probably always constitutes the major fraction of chlorophyll when this is measured spectrophotometrically in red light.†

In the literature, standard factors corresponding to about 3–4 per cent chlorophyll of dry organic matter have often been used (Banse, 1956; Hanaoka, 1958; Humphrey, 1960; Jenkins, 1955; Laevastu, 1957, and others). An average chlorophyll concentration in the phytoplankton in the sea as high as 3–4 per cent is, however, hardly warranted if greater emphasis is laid on the results of determinations of chlorophyll concentrations in natural phytoplankton than in pure cultures of *Chlorella*. Especially the

* Part of the variation in chlorophyll content per cell may, however, be due to diurnal variation in the cell size.

† Errors may also arise from incomplete extraction of chlorophyll (Gardiner, 1943; Nelson, 1960; Ryther and Yentsch, 1957).

results of McAllister *et al.* (1960) strongly indicate that in the temperate oceanic water the average concentration of chlorophyll a in the phytoplankton may be as low as 1 per cent. However, as it is essential in drawing conclusions regarding the relations between the suspension feeders and their food sources that the standing stock of phytoplankton is not systematically overestimated, I have adopted a concentration of 3 per cent chlorophyll of dry organic matter in the calculations of the phytoplankton standing stocks, irrespective of the regions of the sea from which the measurements are derived. When the measurements of chlorophyll are not invalidated due to methodological errors, the calculated values of standing stocks will probably generally derivate from the true value by less than a factor 3, provided that all the particulate chlorophyll found in a sample of water is present within the living phytoplankton cells. However, there are indications that chlorophyll may be present in significant amounts outside the phytoplankton.

2.1114. *Detrital Chlorophyll*

Detrital chlorophyll must be continuously produced from dying and decaying phytoplankton and from faecal matter delivered by zooplankton organisms. This production presumably varies with the season, being small during phases of rapid increase in the standing stock of phytoplankton, and large when zooplankton maxima develop under or immediately after blooms (Harvey, 1950). The rate of decomposition of phytoplankton varies with the temperature and may be rather slow. In fresh water, Skopintzev and Broock (1940) thus found that 18 per cent of chlorophyll remained after 53 days' decomposition of fresh phytoplankton at 16°C, and 48 per cent at 6°C.

A few estimates have been made of the actual amounts of particulate chlorophyll present in sea water outside the living phytoplankton, but the estimates are probably not very accurate. In his calculations of the phytoplankton mass and its chlorophyll content in the Baltic Sea, Gillbricht (1952) concurrently obtained values for the amounts of detritus and detrital chlorophyll. It was always present in the samples in large amounts, even at times of the year (January and March) when the amounts of phytoplankton, as judged from the cell volume, were negligible. The amounts of detrital chlorophyll may, though, have been overestimated. The reliability of the calculations of particulate, extracellular chlorophyll is dependent upon the accuracy of the method of calculating phytoplankton cell volume, a method that may not adequately record the ultraplankton, especially the naked flagellates. In fact, this fraction is not mentioned in Gillbricht's paper in which the phytoplankton organisms are distributed solely between diatoms and peridineans, whereas at a nearby locality, the ultraplankton might constitute the major part of the phytoplankton (Overback, 1959). In the North Sea near Heligoland, in August, Gillbricht (1961) found the correlations between chlorophyll and phytoplankton mass to indicate that 65 per cent of the chlorophyll was present outside the living phytoplankton cells. Further indirect evidence for the presence of significant amounts of detrital chlorophyll was given by Banse (1956). In the North Sea off the English coast in March, more than 10 μg/l. of chlorophyll was measured whereas particulate protein amounted to only 75 μg/1.10 μg/l. chlorophyll as phytoplankton would correspond to more than 100 μg/l. of protein. The disproportion between chlorophyll and protein was local. In other areas of the North Sea, chlorophyll corresponded to only minor fractions of particulate protein.

Contrary to these results, Steele and Baird (1961) in a study of the correlation between particulate chlorophyll and organic production in the inshore locality Aberdeen Bay, Scotland, and the offshore locality Fladen Ground, North Sea, obtained no indication of significant amounts of detrital chlorophyll at any time of the year. This result would suggest a rapid breakdown of chlorophyll after the death of the phytoplankton, in agreement with the finding of Currie (1962) that phytoplankton chlorophyll a is degraded during the passage of the intestinal canal of the copepod *Calanus finmarchicus*. Also McAllister *et al.* (1961) found that under the approximately natural conditions in a large plastic bag containing sea water and kept suspended in the sea (see above) chlorophyll values would rapidly decrease upon the dying off of the phytoplankton.

Evidently, further investigations are necessary before the significance of detrital chlorophyll in the various parts of the sea can be finally established. However, in the classification of potential food sources for the suspension feeders, it is perhaps of minor importance to be able to distinguish between living and dead chlorophyll. So long as the chlorophyll in decaying phytoplankton matter is still present in particles large enough to be retained by filter paper or membrane filters, the decay is likely to be little advanced and the nutritional value therefore still not much reduced; so it is perhaps justified to include also detrital chlorophyll in the calculations of phytoplankton as a potential food source to the suspension feeders.

2.112. PHYTOPLANKTON STANDING STOCK

Standing stocks recalculated from data available in the literature are collated in Tables 2.6, 2.7, 2.8 and 2.9, grouped according to method used in obtaining the data and to type of locality from which the phytoplankton was secured, distinction being made between oceanic, and coastal offshore and inshore localities. In gathering the data, completeness has not been attempted.

2.1121. *Oceanic waters* (*Tables 2.6 and 2.9*)

In oceanic waters, estimates of the standing stock range from 1 to 100 μg organic matter per litre of water. Values from the Pacific Ocean do not exceed 14 μg/l., or 42 μg/l. if the standing stock is calculated under the assumption of 1 per cent of chlorophyll a instead of 3 per cent (McAllister *et al.,* 1960, see p. 165). The high values of 50 to 100 μg/l. are all from the northern parts of the North Atlantic Ocean, values probably associated with special conditions favouring phytoplankton growth, as upwelling of deep water rich in nutrient salts. The euphotic zone of by far the greater parts of the oceans may contain less than about 10–20 μg of phytoplankton organic matter per litre.

It is characteristic for the oceans that the annual variation in the standing stock is small; phytoplankton blooms so typical of coastal waters are little pronounced or absent in the open oceans. In the North Atlantic (35°N 48°W), the standing stock of phytoplankton during the spring was thus only about two times larger than during the rest of the year (Riley, 1957); and a constant level of the standing stock throughout the year is reported from the Bering Sea (Semina, 1958) and the Pacific Ocean (McAllister *et al.,* 1960). Similar results can be deduced from the measurements of phytoplankton production in the oceans (p. 227).

TABLE 2.6. STANDING STOCK OF PHYTOPLANKTON IN OCEANIC WATERS, AS ESTIMATED FROM DATA ON CHLOROPHYLL A*

Locality	Time of year	Phytoplankton, μg dry organic matter per litre	References
N.W. Sargasso Sea, 15 miles S.E. of Bermuda	May–September	2–6	Hulburt *et al.*, 1960
	October–February	8–15	
	March–April	15–33	
Sargasso Sea	May–June	34	Riley, 1939
	July–August	2	Riley and Gorgy, 1948
North-western North Atlantic	May–June	70	Riley, 1939
North Atlantic:			
South East of Greenland	August	*ca.* 50	Hansen, 1959
West of Ireland	August	*ca.* 13	
Between Faroes and Iceland	June	*ca.* 100	
Irminger Sea	August	*ca.* 60	
Eastern Pacific	—	*ca.* 3	Holmes, 1958a
Pacific 50°N, 145°W	July–August	14	McAllister *et al.*, 1960
Western Pacific off Japan	Summer	10	Ichimura and Saijō, 1959

* Chlorophyll a content assumed to be 3 per cent of dry organic matter.

2.1122. *Coastal offshore waters* (*Tables 2.7 and 2.9*)

The standing stock increases when the continental margins of the oceans, or even oceanic islands (Doty and Oguri, 1956), are approached. This is especially convincingly demonstrated from directly comparable investigations where the same technique has been used in measuring the standing stock at different distances from the coast. For example, Gillbricht (1959) found in June a standing stock of phytoplankton corresponding to 49 μg organic matter per litre in the Irminger Sea, but 80 and 140 μg/l. in the waters off Iceland and Greenland; and Semina (1958, 1960) found in the spring 1 μg/l. in oceanic waters of the Bering Sea, but 160 μg/l. in coastal waters. Over the continental shelf off the east coast of the U.S.A., Yentsch and Vaccaro (1958) measured on the same day over a depth of 1,000 fathoms, 0·39; over 100 fathoms, 1·86; and over 25 fathoms, 3·75mg chlorophyll a per m^3. Furthermore, 15 miles off the Portuguese Coast, Currie (1958) found the concentration of total pigments of phytoplankton to range some 20 times higher than 200 miles off the coast.

The standing stock of phytoplankton in offshore waters at the open coast and over the continental shelf, varies almost a thousand-fold in size; the variation is to about equal parts geographical and seasonal. Some especially thoroughly investigated regions may serve as examples. In the English Channel, the concentration of phytoplankton varied throughout the year between *ca.* 100 and 1,000 μg/l.; maximum values were measured in April, minimum values in June and August (Atkins *et al.*, 1954). At the Atlantic Coast of the U.S.A. off Connecticut, annual variation ranged from 10 to 135 μg/l. (Riley, 1959), and off Florida from 6 to 900 μg/l. (Miller and Moore, 1953). In coastal waters south east of Australia, variation was from 10 to 210 μg/l. (Humphrey, 1960), and west of India from 110 to 800 μg/l. (Subrahmanyan, 1959). Much lower values are recorded from coastal waters in the western Mediterranean where the annual variation was from 5 to only 18 μg/l. throughout the years of 1953–58 (Margalef *et al.*, 1955–59). The standing stock was calculated from measurements of plant pigments according to Harvey under the assumption that 1 UPP equalled 0·14 μg chlorophyll as experimentally found by Margalef *et al.* (1959). The use of a factor corresponding to the more commonly observed relation of 0·26 μg chlorophyll to

TABLE 2.7. STANDING STOCK OF PHYTOPLANKTON IN EUPHOTIC ZONE OF COASTAL OFFSHORE WATERS, AS ESTIMATED FROM DATA ON CHLOROPHYLL A*

Locality	Time of year	Phytoplankton, μg dry organic matter per litre	References
English Channel	March–May	870–1,770	Jenkins, 1955
	June–February	150–430	
	Annual variation, 1952	70–790	Atkins *et al.*, 1954
	Annual variation, 1953	170–1,450	
North Sea:			
Fladen Ground	Annual variation	3–230	Steele, 1956; Steele and Baird, 1961
Southern North Sea	March	90	Banse, 1956
Southeast North Sea	—	530	Krey, 1956
Mediterranean:			
Gulf of Naples	April	180	Korringa and Postma, 1957
	October	70	
Western Mediterranean	Annual variation, 1953–58	5–18, av. 9†	Margalef *et al.*, 1955–59
Golfo de Cadiz	Annual variation, 1957–59	8–31, av. 18†	Muños, 1960
Ionian Sea	Summer	11	Sushchenya, 1963
Aegean Sea		16	
Black Sea near Bosphorus		43	

Atlantic Ocean:			
Faroe coastal waters	June	130	Hansen, 1959
Scoresby Sound, East Greenland	Annual variation	0–90‡	Digby, 1953
South of Cape Cod	July–September	60–280	Riley and Gorgy, 1948
Vineyard Sound	July	65	Yentsch and Vaccaro, 1958
Continental shelf off Long Island	September–March, 1956–59	13–100	Vaccaro and Ryther, 1960
Long Island Sound, Entrance	Annual variation	10–135, av. 56	Riley, 1959
Georges Bank	March–September	130–470, av. 240	Riley, 1941a, 1942
	January	30	
Florida current off Miami	Annual variation	6–900	Miller and Moore, 1953
Florida West Coast	—	76	Marshall, 1956
Northern Gulf of Mexico	—	31–93	Riley, 1937
Pacific Ocean:			
Bering Sea	—	*ca.* 170	Holmes, 1958a
Off Mexico and Peru	—	*ca.* 15	
Off California, U.S.A.	Annual variation	0–60	Graham, 1943
Off Japan	December–March	20–180	Hanaoka, 1958
Tokyo and Shimoda Bays, Japan	Summer	*ca.* 100	Ichimura and Saijō, 1959
Off S.E. Australia	Annual variation	*ca.* 10–210, av. *ca.* 40	Humphrey, 1960
West Coast of India	Annual variation	110–800, av. 340‡	Subrahmanyan, 1959
Bransfield Strait, Antarctica	December–January	13	Burkholder and Sieburth, 1961

* Chlorophyll a content assumed to be 3 per cent of dry organic matter. † 1 UPP = 0·14 μg chlorophyll. ‡ 1 UPP = 0·26 μg chlorophyll.

1 UPP only raises the range of the standing stock to 9–33 μg/l. The low values in the western Mediterranean are in contrast to the findings of Korringa and Postma (1957) in the Gulf of Naples of a phytoplankton standing stock amounting to 180 μg/l. in April and 70 μg/l. in October. Low values were also obtained in the eastern Mediterranean by Sushchenya (1963).*

The annual variation of 0–60 μg/l. found by Graham (1943) off La Jolla, California, is probably underestimated because in some samples almost no chlorophyll was found even when fair numbers of algae were counted. As mentioned, without special precautions, extraction of chlorophyll may be incomplete.

2.1123. *Inshore waters* (*Tables 2.8 and 2.9*)

In bays, fjords, estuaries, rivers, landlocked saline lakes, and other types of inshore waters, the variation in size of the standing stock of phytoplankton is even greater than in coastal offshore waters. Especially the geographical variation is pronounced, as might have been expected from the large variations in environmental conditions such as nutrients, salinity, temperature, transparency, pollution, etc., of different inshore waters.

The largest standing stocks recorded from sea water are those of 6,800 μg/l. from the Schlei in Germany (Krey, 1956) and of 6,700 μg/l. from a tidal creek on the east coast of the U.S.A. (Ryther *et al.,* 1958); both values are based on chlorophyll measurements.

Low values are recorded from the mouth of some rivers, in the Elbe and the Weser in Germany, thus only 11–51 μg/l. in May and November (Bursche *et al.,* 1958, 1959); but high in others, e.g. 300–3,000 μg/l. in the Oosterschelde, Netherlands (Drinkwaard, 1959, 1960) and 40–3,300 μg/l. in the York River, Virginia, U.S.A. (Galtsoff *et al.,* 1947). Ría de Vigo, Spain, falls in between

* Further data from the Mediterranean are furnished by Bernard (1958) who measured the cell volume of the phytoplankton algae, almost exclusively coccolithophores, to be about 50–60mm^3/l. in the surface. However, he found only a slightly smaller standing stock of coccolithophores below the euphotic zone at depths down to 1,000–2,500m. His results are therefore difficult to reconcile with those of the great majority of investigators who find a rapid decrease in the phytoplankton below the euphotic zone, and living phytoplankton cells are generally not found at great depths in the oceans.

with 14–250 μg/l. (Vives and Lopez-Benito, 1958). Thoroughly investigated regions as the Baltic Sea (Gillbricht, 1952; Krey, 1939, 1952a; Lohmann, 1908), the Dutch Wadden Sea (Postma, 1954), Long Island Sound (Riley, 1959), Alligator Harbor, Florida (Marshall, 1956) and the coastal saline Lake Macquarie in south east Australia (Davis, 1959) show annual variations that fall within the range encountered in offshore waters.*

In conclusion, the standing stock of phytoplankton is low, apparently of the order of 10 μg organic matter per litre, over the greater part of the open ocean, and annual variation is small or absent. In offshore and inshore waters, the standing stock is generally characterized by blooms in the spring and often also in the autumn or other times of the year. During the blooms, the phytoplankton may increase in amount to values 10–100 times higher than those typical of the open oceans. However, outside the blooms, the standing stock often decreases to the oceanic level or below.

2.1124. *Comparison between volume and chlorophyll methods*

As mentioned, the calculations of the standing stock of phytoplankton from measurements of number and size of cells might lead to underestimates of the standing stock because of difficulties in recording the ultraplankton (p. 153), whereas calculations made from measurements of chlorophyll might result in overestimates because of the presence of significant amounts of detrital chlorophyll (p. 167). It is, therefore, of interest that in a number of cases both methods have been used in the same region of the sea and at the same time of the year. Such comparable data are shown in Table 2.10. It is noteworthy that the two methods did not yield significantly different results, except in the previously mentioned example from the Baltic Sea (Gillbricht, 1952). It would be close at hand to assume that generally only a small fraction of the phytoplankton is not included in the measurements of the cell volume and that extracellular

* In shallow waters, the benthic vegetation, mostly of diatoms, may contribute significantly to the plankton, especially of course under conditions of agitation of the whole water column, due to the wind or tidal waves (Grøntved, 1949, 1960b, 1962; Grøntved and Steemann Nielsen, 1957; Marshall, 1956; Moul and Mason, 1957).

TABLE 2.8. STANDING STOCK OF PHYTOPLANKTON IN EUPHOTIC ZONE OF INSHORE WATERS, AS ESTIMATED FROM DATA ON CHLOROPHYLL CONTENT OF THE WATER*

Locality	Time of year	Phytoplankton, μg dry organic matter per litre	References
Baltic Sea	Annual variation	7–410, av. 100	Krey, 1939, 1952
Schlei, Germany	May	6,800	Krey, 1956
Flensborg Fjord, Germany	May	80	Krey, 1956
Off Elbe River, Germany	May	11	Bursche *et al.*, 1958
Off Elbe River, Germany	November	17	Bursche *et al.*, 1958
Off Weser River, Germany	May	15	Bursche *et al.*, 1959
Off Weser River, Germany	November	51	Bursche *et al.*, 1959
Aberdeen Bay, Scotland	Annual variation	23–200	Steele and Baird, 1961
Dutch Wadden Sea, The Netherlands	Annual variation, 1951–53	100–1,000, av. 430	Postma, 1954
Oosterschelde, The Netherlands	Annual variation, 1957–58	300–3,000, av. 1,200	Drinkwaard, 1959, 1960
Ría de Vigo, Spain	Annual variation, 1954–57	20–230, av. 80†	Vives and López-Benito, 1958
Mouth of Ría de Vigo	Annual variation, 1955–56	14–250†	Vives and López-Benito, 1958
Landlocked saline lakes, Italy	April	300	Korringa and Postma, 1957
Landlocked saline lakes, Italy	October	110	Korringa and Postma, 1957
Palma Harbour, Mallorca, Spain	January–March, 1953–57	7–16, av. 11†	Balle, 1959a, b
Palma Harbour, Mallorca, Spain	April–June, 1953–57	12–32, av. 19†	Balle, 1959a, b
Palma Harbour, Mallorca, Spain	July–September, 1953–57	20–37, av. 27†	Balle, 1959a, b
Palma Harbour, Mallorca, Spain	October–December, 1953–57	18–24, av. 20†	Balle, 1959a, b

Buzzards Bay, Cape Cod, Mass., U.S.A.	July–August	350	Edmondson and Edmondson, 1947
Long Island Sound, U.S.A.			
Eastern part	Annual variation	26–168, av. 82	Riley, 1959
Middle and western part	Annual variation	66–540, av. 230	Riley, 1959
Tidal creek, Long Island, N.Y., U.S.A.	June	6,700	Ryther *et al.*, 1958
York River, Virginia, U.S.A.	Annual variation	40–3,300, av. 760	Galtsoff *et al.*, 1947
Chesapeake Bay, U.S.A.	—	max. 760, av. 135	Marshall, 1956
Tidewaters of Chesapeake Bay	—	max. 1,570, av. 185	Marshall, 1956
Estuarine waters, Georgia, U.S.A.	February–April	90–330, av. 220	Ragotzkie, 1959
Estuarine waters, Georgia, U.S.A.	May–July	510–910, av. 680	Ragotzkie, 1959
Tampa Bay, Florida, U.S.A.	—	max. 510, av. 220	Marshall, 1956
Alligator Harbor, Florida, U.S.A.	Annual variation	33–460, av. 140	Marshall, 1956
Departure Bay, Nanaimo, Canada	April	400	Parsons and Strickland, 1959
Landlocked saline lake, S.E. Australia	Annual variation	33–130, av. 42	Davis, 1959
New Zealand, off west coast beach	August	690	Cassie and Cassie, 1960
Gerlache Strait, Antarctica	December–January	*ca.* 500	Burkholder and Sieburth, 1961

* Chlorophylla content assumed to be 3 per cent of dry organic matter. † 1 UPP = 0·14 μg chlorophyll.

TABLE 2.9. STANDING STOCK OF PHYTOPLANKTON IN EUPHOTIC ZONE OF THE SEA, AS ESTIMATED FROM DATA ON BIOMASS, CELL VOLUME AND PLASMA VOLUME*

Locality	Time of year	Phytoplankton, μg dry organic matter per litre	Method	References
Bering Sea, oceanic waters	Spring	1	Biomass (Sedimentation)	Semina, 1958, 1960
	Summer	1·5		
Bering Sea, coastal waters	Spring	160		
	Summer	25		
North Central Sargasso Sea, 35°N, 48°W	Winter	13	Cell volume (Rough estimates of volume)	Riley, 1957
	Spring	26		
	Summer	12		
	Autumn	14		
North Sea	April–May	24	Cell volume	Cushing, 1957
North Sea off Lowestoft, England	March	9·3	Cell volume	Wimpenny, 1958
	October	69	Cell volume	
Off Weser River, Germany	May	27	Cell volume	Bursche *et al.*, 1959
Breeger Bodden, Baltic Sea	August	210	Cell volume	Overback, 1959
Irminger Sea, oceanic waters	June	49	Plasma volume	Gillbricht, 1959
Iceland and Greenland, coastal waters	June	80, 140	Plasma volume	
Southern North Sea	March	104	Plasma volume	Krey, 1953
Off the Thames, England	March	8	Plasma volume	
Kieler Bight, Germany	Annual variation	5–160, av. 71	Plasma volume	Gillbricht, 1952 (See also Banse, 1956)
Mouth of Kieler Fjord, Germany	Annual variation	10–210, av. 76	Plasma volume	Lohmann, 1908

* Dry organic matter assumed to be 10 per cent of biomass and cell volume, and 20 per cent of plasma volume.

TABLE 2.10. COMPARISON BETWEEN VALUES OF STANDING STOCK OF PHYTOPLANKTON CALCULATED FROM MEASUREMENTS OF CHLOROPHYLL AND OF VOLUME OF THE ALGAL CELLS

Locality	Standing stock, μg dry organic matter per litre, as measured by		Time of year*	References
	volume method	chlorophyll method		
Oceanic waters:				
Sargasso Sea, 35°N, 48°W	12–26	—	Annual variation	Riley, 1957
Sargasso Sea, 15 miles S.E. of Bermuda	—	2–33	Annual variation	Hulburt *et al.*, 1960
Irminger Sea	49	—	June	Gillbricht, 1959
	—	60	August	Hansen, 1959
Coastal offshore waters:				
Southern North Sea	104	—	—	Krey, 1953
	—	90	—	Banse, 1956
Coastal waters off Iceland and Greenland	80, 140	—	—	Gillbricht, 1959
Coastal waters off Faroes	—	130	—	Hansen, 1959
Inshore waters:				
Off Weser River	27	15	—	Bursche *et al.*, 1959
Kieler Bay	5–160	130–400	Annual variation	Gillbricht, 1952

* Measurements have been made at same time of the year when nothing else is stated.

chlorophyll is mostly present in only small amounts. However, such an assumption is not justified without further evidence, because it might be that, owing to the choice of a too high chlorophyll percentage, the chlorophyll method may underestimate the standing stock to no smaller degree than does the volume method.

2.113. PARTITION OF PHYTOPLANKTON MASS ACCORDING TO SIZE OF THE PHYTOPLANKTON ORGANISMS

The size of the phytoplankton organisms may determine whether they are accessible as food to the suspension feeders: the organisms may be either too small or too large to be handled by the feeding organs. It may, therefore, be of importance to know how the phytoplankton mass is distributed on the various size groups of the phytoplankton organisms. A detailed study was made by Goldberg *et al.* (1952) by means of the technique described above (p. 153). After counting and measuring the cells on the molecular filter through which the sample of water was filtered, they calculated the cell volumes of the algae as a function of the cell size. In two samples of 500ml each they found that the cells smaller than 10 μ in diameter accounted for respectively 0·7 and 2 per cent of the total cell volume, cells 10–50 μ in diameter for 32 and 68 per cent, and cells 50–128 μ for 67 and 30 per cent. It is noteworthy that the total volumes of the phytoplankton contained in the two samples were very low, only 0·008 and 0·004mm^3/l., or about 1 μg dry organic matter per litre. The samples were obtained 28th March, probably in the Pacific Ocean off La Jolla, California, but the locality is not stated.

Further available data on the distribution of phytoplankton mass according to cell size are far less detailed and exact, but mainly consist of estimates of the proportions of the standing stock retained by filters of different porosity, such as nets of bolting silk, filter paper or membrane filters.

The microplankton can be estimated separately as the fraction of the total phytoplankton mass that is retained by nets made from the finest bolting silk. From Table 2.11 it may be seen that the microplankton varies from constituting an insignificant to a dominant fraction of the total phytoplankton mass; the variation is both geographical and seasonal. It is perhaps significant that

TABLE 2.11. STANDING STOCK OF MICROPLANKTON IN PER CENT OF TOTAL PHYTOPLANKTON STANDING STOCK

Locality	Percentage microplankton	Total standing stock determined by means of	References
Oceanic waters:			
Northeast Pacific Ocean	Very small fraction	Membrane filter, 0·8 μ pores	McAllister *et al.*, 1960
Coastal waters:			
Kieler Fjord, Germany	*ca.* 5–*ca.* 100	Filter paper or centrifugation	Lohmann, 1908
Baltic Sea, estuarine waters	0·1–2·4	Counting chamber	Overback. 1959
Oosterschelde, The Netherlands	10–33	Membrane filter	Korringa, 1957, 1958
	35	Membrane filter	Drinkwaard, 1960
English Channel	<2–22	Membrane filter	Harvey, 1950
Gulf of Naples, and saline lakes	Small fraction	Membrane filter	Korringa and Postma, 1957
Messina Strait, Italy	23	Filter paper, hard	Lohmann, 1903
Scoresby Sound, Greenland	*ca.* 33–*ca.* 100	Membrane filter, 0·6 μ pores	Digby, 1953
Vineyard Sound, Mass., U.S.A.	2–47	Membrane filter	Yentsch and Ryther, 1959b
Long Island Sound, U.S.A.	2–78	Membrane filter	Riley, 1941b
Gulf of Mexico	<2	Filter paper, Whatman No. 2	Riley, 1938b
Florida Current	0·08	Filter paper, Whatman No. 54	Miller *et al.*, 1953
Off Japan	12	Membrane filter, 0·4 μ pores	Hanaoka, 1958
Off Eastern Australia	0·03–4	Centrifugation	Wood and Davis, 1956
Freshwaters:			
Lake Görväln, Sweden	10–90	Membrane filter	Rodhe, 1958

only at higher latitudes, the microplankton has been recorded to be dominant, and only at certain times of the year; the only region where the average microplankton mass amounted to more than half the total phytoplankton mass was Scoresby Sound in East Greenland (70°N), with an average of 64 per cent (Digby, 1953). At low latitudes the values were generally below a few per cent.

The determinations of the relative amounts of microplankton tend to provide maximum values due to underestimation of total phytoplankton mass. The accuracy of the measurements of total phytoplankton depends upon the efficiency with which the membrane filters, filter papers or centrifuges retain the phytoplankton cells, and filter paper, even the most retentive, let through large parts of the ultraplankton. Even membrane filters may not be completely impermeable to the ultraplankton; up to 45 per cent losses of phytoplankton have been reported through membrane filters with pores as small as 0·8 μ, whereas membrane filters of 0·4 μ pores size secured effective retention (Lasker and Holmes, 1957). Apparently the algae did not pass the 0·8 μ filter because the cells were small enough, but because the very fragile cells of the ultraplankton disintegrated in the filter.

It is not yet possible concisely to evaluate the role played by the smallest nannoplankton and the ultraplankton of cell size below 5 μ. The few data available are contradictory, probably because they are not directly comparable. Most "ultra plankton" determinations have been made by measuring the fraction of the phytoplankton that pass filter paper (the pores of which have no well defined size), but not membrane filters. The fractions of phytoplankton retained by the highly retentive filter papers Whatman No. 2 and No. 42 may vary from a few per cent (Riley and Conover, 1956) to *ca.* 60 per cent (Harvey, 1950). However, the variation in retention may not only be due to differences in the dominating cell size, but perhaps also to differences in degree of cell disintegration in the filters.

In the English Channel, both Ballentine (1953) and Cushing (1955), contrary to Harvey (1950), found the ultraplankton to constitute only a small fraction of the total phytoplankton. Whether this discrepancy is due to differences in techniques or to seasonal variation in the relative amounts of ultraplankton remains to be seen.

In Oosterschelde, Netherlands, which contains important oyster beds, Drinkwaard (1960) followed the annual variation of three fractions of the phytoplankton; the microplankton averaged 36 per cent of the total standing stock as judged from chlorophyll analyses, larger nannoplankton 24 per cent, and smaller nannoplankton 40 per cent. Larger and smaller nannoplankton organisms were separated by filtration through filter paper, assumed to let through cells smaller than 15–20 μ.

In oceanic water from the north east Pacific Ocean, McAllister *et al.* (1960) found during the summer months that on an average 75 per cent of the phytoplankton standing stock passed through a 10 μ filter. Also from the Sargasso Sea evidence has been provided for the dominant role of the 1–10 μ cells in the phytoplankton of the oceanic waters (Hulburt *et al.,* 1960).

From the available data it thus appears that nanno- and ultraplankton mostly constitute the larger and often even the dominant part of the phytoplankton; it even seems that the ultraplankton alone may constitute a significant or dominant part of the phytoplankton standing stock. There are indications that the ultraplankton and smaller nannoplankton organisms are of special and more constant significance in oceanic waters, especially at lower latitudes.

The relative importance of the various fractions of the phytoplankton not only depends upon the size of the standing stock, but also upon the rate of production of phytoplankton. It is therefore of interest to measure the rates of production as a function of the size of the phytoplankton organisms. A few such measurements have been included in Table 2.12. In the inshore water Skibsted Fjord, Grøntved (1958) found that the phytoplankton cells capable of passing through filter paper were responsible for 15–70 per cent of the production of the total phytoplankton population; the percentage varied with the season of the year. The filter paper is stated to retain only the larger of the naked flagellates. In the freshwater Görväln, Sweden, Rodhe (1958) found that the microplankton accounted for 10–90 per cent of the phytoplankton mass, but only for 5–30 per cent of the production. Thus in terms of production per unit mass, the nanno- and ultraplankton is relatively more important than the microplankton. The real importance as a potential food source to suspension feeders of the smaller phytoplankton organisms is therefore even greater than appears solely from the fraction they constitute of the standing stock.

TABLE 2.12. MICROPLANKTON AND "ULTRAPLANKTON" PRODUCTION IN PER CENT OF TOTAL PRODUCTION, DETERMINED BY MEANS OF THE CARBON-14 METHOD

Locality	Percentage microplankton	References
Oceanic waters:		
Various oceans	6	Steemann Nielsen and Jensen, 1957
Pacific Ocean, tropical waters	0–7	Holmes, 1958b
Coastal waters:		
Various coastal waters	18	Steemann Nielsen and Jensen, 1957
Freshwaters:		
Lake Görväln, Sweden	5–30	Rodhe, 1958
	Percentage "ultraplankton"	**References**
Inshore waters:		
Skibsted Fjord, Denmark	15–70	Grøntved, 1958
Various Danish inshore waters	av. 26	Grøntved, 1960b

2.12. ORGANIC DETRITUS AND PARTICULATE INORGANIC MATTER

The amounts of organic detritus suspended in sea water has mainly been obtained indirectly as the difference between total particulate organic matter and phytoplankton + bacteria.*

Fairly accurate methods are available for the determinations of

* Zooplankton may be excluded from the determinations of total particulate organic matter, for instance by prefiltering of the water sample through a net with pores about 0·2mm in diameter.

total organic particulate matter, and as, furthermore, bacterial organic matter often seems to be relatively low in sea water the accuracy with which detritus can be determined may largely depend upon the accuracy of the determinations of the standing stock of phytoplankton (see also Krey, 1961d).

2.121. METHODS OF DETERMINATION

Total particulate matter may be determined directly by filtering water samples through retensive filters and by weighing the retained material after drying (e.g. Krey, 1950). Indirectly, concentrations of particulate matter may be determined by measuring the Tyndall scattering of light by the suspended particles present in the water (Jerlov, 1951, 1955; Jerlov and Kullenberg, 1953), or by measuring the extinction of light, which is caused partly by the absorption of light by the water itself and partly by the absorption and scattering of light by the suspended particles (Joseph, 1954). The optical measurements vary not only with the concentration of suspended particulate matter, but also with the size and other properties of the particles. The indirect methods can therefore provide absolute measurements of concentration of particulate matter only which the sizes and character of the particles are known, which is seldom the case. However, the optical methods are more sensitive and rapid than methods implying filtration and weighing, and they have been of great value in oceanographical research in providing a detailed knowledge of the distribution of particles in the sea (Jerlov, 1953, 1955, 1959; Joseph, 1954, 1959).

Total organic particulate matter has often been determined simultaneously with the direct determination of total particulate matter from the loss of weight following ignition of the dried sample. A correction may be needed for loss of carbonate during the ignition. Organic matter has also been determined by "wet oxidation" in various modifications, for instance with potassium dichromate in concentrated sulphuric acid (Fox *et al.,* 1952; McAllister *et al.,* 1960).

Special interest has been shown in determining particulate organic nitrogen, especially by von Brand and Rakestraw in the U.S.A. and Krey in Germany. Von Brand modified the method of Krogh and Keys for the determination of dissolved organic nitrogen in sea water (p. 201). Particulate matter was centrifuged down with a precipitate formed by addition of potassium hydroxide to the sample of sea water. Colloidal matter in suspension in the water was presumably also carried down with the precipitate and thus included in the analyses. Whereas the method of von Brand measures both proteinaceous and non-proteinaceous organic nitrogen, the method of Krey determines specifically particulate proteinaceous matter by means of the biuret reaction, which is performed after hydrolysis of the proteins in a solution of sodium hydroxide. The particulate material for analysis is recovered by filtration of sea water samples through paper filters (Krey *et al.,* 1957). The Kjeldahl technique has also been used in determining particulate organic nitrogen in sea water (Fox *et al.,* 1952; Fraga, 1957, 1959).

Determinations of protein in particulate form are of great importance in appreciating the potential food available to the suspension feeders, since this material probably contains a large proportion of valuable food. Krey (1961d) assumes that the particulate protein, as determined by means of the biuret reaction, measures total biomass of phytoplankton, zooplankton and the microheterotrophs. Presumably, however, detrital proteins would also be included in the measurements.

A large part of the total organic matter in sea water is present in true solution; most probably this fraction gradually merges with the colloidal and this with the particulate fraction with no pronounced discontinuities in particle size. Since the minimum particle size that is efficiently retained varies with the methods used in isolating the particulate matter, these methods presumably measure different fractions of the organic matter present in the water. Methods applying centrifugations of water samples with precipitated hydroxides (von Brand, 1935) or special adsorbing filters consisting of a pad of, e.g. magnesium oxide and "cellite" that may completely retain protein molecules as haemoglobin (Fox *et al.,* 1953), must include larger parts of the total organic matter in the water sample than centrifugation without a precipitate, or filtration through membrane filters with pores about 1 μ in diameter, or through the even less retensive paper filters.

The quantitative importance of the colloidal fraction of organic matter in sea water has not been systematically investigated.

Available evidence indicates that the fraction is small compared with the amounts of dissolved organic matter (p. 197); but this does not exclude that the colloidal fraction may be significant compared with the particulate fraction retained by filters of about 1 μ pore size. The size of the colloidal fraction is of special interest to us because most suspension feeders seem only inefficiently to retain particles less than one or a few microns. If colloids not retained by 1 μ membrane filters or paper filters are of quantitative importance, an inspection of the data on amounts of particulate organic matter in the various regions of the sea should reveal an influence of the type of method used on the amounts found. No such influence is definitely indicated (Tables 2.13, 2.14, 2.15, 2.16); however, the concentrations of organic matter vary by more than a factor 100; so we can merely conclude that the variation due to differences in the efficiency of recovery of the colloidal fraction of the organic matter is apparently small compared with the variation due to seasonal or geographical differences in content of particulate organic matter in the sea, or to other differences in technique than the efficiency of recovery of the colloidal fraction.

Data on total particulate matter, and its inorganic and organic fractions, in oceanic, coastal offshore and inshore waters are tabulated in Tables 2.13, 2.14 and 2.15, together with comparative data on the standing stock of phytoplankton. A discussion of the relative sizes of the various organic fractions is given below on p. 219.

2.122. CONCENTRATIONS OF PARTICULATE MATTER

2.1221. *Oceanic surface waters* (*Table 2.13*)

In the North Atlantic Ocean, Krey and his colleagues (1959) found that the standing stock of phyto- and zooplankton might constitute from 14 to 60 per cent of the total particulate matter. In the highly productive areas south west of Iceland and in the Denmark Strait between Iceland and Greenland, particulate matter retained on filter paper reached values between 1 and 2mg/l., phyto- and zooplankton accounting for about half the amounts. In the region south of Greenland, concentrations of particulate matter higher than 2mg/l. were encountered, but here

TABLE 2.13. PARTICULATE MATTER IN OCEANIC SURFACE WATERS

Locality	Time of year	Total particulate matter, μg/l. I	Particulate inorganic matter, μg/l. II	Particulate organic matter, μg/l. III	Particulate organic N as protein (N×6·25), μg/l. IV	Phyto-plankton in per cent of I, II, III, or IV	References
North Atlantic Ocean:							
S.W. of Iceland	June	1,380	—	—	—	60	Krey et al., 1959; Krey, 1961a
Denmark Strait	June	1,700	—	—	—	50	
Between Iceland and Faroes	June	570	—	—	—	14	
Between Faroes and Orkneys	June	670	—	—	—	39	
Between Cape Farewell and Flemish Cape	April, September	140, 160	—	—	—	—	
Off Cape Farewell	April, September	300–2,750	—	—	—	—	
[illegible]	Summer	500–800	—	—	—	—	Lisitzin and Zhivago, 1960

				of total			
Pacific Ocean:							
San Clemente Deep off San Diego, California	February	—	1,600	—	—	—	Goldberg *et al.*, 1952
Atlantic Ocean, 43–49°N, 15–19°W	June	—	*ca.* 300	—	—	—	Armstrong, 1958
N.E. Pacific Ocean, 50°N, 145°W	July–August	—	—	320	128	18(III); 12(IV)	McAllister *et al.*, 1960
North Atlantic Ocean:							
Irminger Sea	—	—	—	—	63–380, av. 263	—	von Brand, 1935
N.W. Atlantic Ocean	November	—	—	—	26	70	Yentsch and Vaccaro, 1958
Sargasso Sea, Gulf Stream, Caribbean Sea	March–July	—	—	—	30–112, av. 64	—	von Brand, 1938
Sargasso Sea, 36°N, 68°W	1937–38	—	—	—	*ca.* 60	—	Rakestraw, 1958
	1939	—	—	—	*ca.* 20	—	
Slope water	Annual variation	—	—	—	25–125	—	

the high content of particulate matter apparently mainly rose from terrigenous material set free from melting icebergs. The terrigenous component was still significant more than 400km south of Greenland, but in the remaining part of the 1,500km long section through the North Atlantic Ocean from Cape Farewell to the Flemish Cap, along which measurements were made, particulate matter averaged 0·14mg/l. during the cruise in April, and 0·16 mg/l. in September. These values of total particulate matter may be compared with the value of *ca.* 13 μg/l. for the phytoplankton standing stock available for August from the North Atlantic west of Ireland (Hansen, 1959, see Table 2.7). According to these data, the phytoplankton would amount to 8–9 per cent of the total particulate matter.

The values of total particulate matter provided for the Antarctic and Indian Oceans, 0·5–0·8mg/l. (Lisitzin and Zhivago, 1960), are several times higher than those for the central part of the North Atlantic, but the composition of the matter is not stated. A high average value for inorganic particulate matter, 1·6mg/l., was found in the Pacific Ocean off San Diego (Goldberg *et al.*, 1952).

In the North-eastern Pacific, total particulate organic matter averaged *ca.* 0·3mg/l.; phytoplankton made up 18 per cent of the organic matter determined by wet oxidation, whereas 12 per cent of proteinaceous particulate matter was contained in the phytoplankton (McAllister *et al.*, 1960).

From the North Atlantic Ocean, a number of analyses are available on particulate organic nitrogen. The values are high in the productive Irminger Sea, and low in the unproductive Sargasso Sea, where the difference between the average value of *ca.* 60 μg/l. for 1937–38 and *ca.* 20 μg/l. for 1939 is noteworthy (von Brand, 1935, 1938; Rakestraw, 1958). Phytoplankton values from the Sargasso Sea range from 2 to 33 μg org. matter/l. (Table 2.7), or, assuming a protein content of 50 per cent, 1–17 μg protein/l. In samples from the north-western North Atlantic Ocean in November, Yentsch and Vaccaro (1958) calculated that 70 per cent of the particulate nitrogen was present in the phytoplankton.

2.1222. *Coastal offshore waters* (*Table 2.14*)

In the coastal waters, large variations are found in the total amounts of particulate matter as well as in the ratios between its various fractions. Some areas studied in more detail may serve as examples. In the Long Island Sound on the east coast of the U.S.A., total particulate matter varied annually between 2 and 7·5mg/l., about half of the matter being organic. Of the organic matter, the phytoplankton constituted from 4 to 30 per cent (Riley, 1959). In the English Channel and the North Sea, similar values have generally been found (Armstrong and Atkins, 1950; Atkins *et al.,* 1954; Banse, 1956; Krey, 1953, 1956). However, in parts of the southern North Sea in English coastal waters very low values of phytoplankton have been found, both absolutely and in relation to the particulate organic matter, less than 1 per cent, apparently due to the influence of the River Thames (Krey, 1953; Banse, 1956).

A tendency towards a decrease in the ratio between phytoplankton and total organic particulate matter in the coastal waters is evident from the data of Yentsch and Vaccaro (1958) who found that whereas the total amounts of particulate organic nitrogen increased strongly in coastal waters compared with oceanic waters, the percentage contained in the phytoplankton decreased from 70 in oceanic surface water to *ca.* 50 in shelf water and 20 closer to the coast in Vineyard Sound, Mass.; the latter value, though, was obtained in July, and the two former in November.

2.1223. *Inshore waters* (*Table 2.15*)

In shallow inshore waters, the wind or the tide may agitate the water masses throughout the whole water column and thus stir up bottom material; and rivers may drain large quantities of suspended material into estuarine waters. Suspension feeders living in such areas are therefore exposed to heavier loads of suspended materials than in other habitats in the sea. Suspended matter in inshore waters often amounts to several mg/l., with maximum values above 100mg/l., e.g. in the Wadden Sea along the coast of north-western Europe. The amounts of suspended matter may vary greatly throughout the tidal cycle, and are higher at the bottom than at the surface. In strongly agitated water, a

TABLE 2.14. PARTICULATE MATTER IN COASTAL OFFSHORE WATERS

Locality	Time of year	Total particulate matter, μg/l. I	Particulate inorganic matter, μg/l. II	Particulate organic matter, μg/l. III	Particulate organic N as protein (N×6·25), μg/l. IV	Phytoplankton in per cent of I, II, III, or IV	References
Southern North Sea:							
N.W. of Wadden Sea and Elbe River	August	2,000	—	2,000	203	17 (I); 17 (III)	Krey, 1956
Off Wadden Sea	August	6,800	—	2,500	256	6 (I); 16 (III)	Krey, 1956
Off Elbe River	August	7,400	—	5,000	722	10 (I); 15 (III)	Krey, 1956
Waters from English Chan.	February–March	750	—	—	79	9 (I)	Krey, 1953
English coastal waters	February–March	7,500	—	—	156	0·1 (I)	
52–54°N, 2–5°E	March	1,160	—	—	22	0·9 (I)	Banse, 1956
Off Dutch coast	—	6,000	—	1,500	—	—	Postma, 1954
English Channel:							
Off Plymouth	Annual variation	2,700–5,770, av. 3,820	—	950–2,410, av. 1,800	48–91, av. 75 (Amino–N) 86–226, av. 137 (Kjeldahl–N)	—	Corner, 1961; Cowey and Corner, 1963
Mediterranean:							
Off Monaco	—	400	200	—	—	—	Kufferath, 1957
Gulf of Naples	October	*ca.* 300	—	—	—	27	Korringa & Postma, 1957
Aegean Sea	Summer	207	—	134	—	11·6 (III)	Sushchenya, 1963
South of Rhodes	Summer	219	—	141	—	—	Sushchenya, 1963
Ionian Sea	Summer	245	—	162	—	5·7 (III)	Sushchenya, 1963
Adriatic Sea	Summer	285	—	184	—	—	Sushchenya, 1963

Long Island Sound	Annual variation	2,000–7,500	—	1,200–3,100	—	4–30, av. 12 (III)	Riley, 1959
Off Florida	Annual variation	<20–100	—	—	—	—	Bsharah, 1957
Off Japan	December–March	400–700	—	—	—	*ca.* 17	Hanaoka, 1958
English Channel	Annual variation	—	450–2,770	*ca.* 1,500	—	—	Armstrong and Atkins, 1950
English Channel	June	—	1,000	—	—	7	Atkins *et al.*, 1954
English Channel, 20 miles offshore	Annual variation	—	160–1,200, av. 610	—	—	—	Armstrong, 1958
North Sea, Fladen Ground	Annual variation	—	—	20–600	—	—	Steele and Baird, 1961
Atlantic Ocean:							
Gulf of Maine, 0–40m	August	—	—	—	75–210	—	von Brand, 1937
60–215m	August	—	—	—	63–106	—	
Georges Bank	July	—	—	—	112–293	—	
Shelf waters, 40°N, 68°W							
0m	—	—	—	—	181	—	
50m	—	—	—	—	70	—	
100m	—	—	—	—	44	—	
200m	—	—	—	—	29	—	
300m	—	—	—	—	22	—	
Shelf water, 42°N, 72°W	Annual variation	—	—	—	60–250	—	Rakestraw, 1958
Shelf water 25 fathoms	November	—	—	—	330	59	Yentsch and Vaccaro, 1958
100 fathoms	November	—	—	—	230	43	
Vineyard Sound, Mass., U.S.A.	July	—	—	—	570	20	
	July–August	—	—	—	81–262, av. 160	—	von Brand (Cit. Fuller, 1937)

TABLE 2.15. PARTICULATE MATTER IN INSHORE WATERS

Locality	Time of year	Total particulate matter μg/l. I	Particulate inorganic matter μg/l. II	Particulate organic matter μg/l. III	Particulate organic N as protein (N × 6·25), μg/l. IV	Phytoplankton in per cent of I, II, III, or IV	References
Baltic Sea 5m	Summer	260	—	—	27	10 (I)	Krey, 1957
10m		280	—	—	33	14 (I)	
15m		500	—	—	33	11 (I)	
60m		110	—	—	8	6 (I)	
Kieler Bay, Germany, 0–26m	Annual variation, 1958–60	200–4,200, av. 1,400	—	—	5–300, av. 80	—	Krey, 1961b
Gulmar Fjord, Sweden	Spring	1,400	—	—	—	—	Pettersson, 1934 (Cit. Atkins *et al.*, 1954)
Aberdeen Bay, Scotland:							
Surface	Annual variation	—	0–5,000	200–3,400	—	—	Steele and Baird, 1961
Near bottom			5,000–30,000				
Wadden Sea:							
Dutch Wadden Sea, western part	Annual variation	3,000–100,000, av. 23,000	—	Summer: 4,000, or 17% of I Winter: 1,000, or 4% of I	—	2 (I)	Postma, 1954
German Wadden Sea:							
8–16‰ salinity	—	40,000–90,000	—	—	—	—	Lüneburg, 1952
22–29‰ salinity	—	*ca.* 40,000	—	—	—	—	
Danish Wadden Sea	June, August	20,000–300,000	—	—	—	—	Gry, 1942
Off Weser River	May	4,400	—	1,600	770	0·3 (I); 0·8 (III)	Bursche *et al.*, 1959

	November	17,800	—	6,900	410	0·1 (I); 0·2 (III)	Bursche *et al.*, 1958
Ijsselmeer, The Netherlands	Annual variation	10,000–240,000	—	—	—	—	Havinga, 1954
Oosterschelde, The Netherlands:							
Zieriksee	Annual variation	5,000–35,000, av. 18,400	—	—	—	13 (I), April	Drinkwaard, 1960
Yerseke	Annual variation	3,000–18,000, av. 6,300	—	—	—	27 (I), April; 34 (I), July	Drinkwaard, 1960
Plymouth Sound and Hamoaze, England	May	600–24,900, av. 5,000	—	—	—	—	Jones and Wills, 1956
		3,100	2,300	800	—	—	
Landlocked saline lakes, Italy	October	3,000	—	—	—	3·6	Korringa and Postma, 1957
Departure Bay, west coast of Canada	April	—	—	950	—	42	Parsons and Strickland, 1959
Slick area near San Diego, California, U.S.A:							
Turbid water from surface; slick	—	—	—	13,540	4,440	—	Fox *et al.*, 1952
6″ beneath slick	—	—	—	8,320	2,780	—	
Clear water from surface	—	—	—	11,800	3,940	—	
Slick area in San Diego Channel:							
Surface	—	—	—	29,270	—	—	
Beneath surface	—	—	—	13,420	—	—	
Near bottom, *ca.* 10m	—	—	—	39,280	—	—	
Ría de Vigo, west coast of Spain	Annual variation	—	—	—	270–860, av.600	—	Fraga, 1957
Station E21	Annual average	—	—	—	440	—	Fraga, 1960
Station E51	Annual average	—	—	—	200	—	

substantial part of the suspended matter may consist of sand grains, which easily settle on the bottom, whereas silt forms more stable suspensions (Postma, 1954).

In the Wadden Sea, the organic particulate matter amounted to 4 mg/l. during the summer, or 17 per cent of the total particulate matter, and 1mg/l. during the winter, or 4 per cent of the total particulate matter (Postma, 1954). Off the rivers Weser and Elbe, Germany, values of about 1·6 and 3·5mg/l. were found in May and 6·9mg/l. in November, or about ½ to ⅓ the total suspended matter (Bursche *et al.,* 1958, 1959). In Plymouth Sound, England, organic matter amounted to 0·8mg/l. or about ¼ the total suspended matter (Jones and Wills, 1956). Exceptionally high values of particulate organic matter were found in slick areas near San Diego, California, viz. up to 38mg/l. (Fox *et al.,* 1952).

The fraction of the organic matter made up by the phytoplankton varies widely, not only seasonally, but also geographically. Low values, mostly less than 1 per cent were found, e.g. off the rivers Weser and Elbe (Bursche *et al.,* 1958, 1959), high values in the river Oosterschelde, where the phytoplankton during the blooms was calculated to constitute up to *ca.* ¼ to ⅓ of the total suspended matter. The phytoplankton standing stock was estimated from the concentrations of particulate chlorophyll and may therefore have been overestimated because significant amounts of detrital chlorophyll may have been present in the water (Verwey, 1960).

2.1224. *Deep waters* (*Table 2.16*)

Only few data are available on the absolute amounts and composition of the suspended matter present in the water column from below the euphotic zone to the greatest depth of the sea. In the North Atlantic Ocean, Krey *et al.* (1959) measured total suspended matter down to depths of *ca.* 3,000m along a 1,500km long section from Greenland to the Flemish Cap. They found values around 30–60 μg/l., retained by filter paper, except off Cape Farewell where values above 300 μg/l. have been recorded throughout the water column, probably due to admixture of inorganic particulate material from melting glaciers and icebergs (see above, p. 190). At oceanic stations off northern Spain, Armstrong (1958), using membrane filters of 0·2 μ average pore diameter found values of inorganic particulate matter ranging from 50 to 570 μg/l. at sampling depths of 2,400–5,900m. Some of the figures are assumed to be high owing to precipitation of calcium carbonate during storage. However, in separate analyses of particulate SiO_2, Fe_2O_3 and Al_2O_3, the sum of these components ranged from 50 to 140 μg/l., average 110 μg/l., or significantly higher than the values for total particulate matter recorded by Krey from deep waters further north in the Atlantic Ocean. From the Indian Ocean at 3,000–5,000m, Lisitzin and Zhivago (1960) mostly recorded concentrations of total particulate matter ranging from 100 to 500 μg/l., but values up to 1,500 μg/l. were found. At a depth of 100–600m in the Pacific Ocean off San Diego, California, Fox *et al.* (1952) found values as high as 280–940 μg/l. of particulate organic matter. In the north-eastern part of the Pacific Ocean, Parsons and Strickland (1962a) found about 72 to 117 μg/l. at depths varying from 300 to 3,000m, the minimum values being recorded at a depth of 1,500m. Particulate organic nitrogen, calculated as protein, sometimes accounted for more than the total organic particulate matter, indicating that not all particulate organic nitrogen was present in aminoacids. In fact, aminoacids isolated from composite samples of detritus taken at a depth of 400m amounted only to 60 per cent of the total protein determined by nitrogen analysis.

Lower values were found for particulate organic nitrogen in the North Atlantic Ocean by von Brand (1938) and Rakestraw (1958); calculated as protein, the values ranged from 20 to 50μg/l. at a depth of 200–1000m. No definite change in concentration was observed with increasing depth down to about 5,000m (von Brand, 1938).

The origin and nature of the particulate matter in the aphotic deeper zones of the sea will be discussed below (p. 280).

2.13. DISSOLVED ORGANIC MATTER

In most waters, by far the larger part of the organic matter is not brought down by centrifugation or retained by the filters used to obtain phytoplankton or other particulate matter. This dissolved organic matter is of special interest to us because it has often been considered important as food for aquatic organisms. The matter may be present in true or colloidal solution; the latter

TABLE 2.16. PARTICULATE MATTER IN OCEANIC DEEP WATERS

Locality	Depth in m	Time of year	Total particulate matter, μg/l.	Particulate inorganic matter, μg/l.	Particulate organic matter, μg/l.	Particulate organic N as protein (N × 6·25), μg/l.	References
North Atlantic Ocean:							
Between Cape Farewell and Flemish Cap	100–3,000	April, September	30–60	—	—	—	Krey *et al.*, 1959; Krey, 1961a
Off Cape Farewell	100–3,000	April, September	> 300	—	—	—	
Indian Ocean	3,000–6,000	Summer	100–500	—	—	—	Lisitzin and Zhivago, 1960
			ca. 1,000				Lisitzin, 1959
Equatorial Pacific Ocean	Whole column	—	40–94, av. 56	—	—	—	Kullenberg, 1953
Pacific Ocean:							
San Clemente Deep off [illegible]	0–1,200	February	—	200–6,100, av. 1,600	—	—	Goldberg *et al.*, 1952

15–19°W				av. 110			
Pacific Ocean:							
33°N, 120°W	150	—	—	—	940	—	Fox *et al.*, 1952
33°N, 119°W	100	—	—	—	340	—	
	300	—	—	—	320	—	
	600	—	—	—	280	—	
49–51°N, 133–145°W	300–500	July	—	—	117	114	Parsons and Strickland, 1962a
	1,500	July	—	—	72	90	
	2,700–3,000	July	—	—	110	58	
North Atlantic Ocean:							
Sargasso Sea, Gulf Stream, Caribbean Sea	50–400	March, July	—	—	—	6–41, av. 22	von Brand, 1938
	400–5,000	March, July	—	—	—	0–32, av. 12	
Sargasso Sea, 36°N, 68°W	200–1,000	1937–38	—	—	—	*ca.* 30	Rakestraw, 1958
	200–1,000	1939	—	—	—	*ca.* 20	
Slopewater, 39°N, 71°W	200–1,000	1937–38	—	—	—	*ca.* 50	
Slopewater, 39°N, 71°W	200–1,000	1939	—	—	—	*ca.* 30	

fraction is probably small in unpolluted sea water. Thus, Krogh and Keys (1934) found that water filtered through a bacteriological filter did not lose measurable amounts of organic matter by being filtered through a colloidal filter. Furthermore, Datzko (1939) compared the organic nitrogen content in several samples, before and after removal of particulate and colloidal matter by precipitation with silver sulphate, and obtained no significant change in nitrogen content in water samples from the Greenland Sea, White Sea or Black Sea.*

Thus, the amounts of organic nitrogen contained in the particulate and colloidal fractions were too small to be measured with the technique applied. These and other investigations show that the amounts of organic matter in true solution in the sea is mostly many times the amounts present in colloidal or particulate form. Great difficulties are encountered in measuring the amounts of dissolved organic matter in sea water because of the large quantities of salts present in the water. The first reliable determinations were made by Krogh and Keys (1934) who removed halides from the sea water with thallium sulphate before further analysis. After evaporation, the sea water residue was oxidized with chromic acid and the evolved carbon dioxide determined by titration. The method has been repeatedly modified (Camps, Selga and Arias, 1960a; Datzko and Datzko, 1950; Duursma, 1961; Kay, 1954a). In these later methods, the halides are not precipitated before analysis, but the chlorine developed during the oxidation is removed separately before the final determination of the carbon dioxide.

Besides organic carbon, total or dissolved organic nitrogen has also been determined. Owing to lack of sufficiently pure reagents, the classical Kjeldahl technique for measuring organic nitrogen was previously found unsuitable for analysis of the very dilute solutions met with in unpolluted sea water (Raben, 1910; Krogh and Keys, 1934).†

* In the Caspian Sea and Sea of Azov, which are isolated from the oceans, Datzko found that $\frac{1}{4}$–$\frac{1}{3}$ of the total organic nitrogen was precipitated by addition of silver sulphate.

† Lack of pure reagents does not seem any longer to prevent the use of the Kjeldahl technique in determining dissolved organic nitrogen in sea water, and methods based on this technique have recently been described (Fraga, 1959; Strickland and Parsons, 1960).

Krogh and Keys developed a method according to which samples of sea water are digested at 500°C with sodium hydroxide in an atmosphere of pure hydrogen, followed by titration of the evolved ammonia.

Measurements of dissolved organic carbon and nitrogen in sea water obtained by means of the above mentioned methods are presented in Table 2.17, which includes data for oceanic waters, and Table 2.18, which includes data for coastal offshore and inshore waters.

Oceanic waters

The earlier results showed a rather uniform distribution of dissolved carbon and nitrogen in the oceans. In the North Atlantic, Krogh (1934) obtained average values of 2·34mg C/l. and 0·24mg N/l., with no significant vertical variation from the surface to a depth of 5,400m; in the Greenland Sea, Datzko (1939) recorded 2·1mg C/l. and 0·22mg N/l. Expressing organic carbon in terms of organic matter by multiplying with the factor of 2, and organic nitrogen in terms of protein by multiplying with the factor of 6·25, "protein" is found to constitute about one-third of the total organic matter in solution.

In the Pacific Ocean, measurements revealed larger variations in content of organic carbon (Plunkett and Rakestraw, 1955). In the eastern part of the ocean, off California, an average of 2·3mg C/l. was found for the water column from the surface to a depth of 3,000m or about the same as measured in the Atlantic Ocean by Krogh. However, in the middle of the Pacific Ocean, at 41°N, 170°E, the value was 1·6mg C/l. and in the western part of the Ocean off Japan only 1·2mg C/l. Furthermore, at three of the four stations investigated, the carbon content varied vertically with minimum values encountered at about a depth of 500–1,000m.

Even more irregular patterns in geographical and vertical distribution of dissolved organic carbon and nitrogen resulted from the recent work of Duursma (1961) who investigated the contents of organic matter in various water masses from the region of the Norwegian Sea to the North Atlantic Ocean between Greenland and Western Europe. All his values of organic carbon are lower than those recorded above. The highest values were found in the Norwegian Sea, viz. 1mg/l. in the

Table 2.17. Dissolved Organic Matter in Oceanic Waters

Locality	Depth (m)	Organic C (mg/l.)	Total organic matter (C×2)	Organic N (mg/l.)	"Protein" (N×6·25)	"Protein" in per cent of total organic matter	References
Atlantic Ocean, 30°N, 69°W	0–5,400	2·35±0·09	4·7	0·244±0·010	1·5	32	Krogh, 1934a, b
Oceanic waters (Atlantic Ocean)	—	1·2–2·0	3·2	—	—	—	Keys *et al.*, 1935
Greenland Sea, 80°N, 4°E	0–750	2·0–2·1	4·1	0·21–0·23	1·4	34	Datzko, 1939
Norwegian Sea, 64–66°N, 0–2°E	0–200	0·70–1·38, av. 1·02	2·0	0·02–0·21, av. 0·11	0·69	34	
	200–3,000	0·49–1·17, av. 0·79	1·6	0·02–0·18, av. 0·12	0·75	47	
East Greenland Current, 60–62°N							
April	0–250	0·20–0·38	0·58	0·15–0·20	1·1	190	
September		0·42–0·64	1·06	0·06–0·15	0·66	62	Duursma, 1961
North Atlantic Central Waters	0–800	0·50–0·60	1·1	0·04–0·15	0·59	54	
North Atlantic Deep Waters	300–3,000	0·45–0.55 (max. 0·7)	1·0	0·15–0·23 (max. 0·3)	1·2	120	
Arctic Atlantic Intermediate Water	300–1,800	0·30–0·40	0·7	0·08–0·15	0·72	103	
Atlantic Ocean	Various depths	1·04–1·97	2–4	—	—	—	Skopintzev, 1959
Mediterranean, 43°N, 6°E	0–2,180	0·51–1·05, av. 0·69	1·38	—	—	—	Duursma, 1963
Atlantic Ocean	0–3,000	—	—	0·11–0·24	1·3	—	von Brand and Rakestraw, 1941b
Pacific Ocean:							
29–33°N, 118–121°W	0–3,000	1·41–2·82, av. 2·28	4·6	—	—	—	
32°N, 142°E	0–6,290	0·59–1·72, av. 1·17	2·3	—	—	—	Plunkett and Rakestraw, 1955
41°N, 170°E	0–3,345	1·29–2·72, av. 1·63	3·3	—	—	—	
23°S, 166°E	0–700	0·70–1·32, av. 0·95	1·9	—	—	—	Duursma, 1963

TABLE 2.18. DISSOLVED ORGANIC MATTER IN COASTAL WATERS

Locality	Depth (m)	Organic C (mg/l.)	Total organic matter (C×2)	Organic N (mg/l.)	"Protein" (N×6·25)	"Protein" in per cent of total organic matter	References
White Sea, 66°N, 37°E	25–100	2·5–2·6	5·1	0·30–0·32	1·9	37	Datzko, 1939
Black Sea, 44°N, 37°E	25–300	2·3–2·4	4·7	0·22–0·27	1·5	32	(Datzko, 1939)
Black Sea, 44°N, 37°E	5–2,000	2·4–2·9	5·3	—	—	—	Datzko, 1951
Norwegian Sea, 59°N, 5°E	10–230	0·67–1·37, av. 0·98	2·0	0·05–0·28, av. 0·13	0·81	41	Duursma, 1961
North Sea	—	0·5–1·8	*ca.* 2	0·1–0·5	*ca.* 1·9	*ca.* 95	(Duursma, 1961)
Dutch Wadden Sea, August	—	av. 2 96	6·0	av. 0·27	1·7	28	(Duursma, 1961)
Dutch Wadden Sea, November	—	av. 1·85	3·7	av. 0·29	1·8	49	(Duursma, 1961)
Ijsselmeer, The Netherlands	—	7·5–8·0	16	0·5–0·6	3·4	21	(Duursma, 1961)
North Holland Canal	—	24	48	1·62	10	21	(Duursma, 1961)
Baltic Sea, Kieler Bay, Germany	0–35	2·0–4·6 av. 3·3	6·6	—	—	—	Kay, 1954b
Mediterranean, off Spanish Coast	Surface water	1–3 (max. 5)	4	—	—	—	Camps, Selga, Arias, 1960b
Baffin Bay, Texas, U.S.A.	Shallow estuary	26, 39	65	—	—	—	Wilson, 1961
Ría de Vigo, N.W. Spain	—	—	—	0·11	0·69	—	Fraga, 1960
Off Massachusetts, U.S.A.	—	—	—	0·15	0·9	—	von Brand *et al.*, 1942
Woods Hole Harbor, Mass., U.S.A.	—	—	—	0·3–0·4	2·2	—	(von Brand *et al.*, 1942)

surface waters and 0·8mg/l. from a depth of 200–3,000m. In the North Atlantic proper, down to 0·2mg C/l. was measured; most values fall between 0·3 and 0·6mg C/l. The concentrations varied with the water masses, but not with the depth of sampling.

Also the nitrogen values came out lower than those previously recorded from the Atlantic Ocean by Krogh (1934) and von Brand and Rakestraw (1941b), moreover the ratios between nitrogen and carbon varied much more, and were mostly higher, than found by Krogh and Datzko. According to Duursma, the ratios may vary with the season: in the East Greenland Current the ratios were high in April, mainly due to a low content of organic carbon; and low in September, mainly due to a low content of nitrogen. It might thus seem as if organic dissolved nitrogen and carbon can vary widely and independently of each other.

When the organic nitrogen is expressed as protein, it will be seen (Table 2.17) that it sometimes constitutes more than 100 per cent of the total dissolved organic matter, the highest value being 190 per cent from the East Greenland Current in April. Provided the analyses of nitrogen and carbon are correct, it would therefore follow that the organic nitrogen must be present in compounds much richer in nitrogen than are proteins (amino-acids).

Further investigations probably seem to be needed in order to decide which are the correct values of dissolved organic carbon and nitrogen in oceanic waters. However, large seasonal variations beforehand in the composition and concentrations of dissolved organic matter in oceanic waters do not seem likely or possible, because the amounts of dissolved organic matter present in the oceans represent many years' total phytoplankton production (p. 239). Changes in oceanic dissolved organic matter must therefore generally be slow.

Coastal waters

In coastal offshore waters, organic carbon and nitrogen mostly do not differ much from values recorded from oceanic waters (Table 2.18). However, inshore waters receiving significant drainage from land are enriched in dissolved organic matter.

The effect of drainage from land is especially conspicuous from the investigations of Duursma (1961) in the area of the North Sea and Dutch Wadden Sea. The content of organic carbon and nitrogen in the water gradually increases from the offshore waters through the Wadden Sea into the effluents draining in the Wadden Sea. In the effluents, the Isselmeer and the North Holland Canal, organic carbon amounted to several mg per litre and nitrogen to about 1mg, whereas in the North Sea the values approached those of unpolluted oceanic waters.

2.131. CHEMICAL NATURE OF THE DISSOLVED ORGANIC MATTER

A beginning has been made towards chemically identifying and quantitatively measuring the constituents of the dissolved organic matter in natural waters; the subject was reviewed by Vallentyne (1957).

Carbohydrates

Carbohydrates in the sea water have been extensively determined by means of the anthrone or *N*-ethyl carbazole colorimetrical tests; the method was introduced by Collier *et al.* (1950) and described in detail by Zein-Eldin and May (1958). In the inshore waters of Pensacola Bay at the Mexican Gulf, substances responding to the tests for carbohydrates were found in concentrations that, expressed in arabinose equivalents, ranged from 2 to 24mg/l. (Table 2.19A). In coastal offshore waters and in the open sea of the Mexican Gulf, average amounts of 1–2mg/l. were recorded. Apparently, the values did not vary significantly with depth down to 2,500m, the maximum depth from which samples were obtained (Collier, 1958a, b).

A characteristic feature of the carbohydrate values in the various areas of the Mexican Gulf is their discontinuous distribution; throughout the water column, high and zero values may alternate within distances of some few metres. The same discontinuous distribution of carbohydrate values was found by Wangersky (1959) in Long Island Sound and by Anderson *et al.* (1956a, b) in the Atlantic Ocean, both in coastal offshore waters and over the continental slope at depths from 0 to 1,900m. In Long Island Sound carbohydrate values were low, or even zero

TABLE 2.19A. DISSOLVED " CARBOHYDRATE " IN SEA WATER, DETERMINED AS ARABINOSE OR SUCROSE EQUIVALENTS

Locality	Depth (m)	mg/l.	References
Pacific Ocean:			
100–2,000 miles off Central America to S. California	Various depths	<0·7	Lewis and Rakestraw, 1955
Off Scripps' Pier, California, U.S.A.	0–20	0·15–0·36	Lewis and Rakestraw, 1955
Pier and kelp bed	—	0·16–0·45	Lewis and Rakestraw, 1955
Coastal lagoons, San Diego, California	—	0·3 –7·9	Lewis and Rakestraw, 1955
Gulf of Mexico, continental shelf	—	0–3·0	Collier, 1958a
Gulf of Mexico, western coast	0–32	0–2·6, av. *ca.* 1·3	Collier, 1958b
Gulf of Mexico, northern and eastern coast	0–20	0–3·6, av. *ca.* 0·6	Collier, 1958b
Gulf of Mexico, deep water	0–2,530	0–6·9, av. *ca.* 1·8	Collier, 1958b
Naples Pier, Florida, U.S.A., Annual variation	Surface	0·2–4·1	Dragovich, 1961
Pensacola Bay, Florida, U.S.A.	—	2–25	Collier *et al.*, 1950, 1953
Long Island Sound, U.S.A.			
1. Station	0–10	0	Wangersky, 1959
2. Station	0–30	0–2·5	Wangersky, 1959
3. Station	0–40	0–0·8	Wangersky, 1959
Atlantic Ocean, off southern U.S.A.	Surface	0–15	Anderson *et al.*, 1956a, b
Atlantic Ocean, off southern U.S.A.	0–1,900	0–12	Anderson *et al.*, 1956a, b

Table 2.19b. Dissolved "Protein" in Sea Water, Determined as Tyrosine and Tryptophane Equivalents

Locality	Depth (m)	mg/l.	References
Gulf of Mexico:			
Western Coast	0–32	0–1·5, av. *ca.* 0·5	Collier, 1958b
Northern and eastern Coast	0–20	0–3·8, av. *ca.* 0·9	Collier, 1958b
Deep water	0–2,500	0–3·2, av. *ca.* 0·6	Collier, 1958b
Naples Pier, Florida, U.S.A.	Surface	0–10·5	Dragovich, 1961
Atlantic Ocean, off southern U.S.A.	1–1,900	0–7	Anderson *et al.*, 1956a, b

Table 2.19c. Dissolved Lipids in Sea Water

Locality	Depth (m)	μg/l.	References
Pacific Ocean:			
33°N–46°S, 114–124°W	987–2,427	3·4–89·7, av. 42·5	Williams, personal communication
30°N, 119°W	10	30	Williams, personal communication
30°N, 119°W	91–924	84–112	Williams, personal communication
30°N, 119°W	1,388–2,665	51–33	Williams, personal communication
Gulf of Mexico	10–900	400–500	Slowey *et al.*, 1959

TABLE 2.19D. DISSOLVED ORGANIC ACIDS IN SEA WATER

Locality	Depth (m)	Organic acid	mg/l.	References
N.W. Pacific Ocean, offshore	Surface	Acetic acid	0·09–0·82	Koyama and Thompson, 1959
		Formic acid	0·03–0·38	
		Lactic acid	0·03–0·08	
		Glycollic acid	0·14–1·0	
Pacific Ocean, 31°N, 20°W	1,840	—	0·147	Williams, 1961
33°N, 22°W	2,500	—	0·158	
Bassin d'Arcachon, France	—	Citric acid	0·025–0·145	Creac'h, 1955
	—	Malic acid	0·028–0·277	

for a period of several months; but in the offshore and slope waters of the Atlantic values might exceed 10mg/l.

In the Pacific Ocean, the carbohydrate tests were applied by Lewis and Rakestraw (1955) who, in offshore and oceanic waters, obtained rather constant values well above 1mg/l. of sucrose equivalents, with no significant difference between filtered and unfiltered samples. Only in coastal lagoons do values run as high as 8mg/l. in water filtered through membrane filters of 0·5 μ pore size.

The substances in sea water that react with anthrone or *N*-ethyl carbazole have not yet been finally identified, but Wangersky (1952) isolated a substance which had properties of a rhamnoside and which was present in inshore waters of the Mexican Gulf in concentrations up to 100mg/l. The presence of dehydroascorbic acid was also indicated.

Proteins

The presence of dissolved proteins or aminoacids in sea water is only indirectly indicated. Belser (1959) tested sea water for various aminoacids by adding samples of filtered water to a number of artificially produced mutant strains of the marine bacterium *Serrata marinorubra* requiring the aminoacids for growth. By means of this technique, Californian coastal water has been found capable of stimulating growth of isoleucine, glycine, tryptophan and threonine exacting strains, but not of, e.g. histidine, methionine or arginine exacting strains of the bacterium. Support of growth of the isoleucine exacting strain was not regularly recorded, whereas only isolated occurrences of glycine, tryptophan and threonine were indicated. Other mutant strains indicated the frequent presence of biotin and uracil.

Attempts at quantitative determinations of dissolved protein was made by Collier (1958b) who adapted a photometric test for tyrosine or tyrosine–tryptophan for use on sea water samples; the test is based on the absorption in the ultraviolet part of the spectrum by aromatic aminoacids (Sizer and Peacock, 1947). Expressed as tyrosine equivalents, the values averaged 0·3–1·0 mg/l. in coastal and oceanic waters in the Gulf of Mexico (Collier, 1958b). Similar values have been recorded from the Atlantic Ocean (Anderson *et al.*, 1956a, b), with no systematic variation with depth (Table 2.19B).

As was the case with dissolved carbohydrates, so the proteins were found to vary greatly between samples taken at the same time from various depths. Generally, zero values have been obtained at one or more levels, but within some metres of such levels high concentrations might be measured. Furthermore, in series where carbohydrates and proteins have been determined in the same samples, it can be seen that the two may vary independently of each other. The discontinuity in the distribution of carbohydrates and protein is difficult to reconcile with the much smaller variation found in concentrations of total organic carbon and nitrogen. Moreover, the recorded amounts of carbohydrates, as arabinose equivalents, and tyrosine equivalents often greatly exceed the total amounts of organic matter or protein measured as organic carbon or nitrogen. Evidently, further investigations are needed before the significance of the measurements of carbohydrate and protein can be finally evaluated.

Lipids

A number of saturated and unsaturated fatty acids have been isolated from waters of the Pacific Ocean and their concentrations measured (Williams, 1961). The organic acids were extracted with carbon disulphide and chloroform from filtered samples of sea water. The residues from the extracts were methylated and the fatty acids identified by means of gas chromatographic analysis. Total amounts of fatty acids varied between 3·4 and 89·7 μg/l. in deep sea samples collected at various oceanic stations at depths between 1,000 and 2,400m (Table 2.19c). At one station, the variation in concentration of fatty acids was followed throughout the water column from depths of 10–2,700m. The highest values, 84–112 μg/l., were recorded from depths of 90–900m; in the surface layer and below *ca.* 1,300m, values ranged from 30–51 μg/l.

Organic acids

Williams also isolated organic acids from sea water by co-precipitation with ferric hydroxide. Two samples from depths of 1,840 and 2,500m yielded 147 and 158 μg/l., respectively, of these acids which are most likely derived from carbohydrates (Williams, personal communication).

Koyama and Thompson (1959) extracted quantitatively organic acids from filtered surface samples taken offshore in the north western Pacific Ocean. The acids were determined by partition chromatography on a silica column. The concentrations of the various acids were found to decrease with distance from the shore (Table 2.19D). Of special interest is the finding of up to 1mg/l. of glycollic acid which seems to be an important extra cellular product liberated by the living phytoplankton (Fogg and Nalewajko, 1963).

In the water of the landlocked Bassin d'Arcachon in France, Creac'h (1955) reported the presence of 25–145 μg/l. of citric acid and 28–277 μg/l. of malic acid (Table 2.19D). The acids are stated to represent only a small part of the total dissolved organic matter.

It thus appears that the sum of the chemically identified organic substances dissolved in sea water mostly amounts to only a small fraction of the total amount of dissolved organic matter.

Biologically active compounds

Dissolved organic substances may be of importance to aquatic organisms, including the suspension feeders, in other ways than by constituting potential sources of energy for maintenance and growth. During recent years, an increasing interest is shown in organic factors in natural waters exerting specific biological activities, as for instance vitamins, antibiotics, growth substances and enzymes (Fogg, 1959; Lucas, 1958; Vallentyne, 1957). Mostly, these substances are only known from their physiological or ecological effects (see p. 269), whereas their chemical nature is more or less obscure. Some vitamins have, however, been isolated from lake and sea water. Thus vitamin B_{12} has been demonstrated both in coastal and oceanic waters. In bioassays on vitamin B_{12} exacting micro-organisms, 0·2–2 $\mu g/m^3$ of the vitamin has been found in filtered surface waters from various localities from the North Sea to the North Atlantic, with high levels during the winter and low during the summer (Cowey, 1956). In Long Island Sound, U.S.A., annual variation was from 3 to 16 $\mu g/m^3$ (Vishniac and Riley, 1961). The vitamin is not confined to the surface waters; in the Bay of Biscay, the values at intermediate depths of 200–2,000m, *ca.* 2·3 $\mu g/m^3$, were even generally higher than in the euphotic zone or at the greatest depths (Daisley and Fisher, 1958). As previously mentioned, also biotin (Belser, 1959) and ascorbic acid (Wangersky, 1952) have been recorded from sea water.

2.14. BACTERIA

Bacteria are present in all natural waters from the surface to the greatest depths, and locally in great numbers. They have often been ascribed a role as food of aquatic organisms, and we should therefore attempt to evaluate their possible significance as food of the suspension feeders.

Determinations of standing stocks of bacteria have encountered great difficulties which are not yet finally overcome, so our knowledge of the mass of bacteria present in the various parts of the sea is even more fragmentary and inaccurate than is our knowledge of the mass of phytoplankton and organic detritus.

The methods of enumerating bacteria in water are either direct or indirect; the methods are reviewed by ZoBell (1946). In the indirect methods, the water sample to be analysed for content of bacteria is either mixed with a medium favourable for growth of bacteria, or the sample is passed through a membrane filter, and the filter with the retained bacteria is brought in contact with the medium. The number of colonies developing during the following period of incubation reflects the number of bacteria present in the water sample. In the direct methods, the bacteria are counted individually under the microscope. In polluted water, the content of bacteria may be high enough to allow direct counting of the cells in a water sample, but in most parts of the sea the bacteria are so scarce that they must be concentrated before microscopical enumeration. Concentration is mostly brought about by filtration through membrane filters. The bacteria may then either be transferred from the filter on to a slide (Cholodny, 1929; Jannasch and Jones, 1959) or they may, after fixation and staining, be counted directly in the filter, which is placed on a slide and made transparent (Butkevich and Butkevich, 1936; Jannasch, 1953, 1958; Kriss, 1961).

It has been repeatedly observed that the indirect methods generally yield much lower counts of bacteria than the direct ones. A comparison of seven methods of enumeration, five indirect and two direct, of bacterial populations in sea water was recently made by Jannasch and Jones (1959). Of special interest to us is their findings that not only revealed the direct methods' much higher counts than the plate methods, but also that the two direct methods gave widely differing results. Enumeration of bacteria transferred to slides after concentration by filtration gave on an average values about 14 times higher than direct counts on the membrane filter; the latter method on the other hand gave values about 150 times higher than the culture methods; but the variation was great (Table 2.20).

There are several reasons for the great difference in the results obtained by means of the various methods, and especially of the direct and the indirect methods: 1. Bacteria in the natural waters are often found in aggregates, and these register only as single individuals in the culture methods, whereas under the microscope, the individual cells within the aggregates may be counted.* 2. The culture methods may give lower values than the direct counts due to the presence of dead or inactive cells.

* Many bacteria are easily adsorbed, probably by electrostatic forces, to inert particles (Rubentschick *et al.,* 1936; Wood, 1953). The aquatic bacteria are thus not purely planktonic, but whether they proliferate freely suspended in the water or as aggregates attached to solid particles also depends upon the content of assimilable organic matter in the water. Thus Jannasch (1958) found that at low concentrations of organic matter, such as 0·1–0·5mg peptone per litre, growth of *Bacillus subtilis* is very restricted unless suspended particles are present in the water; but if inert particles are added, a rich growth of bacteria may take place, apparently due to the adsorption and concentration of organic matter on the surface of the particles. However, at higher concentrations of dissolved organic matter, e.g. 5mg/l. of peptone, the bacteria may grow freely in the medium, and addition of extra surface in the form of inert particles did not cause aggregation of bacteria on these particles. Under natural conditions, Jannasch found that bacteria occurred singly in the true planktonic state only in contaminated water with much organic matter in solution. In unpolluted sea water a large part of the bacteria present are attached to solid particles of various kinds (Jannasch, 1958; Jannasch and Jones, 1959; Jones and Jannasch, 1959; Kriss, 1961; Waksman, 1934; Waksman and Vartiovaara, 1938; Wood, 1953; ZoBell, 1946). The importance of solid surfaces for the utilization of the low concentrations of dissolved organic matter in natural sea water appears from the stimulating effect on bacterial growth of transferring samples of sea water to small flasks, or of adding inert glass beads to such samples (ZoBell, 1943).

However, according to Alfimov (1954, cited by Jannasch and Jones, 1959) dead cells usually constitute less than 10 per cent. 3. Any one culture method can only be expected to support growth of a part of the bacterial species present in the water, since no one medium can be expected to provide for the nutrient

TABLE 2.20. RATIOS OF DIRECT BACTERIAL COUNTS TO COLONY COUNTS (After Jannasch and Jones, 1959)

	Direct counts on membrane filter	Cholodny method
Mean	147	2,100
Range	13–840	108–9,700

requirements of all bacteria. Thus, with the direct methods, Jannasch and Jones (1959) observed large spirilli-like organisms at all sampling depths, from the surface to 200m, but none of these forms produced colonies to be observed in the culture methods. 4. Bacteria may be separated only with difficulty from microflagellates (Wood, 1953), or other non-bacterial cells, or even dead organic or inorganic particles (Kriss, 1959a; ZoBell and Morita, 1959).

However, none of these factors explain the large difference in counts obtained by Jannasch and Jones (1959) in the two direct methods. Here the low value obtained by counting directly on the filter is thought to be caused by small and weakly stained cells escaping observation on the filter due to the lower optical contrast of the preparation.

STANDING STOCK OF BACTERIA

From the above, it would appear that reliable estimates of the standing stock of bacteria in natural waters cannot be obtained from enumerations of the bacterial populations by means of the culture methods, but only by means of direct counts and measurements of the bacteria. Unfortunately, by far the greater part of enumerations of marine bacteria have applied culture methods.

Furthermore, direct counts of bacteria have mostly been made directly on the membrane filter, in which cases estimates of bacterial mass would tend to be low. But, it is not possible to correct the estimates by means of the ratio found by Jannasch and Jones between counts made directly in the filter and after transfer to a slide, since with a slight change in technique or with another composition of the bacterial populations, the values obtained by direct counts on the filter might be much closer to the true values than those obtained by Jannasch and Jones.

The various authors have mostly used a standard bacterial size in estimating standing stocks from counts of bacteria. However, Kriss (1959a, b, 1961) based his estimates on actual measurements of the size of the various types of bacteria in the samples counted. In the Black Sea the average bacterial size thus varied from about 0·2 μ^3 to 2 μ^3 (Butkevich, 1938; Kriss, 1959a, b; Wood, 1953; ZoBell and Morita, 1959). In the present paper, an average bacterial size of 0·5 μ^3 has been adopted in the calculations of standing stocks of bacteria from available data on bacterial numbers. This corresponds to the volume of a spherical bacteria, 1 μ in diameter; or of a rod-shaped bacteria 2·5 μ in length and 0·5 μ in width, which is the average size of marine bacteria (ZoBell, 1946, p. 116). In order to make direct comparison possible with the data already tabulated on the size of the standing stock of phytoplankton and other organic fractions in sea water, the standing stock of bacteria has been calculated as μg of organic matter per litre of water by assuming a bacterial water content of 80 per cent and a specific weight of 1.1 μg of organic matter is thus contained in 10^7 bacteria.

Estimates of standing stocks of bacteria in various regions and depths of the seas are recorded in Table 2.21. From oceanic surface waters of the Pacific Ocean two sets of bacterial counts are available; those of Kriss (1959a, b) are some 500 times higher than those of Jannasch and Jones (1959), even when both sets of counts are obtained by means of the same method of direct counts on membrane filter. This difference in results probably tells more about the inaccuracy of the methods available for estimating standing stocks of bacteria than about regional or seasonal variations in the size of the bacterial populations in the Pacific

Ocean.* The values of about 10 μg/l. of bacterial organic matter obtained by Kriss is very high compared with the values found for the standing stock of phytoplankton (Tables 2.6 and 2.9). The value of 0·02 μg calculated from the data of Jannasch and Jones are, according to these authors, low. However, the true value should most probably not have been rated 20 times higher even if application of the Cholodny method should have revealed some 20 times higher counts, because the bacteria that may have escaped recording by counting on the filter tend to be small cells of much smaller average mass than those recorded.

From the region of the North Pole, bacterial counts amounted to about one-tenth the counts from the Pacific Ocean, whereas in the Caspian and Black Sea, counts in the surface waters were similar to those of the oceanic Pacific waters (Kriss, 1959a, b, 1961).

In most waters investigated, the standing stock of bacteria decreases with increasing depth of sampling. In the Pacific Ocean, at 2,000–5,000m, the bacterial counts have fallen to about one-hundredth of that found in the surface waters, and in the abyssal depths of the deep trenches from 5,000 to 9,000m, to only one-thousandth. In the bottom deposits, however, the bacteria get numerous again, amounting to 10^6–10^8 cells per ml. in the top-most layer (Kriss, 1961; ZoBell and Morita, 1959), or 0·1–10mg organic matter per litre mud. At lower depths, in estuarine silt, 10^9 cells may be counted per ml. (Wood, 1953).

The bacterial counts may increase excessively in inshore waters, especially above mud flats and in polluted waters. In the shallow waters of lagunes in Texas, Oppenheimer and Jannasch (1962) recorded a standing stock of bacteria ranging from 300 to 2,900 μg organic matter per litre, whereas on the beach of the Gulf of Mexico outside the lagunes the values were much lower, though still rather high (Table 2.21). Comparable values were obtained by ZoBell and Feltham (1942), who used the plate count method, from the water leaving and entering the mud flats of Mission Bay, California. The tremendous effect of pollution appears, e.g.

* Sorokin (1962) has criticized Kriss for using non-sterile Nansen's metal bathometers in obtaining water samples for the enumeration of bacteria by the culture method. It is not clear to what extent this criticism may invalidate the results obtained by the direct counting method.

Locality	Sampling depth (m)	μg/l. dry organic matter	Method	References
N.W. Pacific Ocean	0–50	13	A	Kriss, 1959a, b, 1961
	50–200	4		
	200–500	0·6		
	500–2,000	0·16		
	2,000–5,000	0·03		
	5,000–9,000	0·005		
Region of North Pole	0–100	0·8		
	100–200	0·3		
	200–2,000	0·05		
	2,000–3,000	0·015		
Caspian Sea	0–50	14		
	50–100	4		
	100–200	1·7		
	200–500	0·6		
	500–750	0·02		
Black Sea	0–50	17		
	50–2,000	6		
Pacific Ocean:				
16 miles off San Diego, California	0–200	0·02	A	Jannasch and Jones, 1959
Off Scripps' Pier, California		0·2	A	
		4	B	
Gulf of Mexico:				
Beach outside surf zone, clear water	1·8	90	B	Oppenheimer and Jannasch, 1962
Surf zone	1·2	160		
Shallow lagune water, Texas	0·6	300–2,900		
Mediterranean Sea:				
Offshore, 1,200 m's depth	1·0	0·03	A	Jannasch, 1958
Naples Harbour	1·0	8,300		

A, Counts on membrane filter. B, Cholodny method.

from the finding of Jannasch (1958), that the standing stock of bacteria in the harbour of Naples was some hundred thousand times larger than in the unpolluted offshore water.

Below (p. 286), an evaluation is attempted of the possible importance of bacterial populations in the various main strata of the water masses and in the bottom sediments of the oceans in utilizing dissolved organic matter in the water. It is, therefore, of interest to compare the size of the populations in these strata and in the upper layer of the bottom sediments from where bacteria may be taken into suspension by the bottom currents. This layer is assumed to be 1cm thick and to contain 10^7 bacteria per cm^2 (ZoBell, 1954; ZoBell and Morita, 1959). In Table 2.22,

TABLE 2.22. RELATIVE ABUNDANCE OF BACTERIA IN THE VARIOUS STRATA OF THE WATER MASSES OF THE PACIFIC OCEAN AND IN THE OCEAN BOTTOM SEDIMENTS (From Kriss, 1952, 1959a, b, 1961, and ZoBell and Morita, 1959)

Depth (m)	Per cent of total standing stock of bacteria from surface to 1cm deep in the bottom sediments
0–100	52
100–1,000	31
1,000–5,000	15
5,000–9,000	1
Topmost cm of ocean floor	1

the number of bacteria present in the various parts of the water column and in the upper layer of the bottom sediments are expressed in per cent of the total number present from the surface of the Pacific Ocean to a depth of 1cm in the bottom sediments. According to these data, half the standing stock of bacteria per unit surface is contained within the upper 100m, which roughly correspond to the euphotic zone; whereas the fraction of the standing stock contained within the uppermost layer of the bottom sediments amounts to only 1 per cent. The fraction contained within the deep-sea zone from 1,000 to 5,000m is some 15 times larger than that found in this bottom layer. Even the zone of the trenches, very poor in bacteria, contains a standing stock that does not seem to be smaller than that of the bottom layer.

2.15 RELATIVE SIZE OF THE VARIOUS FRACTIONS OF ORGANIC MATTER

The previous sections reviewed our knowledge of the content of phytoplankton, organic detritus, dissolved organic matter, and bacteria in sea water. In Table 2.23 are presented estimates of the relative size of these organic fractions in the sea. The estimates are very crude and only suggestive, both due to the inaccuracies of the available methods, and to the great geographic and seasonal variations encountered, especially in the standing stock of phytoplankton and bacteria.

TABLE 2.23. ESTIMATES OF STANDING STOCKS OF PHYTOPLANKTON AND BACTERIA IN PER CENT OF TOTAL PARTICULATE ORGANIC MATTER; AND OF PARTICULATE ORGANIC MATTER IN PER CENT OF TOTAL ORGANIC MATTER IN SEA WATER

Habitat	Phytoplankton in per cent of total particulate organic matter	Bacteria in per cent of total particulate organic matter	Particulate organic matter in per cent of total organic matter
Oceanic surface waters	*ca.* 10–20	0·1–1	*ca.* 10
Coastal offshore waters	*ca.* 10–20	0·1–1	up to *ca.* 50
Coastal inshore waters	*ca.* 10–20	Strongly varying	up to *ca.* 50
Deep Sea	—	0·1–1	1–10

During photosynthetic periods, phytoplankton generally seems to constitute from some few hundredths to about one-third of the total particulate organic matter within the euphotic zone, with no significant difference between oceanic and coastal waters. The bacteria seem to amount to 0·1–1 per cent of the organic particulate matter, both in the surface layers of unpolluted coastal and oceanic waters and at the great depths of the oceans. In the oceans, particulate organic matter seems to constitute up to about one-tenth of the total organic matter. In coastal waters, the particulate organic matter may increase strongly, whereas the dissolved organic matter only increases moderately, so that the particulate organic matter in places amounts to half or more of the total organic matter present in the water.

It is close at hand to speculate as to the reason why the content of dissolved organic matter in sea water is mostly so much higher than the content of particulate organic matter. The two fractions are probably approximately in a steady state (p. 291), so that the rates at which new substance is added to the fractions equal the rates at which substance is subtracted. The concentrations reached will therefore depend upon the efficiency of the processes that subtract substance. Processes which act on both fractions are decomposition by bacteria and retention by suspension feeders and other organisms; but organic matter in particulate form is more readily available to bacterial decomposition and to retention by suspension feeders than is dissolved organic matter (see p. 290 and p. 132). Furthermore, particulate matter is continuously being removed from suspension by sedimentation, whereas dissolved organic matter may be transferred to the particulate fraction by being adsorbed to particles in suspension; significant amounts may also be adsorbed to the particles of the bottom sediments. In the greater part of the sea, the combined effect of these four processes is presumably largely responsible for the ratios established between particulate and dissolved organic matter.

2.2. CONCENTRATIONS OF THE VARIOUS ORGANIC FRACTIONS IN FRESHWATERS

The freshwaters represent an extremely heterogeneous assembly of biotops ranging from ponds to ocean-like lakes, from barren arctic or alpine lakes to eutrophic lowland lakes, from crystal-clear springs to more or less polluted rivers, etc. No attempt is made to characterize all the types of freshwater habitats with respect to their contents of the various fractions of particulate and dissolved organic matter, but examples shall be given, especially from lakes of varying degrees of productivity, in order to provide materials for comparisons with the marine habitats.

Phytoplankton

Data on phytoplankton concentrations are collated in Table 2.24. It may be seen that the concentrations reached during the periods of high productivity vary from 1,000 μg/l. or more organic matter in eutrophic lakes to about 50 μg/l. in an oligotrophic alpine lake. Even lower values have been encountered in extremely oligotrophic lakes of Lapland (Rodhe, 1963, p. 31). The concentrations of phytoplankton in freshwaters thus seem to cover the whole range met with in marine waters.

Bacteria

The data on bacterial organic matter obtained from direct counts (Table 2.25) indicate that in many lakes the bacterial fraction of particulate organic matter is of the same order of magnitude as the phytoplankton organic matter, that is, generally much higher than in unpolluted sea water; however, the figures vary very much between authors. Thus, in the mesotrophic Lake Erken, Sweden, Nauwerck (1963) only encountered about one hundredth the bacterial population of the mesotrophic lakes in U.S.S.R.

It is noteworthy that extreme variation may be obtained within the same body of water; in Seeburger Lake, Jannasch (1958) obtained about 1,000–10,000 times higher counts 10cm above the bottom than at the surface of the lake. The importance of contamination is obvious from the finding in River Fulda of 100 times higher counts in a highly contaminated section of the river compared with a rather clean section. No bacteria were found in the continuous spring of a brook.

Organic detritus

Wright (1959) has reviewed data on the ratios between detritus and total particulate suspended matter obtained by various authors in various lakes and ponds. The estimates range from about 20 to 80 per cent detritus, organic + inorganic. In Lake Erken, Sweden, Nauwerck (1963) found the annual average volume of detritus to be about the same as that of phytoplankton. However, the ratios between the two fractions varied throughout the year: during the phytoplankton blooms in April and August the detritus volume was only about half that of the phytoplankton volume, but 10 times the phytoplankton volume in June. In

Table 2.24. Standing Stock of Phytoplankton in Fresh Waters, as Estimated from Cell Volume or Chlorophyll Content

Locality and type of freshwater	Time of year	Phytoplankton, μg dry organic matter per l.	References
Sweden:			
Lake Erken, mesotrophic	Annual variation	10–1,000	Nauwerck, 1963; Rodhe *et al.*, 1958
	Summer average	200	
Görväln	May–September	100–500	Willén, 1959
Poland:			
Lake of Raigrod, mesotrophic	April–November	*ca.* 800	Czeczuga, 1958
Lake Biale, mesotrophic		*ca.* 400	
Lake Drestwo, mesotrophic		400–1,300	
Lake Krzywe, eutrophic		3,000–7,000	
Lake Slepe, dystrophic		*ca.* 200	
Austria:			
Lunzer Untersee, alpine lake	Annual variation	3–50	Steemann Nielsen, 1959
U.S.A.			
Linsley Pond	Annual variation	20–400	Riley, 1940
Canyon Ferry Reservoir	April–October	100–900	Wright, 1959

Table 2.25. Standing Stock of Bacteria in Fresh Waters

Locality and type of fresh water	Time of year	Bacteria, μg dry organic matter per l.	References
U.S.S.R.			
Oligotrophic lakes: Baikal, Pert and Kontsch	Summer stagnation	15	Kusnezow, 1959
Mesotrophic lakes: Jolowoje, Ilmen	Summer stagnation	100	Kusnezow, 1959
Eutrophic lakes: Beloje, B. Medweshje	Summer stagnation	300	Kusnezow, 1959
Dystrophic water: peat bog	Summer stagnation	200	Kusnezow, 1959
Salt lakes: B. Kriwoje, M. Umreschewo	Summer stagnation	1,200	Kusnezow, 1959
Bittersalt lakes: Balpasch, Orassor, Gorkoje	Summer stagnation	30	Kusnezow, 1959
Lake Naroch, mesotrophic	Annual variation	100–300, av. 180	Beliatskaia, 1958
Lake Batorin, eutrophic	Annual variation	580–2,100, av. 1,260	Beliatskaia, 1958
Lake Miastro, intermediate	Annual variation	120–590, av. 240	Beliatskaia, 1958
Various lakes around Lake Ladoga:			
open lake	—	130–1,800, av. 600	Rodina, 1960
weed beds	—	1,000–5,400, av. 2,600	Rodina, 1960
Germany:			
Seeburger Lake: surface	—	0·05	Jannasch, 1958
10 cm above bottom	—	260	Jannasch, 1958
River Fulda: rather clean section	—	29	Jannasch, 1958
highly contaminated		1,123	Jannasch, 1958
Brook Rase: continuous spring	—	0	Jannasch, 1958
1km after spring	—	0·25	Jannasch, 1958
Sweden:			
Lake Erken, mesotrophic	Annual variation	0·1–4	Nauwerck, 1963
U.S.A.:			
Lake Mendota	Annual variation	2–100	Bere, 1933
Lake Wingra	Annual variation	15–220	Bere, 1933
Lake Trout	Annual variation	3–14	Bere, 1933

rivers much higher proportions of detritus have been found, up to more than 99 per cent (Bursche, 1959). The proportion of detritus thus seems to vary strongly, not only between different types of freshwaters, but also seasonally. At certain times of the year and in certain localities the greater part of particulate organic matter available as food to the suspension feeders may be present in living form in the phytoplankton, at other times or in other localities detritus may dominate the particulate organic fraction; however, the exact proportions of the organic detritus are not known, since the methods of measuring the detritus are inaccurate, and most often no differentiation has been made between organic and inorganic detritus.

Dissolved organic matter

The content of dissolved organic matter in freshwaters varies much more than in marine habitats, but, as in sea water, the fraction of dissolved is generally many times larger than that of particulate organic matter. In a great number of lakes in Wisconsin, U.S.A., Birge and Juday (1934) found the dissolved organic matter to range from a few mg to more than 50mg per litre. Part of this variation is due to variation in the productivity of the lakes, the internally produced dissolved organic matter increases in concentration with increasing productivity. However, in the lakes with high organic contents much of the matter consists of humic and other substances derived from external sources and drained into the lakes with water from the catchment area of the lake or carried with the wind. Measurements from other geographic areas fall within the range indicated by Birge and Juday (Krogh and Lange, 1931; Forsblad, 1955).

A large number of organic substances have been identified in the dissolved organic fraction of various freshwaters, and often their quantities have been measured; but much remains to be learnt about the quantitative composition of the dissolved organic fraction (Hutchinson, 1957; Vallentyne, 1957).

When we know the amounts of the various organic fractions present in the water we can calculate how much water should be cleared of utilizable organic matter in order to cover the food requirements of the suspension feeders inhabiting the water. But we cannot appreciate the absolute and relative importance as food of the various organic fractions until we know their rates of production and of turnover. We should, therefore, next look into how accurately these parameters have been estimated.

2.3. RATES OF PRODUCTION AND TURNOVER TIMES OF ORGANIC FRACTIONS

2.31. PHYTOPLANKTON

2.311. METHODS OF MEASURING PRODUCTION

Phytoplankton production has been calculated from the decreases in concentrations of carbon dioxide, phosphate or nitrate that occur due to the photosynthetic activity of the plankton (Atkins, 1922, 1923; Cooper, 1958; Steele, 1957, 1958). These methods can however only be used under special hydrographical conditions and have therefore been used only locally. Most determinations of phytoplankton production have been made by means of the oxygen and the carbon-14 techniques. The oxygen technique was introduced by Gaarder and Gran (1927); it consists of measuring changes in concentration of oxygen in pairs of water samples enclosed in transparent and black bottles, respectively, the bottles being exposed to the temperature and the light conditions of the place of sampling. If the processes going on in the water samples are not significantly influenced by the transfer to the bottles, the fall in oxygen content in the black bottles will correspond to the natural rates of oxygen consumption of the living phyto- and zooplankton organisms, including bacteria and other micro-heterotrophs, whereas the oxygen content in the transparent bottles will change according to the relation between oxygen consumed due to respiration and oxygen produced due to photosynthesis. From the difference between concentrations of oxygen in pairs of transparent and black bottles, it is thus possible to calculate the oxygen production or gross photosynthesis during the experimental period.

The method has been extensively used and provides reliable results in productive areas, in sea water as well as in freshwater. However, the technique is presumably not sensitive enough to be used in oligotrophic areas such as the greater parts of the oceans, especially within the tropical regions. In these regions the experimental periods needed may be so long, up to some days, that the composition and behaviour of the populations of phyto- and zooplankton and bacteria inside the bottles may change significantly, so that the final changes in oxygen content no longer represent natural rates of respiration and assimilation (Steemann Nielsen, 1958a; Prat and Berkson, 1959; Strickland, 1960).

A great improvement in the technique of measuring the primary production was introduced with the carbon-14 method of Steemann Nielsen (1952), for the first time extensively used during the Danish "Galathea" Expedition around the world 1950–52. The method has been described in detail by Steemann Nielsen and Jensen (1957), Steemann Nielsen (1958a) and Strickland (1960). In the method, incorporation of tracer amounts of carbon-14 in the phytoplankton is used as a measure of production. Samples of sea water or freshwater are transferred to bottles, with known amounts of added $^{14}CO_2$ and exposed to the natural temperature and light conditions, either by being replaced in the original surroundings (*in situ* experiments) or in temperature-regulated tanks under controlled artificial light conditions. After exposure, the samples are filtered through membrane filters and the amounts of carbon-14 assimilated are measured by counting the β-radiation from the plankton retained by the filters.

In experiments of short duration, a few hours, the method measures something between gross and net production (see Strickland, 1960). As respiration represents only some 10 per cent of the light-saturated photosynthesis (Steemann Nielsen and Hansen, 1959b), the measurements deviate only little from either net or gross production (see below p. 232). In experiments of long duration, 24 hr as used, e.g. by Ryther and co-workers, equilibrium may be approached between ^{14}C and ^{12}C in the algae and the medium. The final carbon-14 count therefore approximately represents the total amounts of phytoplankton carbon, that is the carbon originally present in the phytoplankton sample plus the net production of algae during the experimental period. This net production, though, may deviate significantly from the net production under natural conditions, because, as mentioned, the artificial conditions inside the bottles may cause changes in species composition and physiology of the populations. Apparently, oceanic species of the phytoplankton are generally more sensitive to the artificial surroundings inside the bottles than species from coastal waters. However, the carbon-14 method is so sensitive that assimilation periods of some few hours' duration suffice to provide accurate estimates of production even in the least productive areas of the oceans, as the Sargasso Sea; and experimental periods of few hours' duration were found not to influence measurably the rate of assimilation (Steemann Nielsen and Jensen, 1957).

Mostly, measurements of the primary production aim at estimating the total production throughout the photosynthetic zone. Measurements are therefore made in samples of water from several levels within this zone and at the light intensities characteristic of the various levels. The final result is generally expressed in mg of carbon assimilated in 24hr/m^2 of surface of the sea.

2.312. RATES OF PRIMARY PRODUCTION IN VARIOUS REGIONS OF THE SEA

In Tables 2.26 and 2.27 are collated data on gross production of phytoplankton representative of the various marine (and freshwater) biotops. The vast tropical and subtropical regions of the oceans produce some 100–200mg C/m^2/day with little annual variation; but in border regions between currents with upwelling of nutrient-rich deep water the productivity increases, for instance to 200–500mg C/m^2/day at the Equatorial Counter-Currents in the different oceans. In offshore waters on the western sides of the Continents strong upwelling of deep water causes very high production rates, up to 3,000mg C/m^2/day. By contrast, in regions with "old" surface water as e.g. the Sargasso Sea, production rates generally do not exceed some 100mg C/m^2/day. At higher latitudes, rates of production in oceanic waters vary greatly, not only with the amounts of nutrients available, but also with the great seasonal variations in light and temperature.

Generally, nutrients are more plentiful and production therefore greater in the coastal waters over the shelf than in true oceanic waters outside the continental slope. In coastal waters, the production varies with the seasons of the years, especially pronouncedly, of course, at higher latitudes where rates of 1,000mg C/m^2/day may be reached during the blooms of the summer

TABLE 2.26. PHYTOPLANKTON PRIMARY PRODUCTION IN OCEANIC WATERS, MEASURED BY THE C-14 METHOD

Locality	Time of year	Gross production mg C/m²/day	References
Various oceanic waters:			
(1) Areas with "old" surface water, e.g. central part of the Sargasso Sea	Annual variation	*ca.* 50	Steemann Nielsen and Jensen, 1957
(2) Largest parts of the tropical and sub-tropical regions	Annual variation	100–200	Steemann Nielsen and Jensen, 1957
(3) Areas with some upwelling of deeper water, e.g. Equatorial counter-current	Annual variation	200–500	Steemann Nielsen and Jensen, 1957
Northern North Atlantic Ocean:			
S.E. of Greenland	August	*ca.* 250	Hansen, 1959
W. of British Islands	August	*ca.* 75	Hansen, 1959
Sargasso Sea:			
15 miles S.E. of Bermuda	February–April	*ca.* 300–800*	Ryther and Menzel, 1960
15 miles S.E. of Bermuda	May–January	*ca.* 100–200*	Ryther and Menzel, 1960
Mediterranean Sea	July, October	32, 42	Brouardel and Rinck, 1956
N.E. Pacific Ocean, 50°N, 145°W	July–August	205	McAllister *et al.*, 1960

* Net production.

TABLE 2.27. PHYTOPLANKTON PRIMARY PRODUCTION IN COASTAL WATERS, MEASURED BY THE C-14 METHOD

Locality	Time of year	Gross production mg C/m^2/day	References
Tropical and sub-tropical offshore waters with strong upwelling of nutrient-rich deep water	—	500–3,000	Steemann Nielsen and Jensen, 1957
Coastal waters of the Faroes	June	800	Hansen, 1959
Faroe–Iceland Ridge	June	650–2,700	Hansen, 1959
Norwegian Sea	June	400–1,800	Berge, 1958
Godthåb Fjord, W. Greenland	Annual variation	0–1,200, av. 260	Steemann Nielsen, 1958b
Outside Narssak, S.W. Greenland	Annual variation	0–200, av. 80	Steemann Nielsen, 1958b
Great Belt, Denmark	Annual variation	10–670, av. 180	Steemann Nielsen, 1958b
Inshore shallow waters, Denmark:			
Venø Bay, Limfjord	Annual variation	0–600, av. 290*	Grøntved and Steemann Nielsen, 1957
Ulvedybet, Limfjord	April–October	50–200, av. 170*	Grøntved, 1960
Dybsø Fjord	Annual variation	0–70, av. 17†	Grøntved and Steemann Nielsen, 1957

* Macrovegetation poor. † Macrovegetation rich.

half of the year, whereas production is low or nil during the winter.

In shallow inshore waters, competition can apparently take place between the phytoplankton and the macrovegetation of the bottom. Thus in the shallow Dybsø Fjord, Denmark, with a rich underwater macrovegetation, the phytoplankton production was $\frac{1}{10}$ or less of that measured in comparable localities, but with a poor macrovegetation. In Dybsø Fjord, the annual production of macrovegetation was more than 10 times larger than the phytoplankton production (Grøntved and Steemann Nielsen, 1957).

Of more general significance than the macrovegetation in shallow waters is the microbenthos, mainly consisting of diatoms, free living or attached to sandgrains and other particles in the bottom. In shallow Danish fjords, Grøntved (1960, 1962) found that the production of the microbenthos was about four times larger than the production of phytoplankton. In shallow waters, turbulence due to the wave action or tidal currents may stir large amounts of these bottom diatoms into suspension in the water; the microbenthos may thus under circumstances contribute significantly to the potential food sources of the suspension feeders inhabiting the shallow waters.

2.313. RELATIONSHIP BETWEEN PRIMARY PRODUCTION AND STANDING STOCK OF PHYTOPLANKTON (TURNOVER TIME)

It appears from a comparison of the geographical and seasonal variation in the standing stock of phytoplankton (Tables 2.6–2.9) with the corresponding variations in primary production, that the two are correlated. The correlation would have been even closer if production had been expressed per unit volume of the whole photosynthetic zone, because the depth of the zone varies inversely with the rate of production: in waters where light adsorption due to particles is mainly caused by the phytoplankton algae, the photosynthetic layer was found to be 100m deep at maximal photosynthetic rates of *ca.* 0·1g C/m^2/day and 0m at *ca.* 4g C/m^2/day (Steemann Nielsen and Jensen, 1957).

The relationship between production and standing stock of phytoplankton will be dealt with in some detail because of our interest in knowing the rates of turnover of the standing stock. These rates of course ultimately depend upon the rates at which the chlorophyll contained in the algae can support assimilation of carbon dioxide and production of organic matter.

A number of measurements have been made of photosynthetic activity per unit of chlorophyll at optimum light conditions, both in pure cultures of algae and in natural phytoplankton populations. The values range from 1 to 10mg C assimilated per hour and mg of chlorophyll a (Holmes, 1958a; McAllister *et al.,* 1961; Steemann Nielsen and Hansen, 1959a; Ryther and Yentsch, 1957; Shimada, 1958; Steele and Yentsch, 1960, etc.). In natural phytoplankton populations, Steemann Nielsen and Hansen (1959a) showed that the high values characterized phytoplankton grown at high light intensities, sun phytoplankton, and low values phytoplankton grown at low light intensities, shade phytoplankton. Thus the value was close to 8 in tropical surface plankton; about 3–4 in temperate and arctic summer plankton from the surface; and close to 1 in temperate winter plankton from the surface and temperate and arctic summer plankton from low levels of the photosynthetic zone under conditions of stable stratification of the water masses, so that mixing of the phytoplankton was only slight between the upper and lower levels of the photosynthetic zone. In northern areas with only slight stabilization of the water masses near the surface and therefore extensive mixing of phytoplankton within the photosynthetic zone, both plankton from the upper and the lower level of the zone yielded values close to 3.

In order to calculate the rates of turnover of the phytoplankton mass the ratios between chlorophyll and total organic matter in the algae should be known. As mentioned previously the ratio is not accurately known and may furthermore vary with the species and external conditions for growth, especially the light intensity. Steeman Nielsen and Hansen calculated gross photosynthesis per unit of phytoplankton carbon at optimal light conditions under the assumption that chlorophyll constituted 3 per cent of the organic matter in shade plankton, 1·5 per cent in arctic surface plankton and 1 or 0·5 per cent in tropical surface plankton. According to their calculations, 0·08mg C was produced per hour and mg phytoplankton carbon in shade plankton (from temperate and arctic waters), 0·17mg C/hr, mg C in arctic surface water, and 0·2 or 0·1mg C/hr, mg C in tropical surface waters.

From the average value of some 0·15mg C/hr, mg C, a maximal rate of turnover of phytoplankton populations in nature may be estimated. Assuming a day length of 12 hr. and an average production throughout the photosynthetic zone of half the production at optimal light intensities, i.e. *ca.* 0·08C/hr, mg C, we find that the daily production in per cent of the standing stock will amount to $100/(0{\cdot}08 \times 12) = 104$ per cent. This value falls within the range of values calculated from direct estimates of standing stock and rates of production under natural conditions (Table 2.28). Production rates have been determined by means of the carbon-14 technique, except by Ragotzkie (1959) who used the oxygen technique, which, however, in eutrophic waters such as the estuarine waters in which Ragotzkie made his investigations yields results in agreement with those obtained by the carbon-14 technique (Steemann Nielsen and Jensen, 1957; Vinberg and Kaler, 1960). Standing stocks were estimated either from measurements of cell volume or from chlorophyll, under the simplified assumption that chlorophyll a constitutes 3 per cent of the phytoplankton organic matter. The values of daily gross production in per cent of the standing stock vary from 20 to 200. Part of this large variation may be due to the disregard of the natural variation in the chlorophyll content of the phytoplankton. Thus, the value of 200 per cent from estuarine waters in Georgia, U.S.A., is probably too high because the phytoplankton in this area is likely to have been sun plankton containing more likely 1 than 3 per cent chlorophyll, in which case the daily production would have been only some 70 per cent of the standing stock.

Within the accuracy of the measurements, the rates of turnover of the phytoplankton standing stock could not be shown to vary significantly with the latitude from subtropical to arctic regions in the summer half of the year (Corlett, 1958; Currie, 1958; Ragotzkie, 1959), or with the type of the water masses, whether oligotrophic oceanic or eutrophic coastal waters (Currie, 1958; Hansen, 1959).

Part of the gross production is consumed during the respiration of the phytoplankton itself. In order to estimate the net

TABLE 2.28. DAILY GROSS PRODUCTION IN PER CENT OF STANDING STOCK OF PHYTOPLANKTON

Locality	Time of year	Gross production in per cent of standing stock	Method of measuring standing stock	References and sources used for the calculations
N.E. Pacific Ocean	July–August	20 62	Chlorophyll a: 1 per cent Chlorophyll a: 3 per cent	McAllister *et al.*, 1960
Atlantic Ocean:				
Northern oceanic and coastal waters	June–August	*ca.* 30	Chlorophyll a	Hansen, 1959
Oceanic and coastal waters off Portugal and North Africa	September	*ca.* 20	Total pigments: 6 per cent	Currie, 1958
Barents Sea, 75°N, 25°E	May	66	Cell volume	Corlett, 1958
North Sea	April–May	22–83, av. 49	Cell volume	Cushing, 1958b
North Sea	October March	31 55 (5°C) 123 (10°C) 168 (15°C)	Cell volume	Wimpenny, 1958
Estuarine waters, Georgia, U.S.A.	February–July	98–435, av. 206	Chlorophyll a	Ragotzkie, 1959

production, which represents the amount available as food to the phytoplankton feeders, the rate of respiration must be known. In a large number of measurements of the rate of respiration in natural phytoplankton from various regions of the oceans and depths within the euphotic zone, Steemann Nielsen and Hansen (1959b) found the respiration to be approximately 10 per cent of the photosynthesis at optimal light intensities. As the rate of respiration is light independent (Brown, 1953, cit. Ryther, 1956), the average net production throughout the euphotic layer can therefore, at a day length of 12 hr, be estimated to about 60 per cent of the gross production (Steemann Nielsen and Jensen, 1957). Consequently, if the daily gross production amounts to some 20–100 per cent of the standing stock of phytoplankton, the net production will amount to some 12–60 per cent of the standing stock, or, in other words, the standing stock is reproduced about every 2 to 8 days within the euphotic zone over wide areas of the natural waters. Much lower rates of turnover of the standing stock are only to be expected during winter at higher latitudes.

2.32. Organic Detritus

The immediate sources of the organic detritus range from dying and decaying plants and animals, and solid excreta of animals inhabiting the water masses, to organic debris of terrestric origin, drained into the sea or lake by rivers. However, except in some near-coastal areas with significant supplies of organic matter from land or from the benthal vegetation, the organic detritus is ultimately derived from the phytoplankton. In the larger part of the sea and the big lakes, the production of detritus is therefore maximally as large as the production of phytoplankton, but generally smaller since significant amounts of the phytoplankton are assimilated and remineralized by grazing animals, or go into solution in the water to be included in the fraction of dissolved organic matter. Attempts will be made to estimate rates of production and turnover of detritus only in regions that are not significantly influenced by land drainage or benthic vegetation; the euphotic zone and the zone below will be considered separately.

2.321. EUPHOTIC ZONE

The phytoplankton produced within the euphotic zone may either leave the zone by passive sinking or active transport downwards in the intestinal tracts of vertically migrating zooplankton that preferentially feed within the euphotic zone; or it may be ingested and assimilated by the grazing animals or it may die within the euphotic zone. Only in the last case is the phytoplankton transferred without loss into detritus or dissolved organic matter. Part of the phytoplankton eaten and assimilated by the grazers, and thus indirectly by the higher links in the food chain, will eventually enter the detritus pool. The proportions of phytoplankton that leave the euphotic zone, or are eaten or directly transferred into detritus, presumably vary, especially with the relationship between the stocks of phytoplankton and of grazers. In order to evaluate the importance of these relations, two extreme ecological conditions should be considered: (1) phytoplankton and zooplankton productions are approximately constant throughout the years, and (2) phytoplankton and zooplankton production show pronounced maxima and minima, the zooplankton maxima following the phytoplankton maxima in time.

(1) Under the first condition, which prevails in tropic and subtropic seas, the steady state is widely assumed to be due mainly to the controlling activity of the zooplankton, which prevents blooming of the phytoplankton by ingesting the greater part of the daily net primary production (p. 271).

Phytoplankton losses by sinking may somewhat arbitrarily be estimated to be about 10 per cent; estimates vary from 4 to 25 per cent (Datzko, 1955, cit. Duursma, 1961; Hasle, 1959; Riley, 1951; Steemann Nielsen, 1962). No exact values are available of the proportion of phytoplankton that dies off and is thus directly converted into detritus and dissolved organic matter. However, the proportion is presumably small if the constancy of phytoplankton production and standing stock is maintained by the grazing activity of the zooplankton; the proportion is here estimated to be 10 per cent of the net primary production, leaving 80 per cent to be ingested by the population of grazers. How much of these 80 per cent are converted into detritus and dissolved organic matter depends upon the gross growth efficiency of the grazers and later links in the food chain. The part of the ingested phytoplankton that is not used in respiration and thus remineralized will sooner or later be converted into dead particulate and dissolved organic matter, either as faeces and other solid or fluid excreta, or upon death of the animals. Growth efficiency varies, especially with age of the animals, and with the amounts of food available; the efficiency decreases with increasing age, and is highest at intermediate concentrations of food in the surrounding water. In the present calculations, an average of about half the ingested phytoplankton is supposed to be respired (see p. 247), leaving 40 per cent of the net primary production for conversion into detritus and dissolved organic matter. According to these estimates somewhat less than half the primary production should thus pass the pool of dead organic matter within the euphotic zone in areas of constant productivity. This matter is to be partitioned between the particulate, that is detrital, and the dissolved fractions. A reasonable ratio would seem to be 5:1, so that these rough calculations indicate that about one-third of the net primary production is converted into detritus and about one-tenth into dissolved organic matter.

The time of turnover of organic detritus relative to phytoplankton can be obtained by multiplying the ratio between concentration of detritus and phytoplankton by the ratio between rates of production of phytoplankton and detritus, which in the present example equalled about 2·5. From Table 2.23 it is indicated that the ratio between concentrations of organic detritus and phytoplankton may often be about 10:1, also for oceanic regions with constant productivity and standing stock throughout the year. The turnover time of detritus under these conditions would therefore be $10 \times 2{\cdot}5 = 25$ times the turnover of phytoplankton. If the turnover times of phytoplankton are 2–8 days, turnover times of detritus would consequently amount to 2–7 months.

(2) In areas of the sea where the second condition prevails, phytoplankton blooms may develop untaxed by grazing of animals. In such areas larger proportions of the phytoplankton production are probably lost by sinking and by conversion directly into detritus than in areas where steady states are established between the phytoplankton and zooplankton. However,

the order of magnitude of turnover time of detritus is not necessarily changed, since the turnover time is to a large extent determined by the ratio between the concentrations of phytoplankton and detritus, and the variation in this ratio has so far not been found to differ significantly between such oceanic areas in which steady states predominate and such higher latitudinal or coastal areas in which plankton production vary greatly throughout the year (Table 2.23). It would thus appear that within the euphotic zone in the greater part of the sea, turnover times of detritus may exceed those of the phytoplankton by at least an order of magnitude, the reliability of the estimates mainly depending upon the accuracy with which the ratios between the concentrations of phytoplankton and detritus are determined.

2.322. DEEP SEA

Estimates of the rate of turnover of detritus in the deep sea are even more uncertain than the estimates applying to the euphotic zone, mainly due to our ignorance about the proportion of the primary production that reaches the deep sea and to our incomplete knowledge about the concentrations of organic detritus in the various regions of the deep sea. Particulate organic matter is continuously being supplied to the deeper layers of the oceans from above. Phytoplankton and detritus are passively sinking from the euphotic zone, vertically migrating zooplankton and larger organisms contribute to the downward transport of organic matter by feeding preferentially at upper levels, but voiding faeces or dying also at lower levels in the sea. Furthermore, deep currents will locally carry particulate organic matter from the surface layers to the deep sea. However, an estimate of the proportion of the primary production that reaches the deep sea say below 1,000m is little more than a guess; 10 per cent would seem a high guess, 1 per cent is perhaps more likely.

The annual primary production of all seas is about $1{\cdot}2$–$1{\cdot}5 \times 10^{10}$ tons of carbon (Steemann Nielsen, 1960; Steemann Nielsen and Jensen, 1957). The annual supply of detritus to the deep sea below 1,000m may therefore amount to about 10^8–10^9 tons of carbon to be distributed over 10^9 km^3 of water, or $0{\cdot}1$–1 μg/l. on an average. The only values available of particulate organic carbon from the deep sea are those of Parsons and Strickland (1962a), who recorded about 50 μg/l. at depths varying from 300 to 3,000m in the north eastern Pacific Ocean (Table 2.16). This would correspond to a turnover time of about 50–500 years. In the North Atlantic, at depths ranging from 400 to 4,000–5,000m, von Brand measured on an average 12 μg/l. particulate "protein" (Table 2.16). Assuming a "protein" content of 50 per cent of the detritus we obtain some 10 μg/l. of carbon, corresponding to a turnover time of 10–100 years. These values compare fairly well with the turnover times of less than 300 years calculated by Kullenberg (1953) for the total amounts of suspended matter in the oceans. The turnover time of detritus in the deep sea may therefore be counted in scores of years, and perhaps some hundred years, which would indicate that the turnover times are determined mainly by the rate of sedimentation of the detritus on the bottom of the oceans.

2.33. DISSOLVED ORGANIC MATTER

SOURCES

The immediate sources of the dissolved organic matter are partly identical with those of the detritus: autolyzing and disintegrating dead organisms from which organic substances leak into solution in the surrounding water; animal excreta containing water soluble organic substances; and dissolved organic matter drained by rivers into the sea. But in addition to these sources, organic substances are given off from living phytoplankton cells.

The importance of the last source has been much debated ever since Pütter (1909) assumed that the larger part of the organic matter produced by algae passed into solution in the surrounding water, and that this dissolved organic matter constituted a main source of food for aquatic animals (see p. 265). More recent investigations have shown that living and healthy algae can give off organic substances to the medium, but not to the extent assumed by Pütter. Most measurements of the amounts of organic matter lost by algae in cultures vary from insignificant amounts to about half the organic matter produced by the algae (Bishop *et al.,* 1954; Fogg, 1952; Guillard and Wangersky, 1958;

Krogh *et al.,* 1930; Lewin, 1956; Myers and Johnston, 1949; Pritchard, Griffin and Whittingham, 1962; Tolbert and Zill, 1956; Wangersky and Guillard, 1960; Watanabe, 1951; and others). Species differences may be partly responsible for the variation in the amounts of organic matter lost, but differences in experimental conditions are probably also of importance. It has sometimes been difficult to make sure whether the organic substances demonstrated in solution in the culture medium are really given off by intact algal cells or by disintegrating algae, or even by bacteria (Krogh *et al.,* 1930). In order to reduce or eliminate this error, analyses have been performed on cultures during the phase of vigorous growth of the algae.

The observations on algal cultures do not directly reveal the importance of the loss of dissolved organic matter from the living phytoplankton under natural conditions. Measurements of the amounts of extracellular organic matter produced by phytoplankton in nature is not available from the sea, but in two freshwater lakes, one oligotrophic and one eutrophic, Fogg (1958) found that on an average the extracellular products of photosynthesis, measured by means of the carbon-14 method, amounted to 2 per cent of the intracellular products per 24 hr. It is probably not unreasonable to assume that similarly small amounts of organic matter are generally lost extracellularly from the living phytoplankton in the sea.*

Rate of production and turnover

As mentioned above, in the larger parts of the oceans about half of the primary production was estimated to pass the stages of detritus and dissolved matter, about $\frac{4}{5}$ entering the former and $\frac{1}{5}$ the latter state. Thus, 10 per cent of the net primary production should go into solution in the sea water, to which amount should be added the probably smaller amounts given off by the living phytoplankton and not included in the determination of

* Recently Fogg and his co-workers found that from 5 to 95 per cent of the carbon-14 fixed by freshwater or marine phytoplankton samples suspended at various depths in a lake or in the sea is present in the filtrates. The organic matter liberated by the algae seems mainly to be glycollic acid (Fogg, 1962; Fogg and Nalewajko *et al.*, 1963).

the primary production. A significant part of the newly formed dissolved organic matter in the euphotic zone is probably oxidized and remineralized within the surface layers, because of the high concentration of bacteria here and because of the generally slow exchange of organic matter between the surface and the deeper layers: in contrast to the particulate, the dissolved organic matter has no downward directed movement of its own in relation to the surrounding water. However, in order to obtain an estimate of the turnover time for the whole ocean we assume the 10 per cent of the net primary production to enter the pool of dissolved organic matter, that is $1{\cdot}3 \times 10^9$ tons C, or 1 μg/l. if uniformly distributed throughout the sea (p. 237). As the most reliable measurements of the concentration of dissolved organic carbon in oceanic water vary between 0·5 and 2·5mg/l., the average turnover time of the dissolved organic matter is probably some 500–2,500 years; the higher figure is perhaps the more likely. It is of interest to note that the turnover times of dissolved organic matter in the oceans are of the same order as the turnover times estimated for the oceanic water itself, viz. the order of magnitude of 1,000 years, the age of the deep water of the Atlantic Ocean being less than that of the Pacific Ocean (Broeker *et al.,* 1961; Cooper, 1961; Fergusson, 1959; Stommel and Arons, 1960; Suess *et al.,* 1959).

2.34. BACTERIA AND OTHER HETEROTROPH MICRO-ORGANISMS

Only little is known about the rates of reproduction of the micro-heterotrophs in the sea (Kriss, 1961; Wood, 1963). Parsons and Strickland (1962b) suggest that the total heterotroph production beneath a unit area of the open ocean could well be of the same order as the photosynthetic production, a result which is a consequence of assuming a mean standing stock of the micro-heterotrophs in the oceans of about 0·1mg C/m^3, and a growth rate that is comparable with that of the phytoplankton. Parsons and Strickland also devised a method of measuring heterotroph productivity by means of carbon-14 labelled organic substances added to the water. However, until more data are obtained it does not seem possible to arrive at realistic estimates of rates of production of bacteria and other heterotrophs in the sea. Furthermore, as the estimates of the standing stocks of the heterotrophs seem to be more inaccurate than of the other organic fractions, we will refrain from attempting to estimate turnover rates of the micro-heterotrophs.

PART 3

FOOD OF THE SUSPENSION FEEDERS

Standard methods of determining diet or food preferences, as analysis of intestinal content and feeding experiments, have also been applied on suspension feeders. An extensive literature deals with the results of microscopical examination of the content of the digestive tract in various suspension feeders, and many attempts have been made to estimate the relative significance as food of the various types of organic matter present, especially in bivalves (e.g. Allen, 1921; Blegvad, 1914; Coe, 1948; Coe and Fox, 1944; Savage, 1925; Verwey, 1952) and in the partly selective copepods (e.g. Beklemishev, 1954; Marshall and Orr, 1955a). But whereas in typically selective feeders, analysis of intestinal content may allow fairly safe conclusions to be drawn concerning food, this is not so in non-selective feeders such as most suspension feeders; in these the composition of the ingested material is roughly similar to that of the suspended material in the surrounding water. However, naked flagellates and other delicate members of the phytoplankton mostly disintegrate shortly after ingestion and therefore easily escape being recognized in the intestinal content, even when they constitute an important part of the plankton. In the digestive tract they may appear as "detritus". Therefore, in a non-selective suspension feeder, analysis alone of the intestinal content does not reveal which fractions of organic suspended matter in the surrounding water are utilized as food; however, some information can be obtained about constituents that are not digested, but appear unchanged in the faeces. Often intact and even viable phytoplankton cells have been found in the faeces of various suspension feeders, a finding which has often led to the suggestion that phytoplankton was only poorly utilized, especially when present at high concentrations in the surrounding water, or when consisting of species equipped with protecting cells walls (see also p. 246).

More promising than analyses of contents of the digestive tract has been the direct testing of the food value of the various components of organic matter in natural waters. However, experiments have been made on only a restricted number of suspension feeders, especially of such representatives of the zooplankton that have been found able to thrive and multiply under laboratory conditions.

In forms that have not, or cannot, be maintained for longer periods in the laboratory, mostly indirect means have been available in evaluating the significance of the various natural food sources: food requirements have been estimated from rates of oxygen consumption and growth, and the estimates compared with the amounts calculated to be ingested of the various organic fractions in the surrounding water. It is, therefore, reasonable to discuss food conditions as a function of the habitat as characterized by the organic particulate matter in suspension. We shall restrict the discussion to a few main types of habitats.

The central position of the phytoplankton as the ultimate food source of most suspension feeders makes it natural to characterize the habitats by their phytoplankton production or content. We may consequently distinguish two main types of habitats, the euphotic zone and the aphotic regions of the oceans and larger lakes. Within the euphotic zone we shall further treat separately coastal waters with a high production of phytoplankton and oceanic waters of low productivity. Within these three main groups of habitats the possible significance as food of phytoplankton organic detritus, dissolved organic matter, and bacteria will be considered separately.

Practically all experimental work on food and feeding in suspension feeders has been made on forms from fresh waters or from coastal waters; and within these areas again mostly on forms from the shallow waters of the eulittoral zone, where photosynthesis extends to the bottom, and where food conditions are similar to pelagic and such benthic suspension feeders that collect their food from the free water masses. Consequently, the most detailed (and fertile) discussion of food of the suspension feeders concerns forms from near-coastal shallow-water habitats. In the two other main habitats selected for discussion the role of the various organic fractions as food is necessarily more theoretically treated, with emphasis on special problems.

3.1. COASTAL WATERS

3.11. PHYTOPLANKTON

3.111. FEEDING EXPERIMENTS

Quite naturally, most experiments have aimed at determining the food value of phytoplankton algae, which have generally been grown in unialgal cultures. Algal diets have been found capable of supporting growth in a number of suspension feeders: e.g. from freshwater, Rotatoria (Pourriot, 1957), Cladocera (Lefèvre, 1942; Ryther, 1954; van Heyningen, 1954) and Anostraca (Kallinowsky, 1955; Moore, 1957); from salt lakes, *Artemia* and the copepod *Tigriopus* (Gibor, 1956, 1957; Provasoli *et al.,* 1959); from sea water, *Balanus* (Costlow and Bookhout, 1953, 1957), copepods (Clarke and Gellis, 1935; Conover, 1960; Fuller, 1937; Marshall and Orr, 1955b, c, 1956, 1961; Raymont and Gross, 1942), *Euphausia* (Lasker, 1960), bivalves: veligers (Bruce *et al.,* 1940; Cole, 1937; Creek, 1960; Davis and Guillard, 1958; Loosanoff and Davis, 1963; Loosanoff *et al.,* 1955; Walne, 1956, 1959, 1963), and metamorphosed (Dean, 1957, cit. Provasoli *et al.,* 1959; Martin, 1928; Wernstedt, 1942), gastropod veligers (Lebour, 1933). In several of the experiments, the algal cultures fed have been bacteria-free (e.g. Davis and Guillard, 1958; Lasker, 1960; Provasoli *et al.,* 1959), so that bacteria cannot have served as supplementary food.

Of special interest is the demonstration that the various algae may vary highly in food value. Comparing the food value to veligers of *Ostrea edulis* of 19 species of algae, Walne (1956) obtained growth with 14 of the algal species, and spat fall with 10 of the species; but sometimes no growth was obtained even with algae that in other experiments proved to be good food.

The ability to utilize pure cultures of phytoplankton as food may vary even between related species of consumers. In extensive studies on the food value of various algae as the only food to about 20 species of bivalve veligers, Loosanoff and his colleagues found that veligers of some species were more exacting than veligers of other species. Thus, young larvae of oysters of the genus *Crassostrea* were able to grow only on few of the algae tested, whereas the larvae of, e.g. *Mercenaria mercenaria* and *Mytilus edulis* utilized most of the algae, provided they were small enough. Other veligers fell in between these extremes (Loosanoff and Davis, 1963).

Apparently, the possession of thick cell walls generally reduces the food value of algae by making the cell contents unavailable to the consumers. Algae as *Chlorella* and especially *Stichococcus* species with thick cellulose walls often proved to be inferior food when compared with naked flagellates or diatoms, at least when used as the only food (Lefèvre, 1942, *Daphnia*; Gibor, 1956 and Provasoli *et al.,* 1959, *Artemia*; Marshall and Orr, 1955b, *Calanus*; Walne, 1963, *Ostrea edulis* veligers, and others). So far, no suspension feeder has apparently been found able to utilize species of *Stichococcus* as food; yet the cells were found to be ingested and digested by the ciliate *Fabrea salina* which may in turn be used as food for fish larvae (Morris, 1956).

Algae may further be of reduced value as food because they may be disliked by or even poisonous to the experimental animal. A few species of algae were thus not at all or only to a small extent ingested by *Calanus* (Marshall and Orr, 1955b). And both *Chlorella* and *Stichococcus,* when added in sufficiently high concentrations, were found to inhibit growth of or kill *Daphnia* (Ryther, 1954), or veligers of oysters and clams (Davis and Guillard, 1958). Also the naked flagellate *Prymnesium* was found to be toxic to the veligers when used as the only food, whereas the closely related *Isochrysis* and *Monochrysis* are among the best food so far available to bivalve veligers (Davis and Guillard, 1958; Walne, 1956). Fatal toxic effects of *Prymnesium* under natural conditions are further indicated by the observation of excessive mortality within a population of newly liberated oyster veligers, coinciding with a high concentration of *Prymnesium* in the phytoplankton; 98 per cent of the veligers died within 24 hr (Waugh, 1960).

Bacteria-free, unialgal cultures may be deficient as food when given alone, but satisfactory when combined. In experiments on *Artemia* and *Tigriopus,* Provasoli *et al.* (1959) found that algae that might serve as the only food throughout the whole life of the parent generation often resulted in infertility or high larval mortality in later generations. Thus both *Rhodomonas* and *Isochrysis* were able to support only a restricted number of generations, but when the two algae were fed together they indefinitely maintained normal fertility and survival. Davis and Guillard (1958) found that mixtures of good food organisms provided faster growth in bivalve veligers than did equal quantities of the single algae. Algae that provide a nutritionally deficient food when used alone may thus be of high value when supplemented.

It has often been assumed that ingested algae may be poorly utilized by the suspension feeders, because the faecal pellets often contain many undigested cells. Experimental evidence, however, points towards a high utilization of most algae. Feeding a great number of different algae labelled with ^{32}P or ^{14}C to *Calanus finmarchicus,* Marshall and Orr (1955b, c) found that most of the algae, even when added in high concentration, were generally utilized to 80 per cent or more; in many cases a utilization of up to 99 per cent was determined; a lower utilization, 47–73 per cent, was found for instance of the dinoflagellate *Prorocentrum triestinum,* and only 15–20 per cent of *Chlorella stigmatophora.* In *Euphausia pacifica* fed on bacteria-free cultures of the flagellate *Dunaliella primolecta* labelled with ^{14}C, Lasker (1960) found that 96 per cent of the ingested carbon was assimilated, whereas in *Daphnia pulex*, Richman (1958) found only about 25 per cent of ingested *Chlamydomonas* cells to be assimilated at a concentration in the medium of *ca.* 6mg dry algal matter per litre. The percentage of assimilated cells decreased with increasing concentration of cells, only 7 per cent being assimilated at a concentration of 25mg/l. However, these concentrations are presumably above natural levels. In experiments in bivalves, Allen (1962) found *Ostrea, Mya* and *Venus* to assimilate from 77 to 95 per cent of ^{32}P-labelled *Nitzschia* cells retained by the feeding organs, that is, only 23 to 5 per cent of the ^{32}P passed out with the pseudofaeces and true faeces; the percentage of ^{32}P assimilated appeared to decrease with increasing concentration of algae in the water. By contrast, veligers of *O. edulis* utilized ^{32}P-labelled algae less efficiently (Walne, 1959). In preliminary experiments in an ascidian, *Ciona intestinalis,* a high utilization of ingested diatoms,

Asterionella japonica, was also found (E. Goldberg and C. B. Jørgensen, unpublished results).

Thus, most suspension feeders have been found to digest and absorb almost completely naked phytoplankton algae, whereas species possessing thick cellulose walls are generally little accessible. As might have been expected, the degree of utilization tends to decrease with increasing concentration of algae in the medium, but even at concentrations much higher than those normally encountered in nature by far the larger part of ingested algae may be assimilated. Additional evidence for the high value of most plankton algae as food for suspension feeders is offered by high gross growth efficiencies obtained in some feeding experiments. In *Artemia* fed optimal concentration of *Duniella,* Gibor (1957) calculated that 53 per cent of the algal mass was utilized in growth. In *Euphausia* fed another species of *Duniella,* the percentage varied between 11 and 74, average 35 (Lasker, 1960). In nature, net growth efficiencies of copepods and of veligers and young stages of gastropods and bivalves have been estimated to be *ca.* 70 per cent (Harvey, 1950; Jørgensen, 1952b, 1955).

3.112. PHYTOPLANKTON AS FOOD IN NATURE

The feeding experiments indicated that most phytoplankton algae are utilized as food by the various suspension feeders investigated, and that mixtures of phytoplankton algae are often better food than monocultures of algae. In nature, phytoplankton populations mostly comprise several species and types of algae, and only exceptionally and locally a single, sometimes toxic, species will dominate, causing so-called red-tides (Ballentine and Abbott, 1957; Collier, 1958a). Presumably, therefore, the importance of the phytoplankton as a food source will generally depend only upon its concentration and rate of production, and upon its availability to the various suspension feeders.

3.1121. *Relationships between food requirements, rates of water transport, and phytoplankton concentrations*

A gap has often been found between the amounts of phytoplankton needed to cover the food requirements of the suspension feeders living in the water and the amounts available, even within the euphotic zone of highly productive coastal waters; a gap was thus found for *Mytilus californianus* (Fox and Coe, 1943) and for *Calanus finmarchicus* and other copepods (Marshall and Orr, 1955b, 1962; and others). However, before we conclude that other organic fractions in the water must therefore be of importance as an energy source we should look into the reliability of the estimates of the phytoplankton concentrations needed to cover the food requirements of the various suspension feeders.

In order to obtain accurate estimates of the phytoplankton concentrations needed we must know: (1) the rate of water transport, (2) the efficiency of retention of the various components of the phytoplankton in the feeding organs, (3) the efficiency of digestion and absorption of the ingested phytoplankton, (4) the calorific value of the absorbed phytoplankton organic matter, and (5) the food requirements.

As already mentioned, most suspension feeders investigated efficiently retain even the smallest phytoplankton organisms; important exceptions are older stages of large copepods and perhaps other crustaceans. Very large or spiny or colonial diatoms and other algae are presumably excluded from the filters or feeding organs of many suspension feeders, but these large phytoplankton most often seem to be of minor quantitative importance (p. 180). We may, therefore, consider almost the total phytoplankton standing stock generally to be available for ingestion to most groups of suspension feeders.

The proportion of the ingested phytoplankton that is digested and absorbed will depend upon the amounts and species composition of the phytoplankton present in the water. However, in feeding experiments with various algae most species have been found to be efficiently utilized when fed in natural concentrations, and the species of algae that were resistant to digestion in some species are mostly of little quantitative importance in natural phytoplankton populations. The high degree of assimilation mostly found of ^{14}C labelled organic matter of phytoplankton indicates that most species of algae generally accumulate only small amounts of organic substances that cannot be broken down and absorbed in the digestive tract of the suspension feeders investigated. If we therefore in our calculations assume half of the organic matter of the phytoplankton collected in the feeding organs to be ultimately absorbed we shall probably more often underestimate than overestimate the true amounts absorbed.

We are probably justified in assuming that the calorific values of the various organic substances metabolized in the suspension feeders are of similar sizes as in other metazoans with predominantly aerobic metabolism, that is, about 4·1kcal/g of carbohydrate or protein, and 9·3kcal/g of fat.

The relative importance of carbohydrate, fat and protein in the phytoplankton varies with its species composition and with the physiological condition of the algae. Some algae store carbohydrates as starch or leucosin as reserve products, e.g. green algae (Chlorophyceae), several groups of flagellates (Chrysophyceae, Chryptophyceae, Dinophyceae, Euglenineae) and, perhaps, diatoms. Other diatoms, green algae, and flagellates (Xanthophyceae, Chrysophyceae, Dinophyceae) store fat, exclusively or in addition to carbohydrate (Fogg, 1953). The relative amounts of reserve products vary with the conditions of growth and the age of the culture, increasing with age. In extreme cases, as in an ageing culture of *Chlorella* lacking in nitrogen and aerated with high concentrations (5 per cent) of CO_2, the fat content may reach 70 per cent of the dry weight of the algae (Aach, 1952). More normal values of fat both for cultures of algae and for natural phytoplankton populations range from less than 10 to 25 per cent of total organic matter (Collyer, 1962; Ketchum and Redfield, 1949; Parsons *et al.,* 1961; Strickland, 1960). Assuming fat to constitute 10–15 per cent, and carbohydrate and protein 85–90 per cent, we obtain a calorific value of 4·6–4·9kcal/g of assimilated phytoplankton organic matter, which probably represents a conservative estimate (compare Richman, 1958).

Metabolic rates are usually determined as rates of oxygen uptake. The calorific value of one litre of oxygen varies from 4·6 to 5·1kcal, with the low values resulting from combustion of protein and fat, and the high value from combustion of carbohydrate. Consumption of one litre oxygen therefore corresponds to the combustion of about 1g of phytoplankton organic matter. In other words, at least 1g of phytoplankton organic matter should be absorbed and metabolized for each litre of oxygen consumed in order to cover the maintenance requirements of the suspension feeder in question; and 2g should consequently be collected in the feeding organs, if only half of the phytoplankton organic matter collected is digested and absorbed.

Generally, little attention has been paid to the possible role of the laboratory conditions in influencing the metabolic rates measured. In animals that can be maintained for long periods in the laboratory in a healthy and apparently normal condition, the rates of oxygen uptake are probably close to normal values, when the animals are adapted to the laboratory conditions. However, in animals that can survive in the laboratory only for a restricted period of time and therefore cannot be considered to be adapted, it is more difficult to ascertain whether a measured rate of oxygen uptake deviates from the natural level of metabolic rate. As suspension feeders have been especially difficult to maintain in the laboratory, the question of the reliability of the available data on oxygen consumption pertains not the least to this group of aquatic organisms. In some cases, oxygen consumption measured on healthy animals has been found to decline rapidly with time. For instance, in the copepod *Temora*, Berner (1962) measured an average uptake of oxygen of about 0·15 μl/hr during the first 6hr, whereas the average of the first 40–72 hr was as low as about 0·02 μl/hr. Berner finds it more likely that the high metabolic rate at the beginning of the experimental period was due to excitement, and thus higher than normal, rather than the later values being low due to depression. A rapidly declining rate of oxygen uptake was also measured in *Mytilus edulis* freshly brought into the laboratory, but excitement, as resulting from tearing off the newly formed byssus threads, caused a secondary transient increase of about 100 per cent in oxygen consumption (Jørgensen, unpublished). On the other hand, suspension feeders that only survived for a restricted period in the laboratory, as *Pecten latiauratus* and *Molgula manhattensis,* showed decreasing rates of oxygen uptake for a week or more (Jørgensen, 1952, 1960). This persistent fall in metabolic rate may represent depressed, and thus abnormally, low rates. The metabolic rates measured under laboratory conditions may therefore under circumstances deviate to either side of the normal range.

In order to secure growth and reproduction, larger amounts of phytoplankton are needed than merely for maintenance. Under optimal conditions in nature and in the laboratory, the proportion of energy that can be utilized for growth or production of eggs in various suspension feeders may amount to $\frac{2}{3}$–$\frac{3}{4}$ of the total absorbed (Jørgensen, 1955; Richman, 1958; Conover, 1962). During rapid growth and reproduction the food requirements are therefore up to 3–4 times greater than indicated by the metabolic rates; or 6–8g of phytoplankton organic matter for each litre of oxygen consumed.

The concentrations at which phytoplankton should be present in the surrounding water in order to cover the food requirements of the various suspension feeders can now be directly calculated from the relationships between the rates of water transport and oxygen consumption given in Table 1.6. A number of such calculations representative for various groups of suspension feeders, pelagic as epifaunal, are presented in Table 3.1. It appears that in typical epifaunal suspension feeders as *Ciona* and several bivalves and sponges, which obtain their food currents from the free water masses and filter about 15 litres for each of ml oxygen consumed, minimal food requirements may be covered at a concentration of about 0·1mg dry organic matter of phytoplankton per litre of water, whereas about 0·5mg/l. is needed for maximal growth or reproduction. It may be seen from Tables 2.7 and 2.8 that phytoplankton concentrations of this order of magnitude are very often indicated as being present in coastal waters from most parts of the world, at least during the periods of high photosynthetic activity; from some areas significantly higher values are recorded, from others lower. As mentioned, the methods of determining standing stocks of phytoplankton are beset with many sources of error, which may under circumstances result in either underestimates or overestimates. The fact that in so many localities estimates of phytoplankton concentrations, at least during longer periods of the year, fall within the range needed to cover requirements of several typical epifaunal suspension feeders strongly suggests that during these periods the phytoplankton can and does serve as the main source of food for these suspension feeders.

Many copepods, which also obtain their food from the free water masses, have been found to clear only about 8 litres of water for each ml of oxygen consumed. As, moreover, later stages of copepods only insufficiently retain phytoplankton cells smaller than 5–10 μ in diameter, the natural phytoplankton concentrations may therefore have to be higher than 1mg organic matter per litre in order to provide for optimal growth. More than 1mg phytoplankton organic matter per litre is sometimes met with in estuaries and other inshore waters (Table 2.8), but seldom in offshore waters. It is therefore possible that phytoplankton is usually present in insufficient amounts to allow enough food to be collected by filtration in copepods. However, this does not necessarily imply that the copepods must therefore utilize other organic fractions besides phytoplankton as food, because copepods seem capable of utilizing the large phytoplankton by direct seizing. Furthermore, as mentioned it cannot be overlooked that the rates of water transport used in the calculations presented in Table 3.1 may be lower than normal.

Other copepods, *Acartia* and especially *Metridia* and *Oithona,* have been found to clear the water of suspended phytoplankton algae at so low rates that, if they represent normal values, phytoplankton collected by filtration can only play a minor role as food.

Many benthic suspension feeders belonging to the infauna, for instance, bivalves, polychaetes, amphioxus, obtain their food from the water layer immediately above the bottom. Filatova (1957), in an attempt to characterize various benthic feeding types, distinguishes between the suspension feeders that take their feeding current from this layer, and those that feed in the free water masses. The forms included in the first group she refers to as "filterers-A", and in the second as "filterers-B". It may be misleading to compare phytoplankton requirements of suspension feeders belonging to the group "A" with data on the standing stock of phytoplankton in the free water masses above the habitat of these suspension feeders, because the concentration of suspended algae in the water layer in contact with the bottom may contain significant amounts of bottom diatoms stirred up by water currents. The available algal organic matter may therefore be considerably higher than could be predicted from estimates of the standing stock of phytoplankton in the free water masses. Not only suspended algae, but also organic detritus and bacteria may increase in concentration close to the bottom. In most habitats the feeding currents of the suspension feeders of group "A" are therefore likely generally to contain higher concentrations of

TABLE 3.1. RELATIONSHIPS BETWEEN RATES OF WATER TRANSPORT AND CONCENTRATIONS OF PHYTOPLANKTON NEEDED TO COVER MAINTENANCE REQUIREMENTS IN VARIOUS SUSPENSION FEEDERS

Species	Litres of water transported for each ml O_2 consumed	Phytoplankton, mg/l. of dry organic matter needed for covering maintenance requirements
Marine forms: Sponges, *Ciona* (ascidian), *several* bivalves (oyster, *Pecten*, *Mytilus*) Freshwater form: *Simocephalus* (cladoceran)	15	0·13
Marine forms: Most suspension feeding copepods, *Sabella* (polychaet)	8	0·25
Marine forms: *Lasaea* (bivalve), some polychaetes Freshwater form: *Bithynia* (gastropod)	2	1
Acartia (marine copepod)	1	2
Metridia (marine copepod)	0·2	10
Oithona (marine copepod)	0·04	50

algae, or other particulate fractions, than the feeding currents of the suspension feeders of group "B". The importance of stirred-up benthic diatoms, alone or together with other particulate organic material from the surface layer of the bottom, has been stressed by Grøntved (1960, 1962), Waugh (1960), and others.

If the rate at which water is cleared by a suspension feeder is adapted to the concentration of food in the water, one might therefore expect generally to find lower clearances in suspension feeders of group "A" than in those of group "B". Only few data are at hand to test this assumption, but the measurements of clearances in a number of bivalves representing both groups do indicate that forms with the inhalant siphon opening at, or very close to the surface of the bottom, transport water at significantly lower rates than forms in which the siphon opens into the free water masses. Thus *Mya arenaria, Venus striatula, Amphidesma ventricosum* and *Lasaea rubra* of group "A" were found to transport water at low rates compared with *Mytilus edulis, Ostrea edulis, Crassostrea virginica* and *Pecten irradians* (Fig. 1.40). In most of these examples, clearances were not expressed in relation to oxygen consumption, but to body weight. However, bivalves of equal weight of the soft tissues probably utilize oxygen at about equal rates. The data therefore suggest that bivalves feeding on the bottom layer of water are transporting water at relatively low rates; but it remains to be ascertained that the rates of water transport measured in the burrowing forms are typical of undisturbed animals (see p. 100).

Evidently, the criterion to be used in deciding whether a suspension feeder should be referred to Filatova's group "A" or "B" must be the extent to which bottom materials have been added to the water from which the feeding currents are derived. It is therefore not always possible definitely to decide to which group a suspension feeder belongs, because the thickness of the bottom layer that becomes enriched with suspended bottom diatoms and other material will depend upon the degree and type of agitation in the water above the bottom, as well as upon the character of the bottom and its flora. Under highly turbulent conditions in an area of shallow water with a high production of bottom diatoms, as for instance in the Wadden Sea along the coast of north western Europe (Grøntved, 1962), these diatoms and other bottom material may contribute significantly to the suspended organic matter of the whole water column. Under these conditions all suspension feeders in the area presumably belong to group "A", even such forms that have the inhalant siphons or other structures of water intake raised to a substantial degree above the bottom. Under conditions of less agitated water these forms may, however, have their area of water intake outside the region that is significantly influenced by the bottom materials, and thus belong to group "B".

Intermediate forms with the region of intake raised only little above the bottom are especially likely to change repeatedly between membership of groups "A" and "B", according to the changing conditions of their habitats, for instance oysters living in tidal areas. The importance of bottom deposits as food for oysters in such habitats has been emphasized, e.g. by Rochford (1952) and Waugh (1960). However, as mentioned above, oysters have been found to transport water at the high rates that are characteristic of suspension feeders of group "B" and of pelagic suspension feeders, which are capable of obtaining enough food at the concentrations of phytoplankton often met with in the free water masses.

3.1122. *Food during seasons with low productivity*

In coastal waters at higher latitudes, the production of phytoplankton is greatly diminished or even stopped during periods of the year, resulting in corresponding decreases in the concentrations of phytoplankton. During such periods, the suspension feeders may not be able to collect enough phytoplankton in the feeding organs to cover maintenance requirements, so that they must starve if other food sources are not available in sufficient quantities.

Many suspension feeders have been found capable of starving for long periods, and reproduction may even take place during the period of starvation. Thus the arctic copepod *Calanus hyperboreus* seems to feed only for a few months in spring during which period it grows to the adult stage, but does not mature sexually. Development of the gonads takes place during the following winter apparently exclusively at cost of other body tissues;

large deposits of fat are present at the beginning of the starvation period (Conover, 1962).

The ability to survive periods of starvation in this and other copepods may be supported by a reduction in metabolic rate (Ussing, 1938; Conover, 1962). Reduction of metabolic rate during starvation has been recorded in a number of other animal groups, e.g. the protozoan *Pelomyxa* (Pace and Belda, 1944), Planaria (Allen, 1919; Hyman, 1919), some insects (Bodine, 1921; Nielsen, 1935), goldfish (Dakin and Dakin, 1925), frog (Hill, 1911–12). Generally, the metabolic rate was found to settle at above half the value of fed animals, but in some species of pulmonate snails, starvation resulted in excessive reduction of metabolic rate, which decreased to 10–20 per cent of the original, within two weeks or more of starvation (von Brand *et al.,* 1948). Reduction of food requirements by reduction of rate of oxygen uptake in suspension feeders may of course improve feeding conditions under periods of low levels of food concentration in the water only if the reduction in oxygen uptake is not followed by a corresponding reduction in the rate of water transport. However, in suspension feeders as *Molgula, Pecten* and *Mytilus* it has been found that the rates of water transport may be kept constant even under conditions where the rates of oxygen uptake were declining to about one-half or one-third of the initial value (Jørgensen, 1952a, 1960, and unpublished). It remains to be found out to what extent values were decreasing with time as a result of starvation and to what extent the initial values were abnormally high because of excitation resulting from the transfer of the animals from their normal environments into the laboratory (see above). However, it is noteworthy that at least some suspension feeders can reduce their food requirements without concomitantly changing the rate of water transport, and it might be worth investigating whether the principle is of importance in nature as an adaptation to declining concentrations of food particles.

The decrease taking place in the concentration of phytoplankton during winter is generally greater than in the concentration of organic detritus. The feeding conditions therefore do not necessarily deteriorate proportionally with the decrease in phytoplankton to such suspension feeders that may utilize detritus, the role of which as food will now be considered.

3.12. DETRITUS AND BACTERIA

The phytoplankton cells are well-defined units with surfaces that are free of bacteria or other adsorbed organic particles as long as the cells are living (Steemann Nielsen, 1955; Oppenheimer and Vance, 1960). The phytoplankton therefore constitutes a well-defined fraction within the particulate organic matter in suspension. By contrast, detritus particles are generally associated with bacteria decomposing the organic substrate. It is therefore natural to discuss together the possible importance of detritus and bacteria as food, especially as bacteria in most waters seem mainly to occur attached to particles, which may, however, also be inorganic (p. 292).

3.121. FEEDING EXPERIMENTS

Detritus

Direct experiments aimed at ascertaining the value of organic detritus as food are few, and the results of the experiments are difficult to interpret, because they have been made with more or less unnatural sources of detritus. So far, no methods are available by which natural organic detritus can be separated on a mass scale from the phytoplankton and other organic constituents of similar particle size suspended in the water; and the artificial detritus used in the experiments may deviate in composition and food value from natural detritus. It would appear reasonable to assume that the food value of natural detritus will vary strongly, especially with its age: the nutritional value of the remains of recently dead organisms probably decreases with time with the progress of bacterial decomposition. This assumption is supported by experimental evidence. In *Daphnia*, van Heyningen (1954) found that whereas freshly boiled phytoplankton still had nutritional value, the value was ultimately lost after exposure to bacterial degradation of the phytoplankton; even the freshly killed phytoplankton was much inferior to living phytoplankton as food. In *Artemia,* Bond (1933) likewise has found that the nutritional value of algae (and bacteria) was much reduced upon killing of the cells by heating. Artificial detritus made from animal and plant material is stated to provide normal growth in oysters (Gavard, 1927, cit. Korringa, 1949), whereas

Loosanoff *et al.* (1951) obtained only restricted growth in veligers of the clam fed on "detritus" as sampled from the bottom of tidal pools or from the bottom of a large tank in which plankton was grown for eight months. In the freshwater copepod *Eudiaptomus gracilis,* Nauwerck (1962) found some value as food of homogenized zooplankton, as judged from its ability to sustain survival and production of eggs in copepods kept in membrane-filtered lake water. In these experiments, except those of Bond, the bacterial content was not controlled, so that part of positive responses may have been due to the presence of bacteria. The importance of detritus as food supplementary to living food organisms seems not to have been experimentally investigated.

Bacteria and other heterotrophic micro-organisms

They have been found capable of sustaining growth when used as the only food of a number of suspension feeders, but the different species of bacteria may vary in nutritional value. Feeding living cultures of 11 species of bacteria to the freshwater cladoceran *Moina macrocopa,* Stuart *et al.* (1931) found that some were able to support growth for several generations of *Moina,* whereas no growth or release of young was obtained with the sporogen *Bacillus subtilis*; other bacteria were directly toxic. Similar results were obtained by Salimovskaya-Rodina (1940, cit. Rodina, 1948b). Also in nature, bacteria may apparently constitute important food for freshwater cladocerans. Thus, Manuilova (1958) found that in natural river water filtered free of algae and detritus, and containing 434,000–1,200,000 bacteria per ml. as directly counted, survival and reproduction were normal in a number of species (e.g. *Daphnia longispina* and *Simocephalus vetulus*), and survival time greatly increased in others (*Sida crystallina*); but if the number of bacteria fell below some 400,000/ml, the cladocerans perished within 2–14 days. Also Sorokin (1959) and Monakov and Sorokin (1961a) found natural bacterial populations to be efficiently utilized by various freshwater cladocerans. Fredeen (1960) reared larvae of the black-fly, *Simulium* sp., in distilled water containing only washed suspensions of *Bacillus subtilis*; at a concentration of *ca.* 160,000 viable cells/ml, growth rate was normal.

Also in freshwater molluscs, normal growth rates have been

obtained on a diet of bacteria (Rodina, 1948b); over periods of about 1 month, adult specimens of the bivalves *Sphaerium, Musculium* and *Pisidium,* and the gastropods *Bithynia* and *Lymnaea* increased in weight by up to 60 per cent when fed pure cultures of bacteria or yeasts isolated from the lake where the experimental animals were sampled. Newly hatched larvae might increase up to 2–3 times in weight. The micro-organisms were fed in amounts corresponding to normal contents of bacteria in eutrophic and mesotrophic lakes. Control animals kept in membrane filtered water with no food added died or lost weight. In all experiments, *Bacillus subtilis* was found to be of value as food, but it was inferior to *Azobacter* or the two species of yeast. The food value of one bacterial species may thus vary significantly in different species of consumers. The low food value often found of *B. subtilis* may at least partly be due to its spore forming habit; in experiments on *Calanus,* Marshall and Orr (1955b) found the spores to be practically indigestible.

Growth on a diet of bacteria has further been obtained in the marine suspension feeders *Mytilus* (ZoBell and Landon, 1937; ZoBell and Feltham, 1938) and *Urechis* (MacGinitie, 1932).

Bacteria have also been found capable of making algal foods nutritionally adequate, when the latter alone did not secure optimal growth in *Artemia* and *Tigriopus* (Provasoli *et al.,* 1959; Shiraishi and Provasoli, 1959). Furthermore, Wood (1953) observed the addition of bacteria to increase the nutritive value of the diet in the Australian oyster.

In other suspension feeders, less positive results have been obtained in feeding experiments with bacteria. In sponges, Kilian (1952) found that addition of suspensions of bacteria influenced the sponges adversely, but Rasmont (1961) obtained good growth in gemmula sponges fed killed bacteria, *Escherichia coli* or *Staphylococcus aureus.* However, high concentrations of bacteria were used in order to secure growth; killed bacteria corresponding to at least 2mg N/l. had to be present in the medium, or about 20mg dry organic matter per litre. In the filter feeding copepods *Diaptomus gracilis* and *D. graciloides,* natural concentrations of bacteria were only poorly utilized, not much better than in the non-suspension feeding *Cyclops.* However, natural concentrations of bacteria supported growth in the nauplial stages of both

Diaptomus and *Cyclops*; and the negative results in adult *Diaptomus* apparently arose from the inability to retain efficiently suspended bacteria, since aggregated bacteria were found to be effectively assimilated (Monakov and Sorokin, 1960; Malovitskaya and Sorokin, 1961a).

In bivalves, bacteria were found unsuitable as food for oyster and other veligers. Thus, none of 13 species of marine bacteria was found to be of value as food to veligers of the oyster *Crassostrea virginica* (Davis, 1953). Bacteria may even cause heavy mortality in cultures of oyster veligers (Walne, 1958), but the mortality is apparently caused by only a few of the species of marine bacteria (Guillard, 1959). Even when a bacterial species is not fatally toxic, its adverse effects may be large enough to mask a possible nutritional value when it is given as the only food. Negative results of feeding experiments using monocultures of bacteria may result from lack of specific nutritional factors in the bacteria, and not necessarily from an inability of the veligers to ingest or to assimilate bacteria. Of other heterotrophic micro-organisms, *Monas* sp. has been found by Imai and Hatanaka (1950) to support normal growth and development of oyster veligers, but these results could not be confirmed by others (Loosanoff and Davis, 1963). It is probably relevant in connection with the results of feeding experiments with bacteria, etc., also to recall that even in experiments feeding algae of well established high nutritional value to bivalve veligers unpredictable failures have often been encountered (Carriker, 1961).

3.122. DETRITUS AND BACTERIA AS FOOD IN NATURE

Several biologists have stressed the importance of detritus in the diet of suspension feeders, also of forms living within the euphotic zone, as appears from the following examples.

In Danish waters, detritus was assumed to be a main source of food to suspension feeders, since it was found to dominate the contents of the digestive tract, as well as the organic matter in suspension in the water (Blegvad, 1914; Boysen Jensen, 1915; Boysen Jensen and Petersen, 1911). In land-near localities, as the Limfjord and along several open coasts, the detritus appeared to be derived principally from the rich *Zostera* beds flourishing at that time; farther off-shore, plankton contributed significantly. Boysen Jensen (1915) attempted to demonstrate that detritus derived from *Zostera* may be utilized as food by suspension feeders by analysing samples of faeces of oysters and of detritus from the locality where the oysters were living for total organic carbon and pentosan. He found the ratio C:pentosan to be 7·45 in the detritus sample, which is close to the values for fresh or decaying *Zostera,* but 17·6 in the sample of faeces collected from 94 oysters. It was thus indicated that the oysters had been digesting and absorbing the pentosan fraction of the detritus to a higher degree than at least some of the other organic components. However, the detritus sample was obtained from the surface of the bottom of the locality where the oysters were living, and not directly from the surrounding water from which the feeding currents were obtained. The ratios C : pentosan of the detritus sample and the faeces sample were therefore not directly comparable and the results are consequently not conclusive.

In the Barents Sea, at depths below 100m, Filatova (1957) assumes the large population of suspension feeding bivalves, e.g. *Cardium,* to be supported by detritus entering from the River Pechora, or originating from the extensive beds of macrophytes on the shores of Novaya Semlya. Detritus was also assumed to constitute a main food to a suspension feeder as *Mytilus californianus* living at the open coast of California exposed to the Pacific Ocean (Fox and Coe, 1943; Coe, 1948), and this assumption was extended to include marine suspension feeders generally (Fox, 1950, 1957). Fox and Coe (1943), put up a balance sheet for the turnover of total organic particulate material by *M. californianus*; but the calculations are not realistic since they are based on the assumption that *Mytilus* can efficiently retain and utilize the total content of 5·5mg organic matter presumably present per litre of sea water. However, as already mentioned, only a small fraction is of particle sizes that do not easily pass the feeding organs of *Mytilus* and other suspension feeders.

In all of the above mentioned examples, as in others not mentioned, no definite evidence was provided in support of the assumption that true organic detritus plays a dominant role as a food source to the suspension feeders. However, the amounts of detritus indicated to be present in coastal waters are often large enough to cover maintenance requirements, or even requirements for growth, at the rates of water transport measured in a number of suspension feeders. Its importance as food will under such conditions depend upon the amounts of digestible, or in other ways utilizable, organic matter it contains. However, organic matter that cannot be digested by most animal enzymes is likely to accumulate in the detritus particles, simultaneously with the progressing decomposition by bacterial action. Most probably, the bacterial moiety of the detritus-bacteria complex generally has the highest food value, as indicated by the feeding experiments which showed that generally a pure bacterial diet provided better results than feeding with detritus. The value of the detritus-bacteria complex as food may thus often largely depend upon the size of the bacterial fraction. It would therefore be of interest to know the ratio between bacterial and detrital organic matter in the various natural waters, and how this ratio may vary.

In the larger part of the sea, coastal waters included, the available data indicate that the bacterial mass amounts to only a small fraction, 1 per cent or less, of the detrital mass (Table 2.23). However, in special localities, e.g. in shallow estuaries, the figures given for organic matter present in the standing stock of bacteria are very high and may even reach the estimates of total particulate organic matter. If the figures provide the correct order of magnitude, bacterial mass may thus in places dominate over that of detrital mass. Direct observations offer support for this possibility in showing that the mass of bacteria found attached to a particle of organic matter not infrequently exceeds that of the substrate itself (ZoBell and Feltham, 1942).

It is thus indicated that in most coastal waters the standing stock of bacteria is too small to be of any quantitative significance as food to the suspension feeders, at least at the rates at which these have been found to transport water through their feeding organs. In order to cover optimal food requirements exclusively on a bacterial diet, a suspension feeder must clear several hundreds of litres of water free of bacteria for each ml of oxygen consumed, if the water contains about 10 μg/l. of bacterial organic matter. However, in some estuaries and other inshore waters bacterial concentrations as high as the phytoplankton concentrations have been recorded, and may be of corresponding significance as food. The importance of bacteria as food to suspension feeders living on the mudflats of estuaries have been emphasized by ZoBell and Feltham (1942), and by MacGinitie and MacGinitie (1949) who found the water over the mudflats along the Californian coast to contain few diatoms, but to be rich in detritus with attached bacteria, which also dominated the stomach content of the suspension feeders living on the mudflats.

Also benthic suspension feeders from somewhat deeper water which obtain their feeding currents from the bottom layer of water may inhale water much richer in bacteria than indicated by the counts from samples of the free water masses. As mentioned, the content of bacteria in the uppermost layer of mud bottoms may contain 10^8 bacteria, or more, per ml, corresponding to about 10mg organic matter per litre. This easily stirred-up material therefore constitutes a rich potential source of food to the suspension feeders dwelling in the soft bottom. Presumably, the concentration of bacteria in the layer of water immediately above the bottom and the thickness of the water layer enriched with bacteria, and detritus, from the bottom must vary widely with local conditions, but few exact figures are available. It may be mentioned that in Lake Seeburger, Jannasch (1958) found the standing stock of bacteria 10cm above the bottom to be about 5,000 times larger than in the surface water (Table 2.25).

3.13. DISSOLVED ORGANIC MATTER

3.131. FEEDING EXPERIMENTS

The question whether the dissolved organic substances in natural waters are of importance as food to the suspension feeders and other aquatic organisms has been much debated, but hardly directly experimentally approached. Most experiments that have aimed at disclosing the possible significance of dissolved organic substances as food have been made by measuring survival of the experimental animals, or their uptake of the dissolved organic matter, in organic solutions differing widely in composition from natural waters. Many experiments of this type are available, but only some examples should be given. In various cladocerans, some growth has been obtained in solutions containing organic colloids, e.g. pond water, extracts of manure and soil, solutions of proteins, but after removal of the colloids by filtration, growth was inhibited (Gellis and Clarke, 1935; Rodina, 1948a; Stuart *et al.,* 1931; van Heyningen, 1954). *Daphnia* failed to attain the second moult in solutions containing 67mg/l. glucose or 33mg/l. peptone (Gellis and Clarke, 1935), whereas Rodina (1948a) obtained limited growth and survival for 12 days of bacteria-free *Daphnia* in sterile filtered solutions containing up to 1,100mg/l. of 11 different amino acids; the results were not improved by addition of 1,100mg/l. of glucose, arabinose or saccharose. Similar results have been obtained on larvae of *Artemia* (Bond, 1933) and *Aedes aegypti* (Trager, 1936). Very high concentrations of substances of high nutritional value were thus needed merely to increase time of survival or to provide a restricted growth in the experimental animals. However, vitamin deficiency or other imbalance of the food may be responsible for the poor results of some of the feeding experiments using dissolved organic substances as the only food, since Singh and Brown (1957) succeeded in rearing larvae of *Aedes aegypti* on a purely synthetic diet of organic and inorganic substances in solution. The solution included glucose, lipids, 17 aminoacids of which 10 were essential to the larvae, RNA, 13 vitamins of which nine were essential, and inorganic salts; in all 18·45g of organic matter per litre, or several thousand times the concentration of dissolved organic matter in natural waters. Still, growth was strongly enhanced in cultures that turned out to have been contaminated with bacteria.

In the experiments with high concentrations of dissolved organic matter it is not known to what extent uptake is via the intestinal tract or through the body surface. However, other experiments have shown that many aquatic organisms, suspension feeders included, are able to take up significant amounts of various organic substances from such dilute solutions that the uptake must presumably have taken place largely through the external surface; otherwise, improbably great quantities of the medium should have passed the intestinal tract. Krogh (1930) found that *Daphnia,* the freshwater mussel *Dreissensia*, fish and tadpoles of frogs utilized glucose even at concentrations as low as 5mg/l. Recently, Stephens and Schinske (1961) have demonstrated uptake of glycine and other aminoacids from dilute solutions in the greater number of 35 genera of marine animals representing 11 phyla. Only the arthropods were consistently found unable to reduce the concentration of aminoacids in the surrounding medium. Occasionally other animals, in particular the suspension feeders *Mytilus, Spisula, Crepidula* and *Ciona* failed to demonstrate any uptake. Glycine uptake could be demonstrated at concentrations as low as 1·5mg/l. The solitary coral *Fungia* took up glucose from solutions containing even less than one mg per litre, but no uptake of galactose, ribose, arabinose or sucrose was seen even at a concentration of 500mg/l. (Stephens, 1960a, b, 1962b). The uptake of aminoacids was studied in more details in the bamboo worm *Clymenella torquata* (Stephens, 1962a). The rate of uptake of ^{14}C labelled phenylalanine at concentrations varying from 0·2 to 8mg/l. was a function of the surface and was independent of the gut. Further increase in concentration did not enhance the rate of uptake correspondingly. Lysine and valine were absorbed at about one-third the rate of phenylalanine and glycine. As regards the amounts of organic matter taken up, *Fungia* was found to absorb enough glucose from a solution containing about 15mg/l. to cover maintenance requirements (Stephens, 1962b). In the experiments of Krogh (1930) about one-fourth of the oxygen consumption was covered by the uptake of glucose from solutions containing 5–40mg/l. However, the loss of organic matter to the medium turned out to exceed that taken up as glucose. The amounts of organic matter given off by the animals investigated by Stephens have not been measured. It is therefore not known whether net uptake of organic matter occurred in his experiments.

It is not known whether natural waters contain significant concentrations of dissolved glucose or aminoacids (see p. 205). At any rate, demonstration of uptake of these substances from dilute solutions does not permit conclusions to be drawn concerning the importance of naturally occurring dissolved organic matter as food for aquatic organisms, as is also stressed by Stephens (1962).

3.132. DISSOLVED ORGANIC MATTER AS FOOD IN NATURE

The discussion on the importance of the dissolved organic matter as a direct source of food to aquatic organisms is closely linked to the name of Pütter who at the beginning of this century put forward the theory that aquatic animals feed on the organic matter in solution in the surrounding water. Pütter (1909) assumed that the dissolved matter was largely derived from the synthesizing phytoplankton organisms which were believed to lose the greater part of the production of organic matter through the cell surfaces into the surrounding water. By far the greater part of the primary production in natural waters should accordingly be offered the consumers in the dissolved state.

Pütter, and others, mainly provided indirect evidence for the importance of the dissolved organic matter as a direct food source. They observed that phytoplankton was present in so small concentrations that aquatic animals in order to cover their energy requirements from this source would have to clear impossibly large volumes of water, whereas dissolved organic matter was present in ample amounts. To mention an example, Pütter estimated that the radiolarian *Collozoum inerne* weighing about 100mg would have to clear 9·4l. of the surrounding water free of phytoplankton algae per hour in order to cover its food requirements, but only 0·55ml of dissolved organic matter. The incorrectness of this first estimate soon became evident when it was shown that the concentrations of phytoplankton had been underrated, and of dissolved organic matter overrated. According to a revised estimate the volume needed to be cleared of algae was reduced to 760ml, and that needed to be cleared of dissolved organic matter increased to 6ml; but still the suggestion seemed justified that a small animal as *Collozoum* would not be capable of clearing the volume needed for algae, but well capable of clearing a volume of 6 ml. Similar estimates were made on a number of other organisms with similar results. Apart from the indirect evidence, support for the theory was also found in the demonstration of uptake of dissolved organic substances by a number of aquatic organisms.

It soon became clear that the methods used by Pütter in measuring the organic fractions of the water were not reliable (Henze, 1908; Raben, 1910). Furthermore, the experiments indicating significant utilization of dissolved organic substances were seriously criticized by Krogh (1931) who found that higher aquatic animals, including suspension feeders as Daphnia and the bivalve *Dreissensia polymorpha,* absorbed smaller quantities of dissolved organic matter than were simultaneously excreted (see above, p. 265). Also Moore *et al.* (1912) found that suspension feeders as sponges and ascidians were unable to reduce the content of organic carbon or nitrogen in filtered sea water. On the contrary, a net loss always occurred of oxidizable organic matter and "albuminoid" nitrogen.

For some time after the severe criticism of Pütter's theory little mention was made of dissolved organic matter as a bulk food of aquatic animals, including the suspension feeders. Recently, however, an inclination has been shown not to accept the rejection of Pütter's theory as final, but to reconsider the possible role of dissolved organic substances as an important energy source to aquatic metazoa. Thus, dissolved organic matter has again been suggested as a possible important food source for various aquatic animals, e.g. by Lucas (1961), Morris (1954) and Stephens (1962). As regards suspension feeders, Korringa (1949) suggests taking up again the problem of the role played by dissolved colloids as food for forms feeding by means of mucus filters, which according to MacGinitie retain large colloidal molecules (p. 11). Fox (1957) considered practically all the dissolved organic matter in the sea to be present in the colloidal state and to be available to suspension feeders as, e.g. bivalves. Furthermore, Collier (1959) and Ricci (1956) assume bivalves as oysters and mussels to be able to utilize as food dissolved organic substances in the sea. Collier and his co-workers obtained experimental evidence for a significant uptake in the oyster *Crassostrea virginica* of the carbohydrate factor found to be present in considerable amounts in estuarine waters of the Gulf of Mexico (see p. 205). As mentioned above, 5–15 per cent of the dissolved carbohydrate was stated to be removed from the water passing the gills of the oyster. The rate of uptake of the carbohydrate increased rapidly with increasing concentrations of the carbohydrate in the water, due to the stimulating effect of the carbohydrate on the rate of water transport (see p. 108). However, the metabolic rate, and thus food requirements, was also found to increase with increasing rates of water transport, closely corresponding to the increase in the rate of carbohydrate uptake. At the moment it does not seem possible to reconcile these results with those of most other investigators.

The suggestion that suspension feeders using mucus filters, e.g. ascidians and perhaps bivalves, might be able to feed on the naturally occurring dissolved organic matter disagrees with the above mentioned findings of Krogh and Moore *et al.* However, in these experiments no precautions were taken to ensure that the animals were feeding normally. The experiments are therefore not entirely conclusive. But in experiments on presumably normally feeding ascidians and bivalves, colloids were found to be only inefficiently retained (pp. 96, 116). Furthermore, colloids presumably constitute only a small fraction of the total dissolved organic matter of most natural waters (p. 200).

In conclusion, it still seems unlikely that the primary production is mainly utilized by the consumers after it has gone into solution in the water. Perhaps only the smaller part of the organic production is generally lost from the living cells of most phytoplankton organisms into the surroundings and no convincing evidence has as yet been provided for the importance of the dissolved organic matter in unpolluted waters as a direct energy source for any metazoan. It has not been proven that the dissolved organic substances are mainly composed of substances that are easily utilized as food by metazoans; and the easily utilizable compounds that may be present are probably not easily available (p. 134).

If aquatic organisms subsist on dissolved food, as do endoparasites as Cestoda, parasitic cirripeds and copepods, etc., their alimentary organs might be expected to be more or less reduced; but they are not. The wide diversity of well-developed feeding mechanisms adapted to dealing with food of sizes ranging from bacteria upwards strongly suggests the general importance of non-dissolved food for aquatic metazoans.

Even when it seems unlikely that suspension feeders, or other aquatic metazoans, utilize dissolved organic matter as their main food, it remains unsettled to what extent such food may serve as a subsidiary source of energy. The final answer to this question is probably not possible until the organic matter in solution in natural waters has been analyzed in its constituents, and the efficiency with which these constituents are retained and utilized by the various suspension feeders has been determined.*

The organic substances in solution in natural waters may not merely serve as potential sources of energy. Some of the substances may play important roles by exerting specific effects on the organisms living in the water, effects that may be produced even at the very low concentrations at which most of the substances are present. The relationships that may exist between such "biologically active" substances and the aquatic organisms are often only very incompletely and vaguely understood, and mostly the substances are not chemically identified, but only known from their more or less firmly established effects. However, examples on specific effects of certain constituents of the dissolved organic fraction in natural waters, especially on phytoplankton organisms, are rapidly accumulating (Fogg, 1962; Johnston, 1963; Lucas, 1947, 1958, 1961; Nalewajko *et al.*, 1963; Pringsheim, 1953; Saunders, 1957). The relationships are of a varied nature and mainly outside the scope of this treatise, but a few examples should be mentioned. The dissolved constituents may be vitamins, e.g. B_{12} and thiamin, which are required for growth by a number of phytoplankters or bacteria (Droop, 1957, 1958). In other instances the identity of the growth stimulating factor is not known (Kain and Fogg, 1958). Further, antibacterial substances liberated from the phytoplankton seem to

* In special environments the water may contain easily utilizable dissolved organic compounds, as aminoacids, etc., in such high concentrations that the compounds may play an important role as food to the animals living in the water. Over mudflats, Stephens (1963) found that aminoacids may amount to about 10mg/l., or about the concentration needed to cover the food requirements of the non-suspension feeding polychaet *Clymenella torquata,* which lives on the mudflats, if the worm was absorbing the aminoacids at the same rates as measured in the laboratory.

An interesting mechanism by which dissolved organic matter in sea water may be made available to suspension feeders was recently demonstrated by Baylor and Sutcliffe (1963). They found that dissolved organic matter is transformed into particulate by bubbling air through filtered sea water, and that this particulate matter could support growth when fed to young *Artemia.* It is suggested that the mechanism may be of importance in producing food to suspension feeders in nature through the air-bubbling arising from wave action.

be present in the water (Steemann Nielsen, 1955; Sieburth, 1959). Substances toxic to a number of aquatic metazoans are also known to be produced by certain algae and to be liberated into the water in harmful amounts during "blooming" of the algae (p. 247).

Of more immediate interest to the biology of suspension feeders are the suggestions of organic constituents necessary for normal growth of bivalve veligers. The common experience of unpredictable results in the culturing of oyster and other bivalve veligers under apparently similar conditions has led to the assumption that unidentified substances of sporadic distribution in the sea may be of importance in stimulating or inhibiting growth (Carriker, 1961; Loosanoff, 1958b). Further, larvae of the sea urchin *Echinus esculentus* and of some annelid worms may develop differently in sea waters of different origin, but indistinguishable from one another by standard physical and chemical procedures. Thus, eggs of *Echinus* often developed normally in water sampled from a water mass that was biologically characterized by the presence of *Sagitta elegans*, whereas under similar conditions abnormal development and death resulted in water characterized by the presence of *Sagitta setosa* (Wilson, 1958). See also p. 108 on dissolved substances stimulating rate of water transport in the oyster.

Most likely therefore constituents of the dissolved organic fraction in natural waters specifically influence various aquatic metazoans, suspension feeders included, but definite details are scarce. In order to exert their activities the substances are presumably generally taken up by the organisms. It is conceivable that the widespread ability of aquatic organisms to take up various organic molecules through their outer surfaces is of importance in securing the uptake of the various biologically active substances.

3.2. EUPHOTIC ZONE OF OCEANIC WATERS

In the open ocean outside areas influenced by upwelling nutrient-rich deep water the phytoplankton concentrations are generally much lower than in coastal waters. It may be seen from Tables 2.6 and 2.9 that over wide areas of the Pacific and Atlantic Oceans concentrations of about 0·01mg/l. have been estimated, with little annual variation. At a phytoplankton concentration of

0·01mg/l. a suspension feeder would have to clear some 150l. of the surrounding water free of phytoplankton for each ml. of oxygen consumed in order to cover maintenance requirements by feeding on the standing stock of phytoplankton. This value is estimated partly under the assumption that chlorophyll a constitutes 3 per cent of the phytoplankton organic matter. As mentioned this percentage may be too high; however, even if we assume oceanic phytoplankton on an average only to contain 1 per cent chlorophyll, the suspension feeders in the surface waters over vast areas of the oceans would seemingly have to clear at least 50l. of water for each ml of oxygen consumed in order to cover their basic energy requirements from the phytoplankton, and more to allow for growth and reproduction. The clearances needed are thus higher than have been measured in any coastal suspension feeders. Unfortunately, determinations are lacking of rates of water transport in most oceanic suspension feeders and no determinations seem to have been made at natural concentrations of food in the water. However, *Euphausia pasifica* was found to clear a suspension of flagellates at the low average rate of 1·2l. for each ml oxygen consumed.

The apparent discrepancy between food requirements and the amounts of algal food that appear to be available, which presumably no longer exist for a number of suspension feeders from coastal and fresh waters, therefore still seems to apply to oceanic suspension feeders. It therefore has been repeatedly suggested, that the phytoplankton is not directly utilized as food, but indirectly in form of organic detritus.* The consequences of such suggestion should therefore be considered.

It is a widely accepted theory that grazing by the zooplankton generally plays an important role in controlling the phytoplankton population, both in areas with great and with small annual variations in the productivity. The quantitative interrelations between the zooplankton and the phytoplankton were first studied by Lohmann (1902, 1908), and have in recent years occupied a central position in the ecology of marine and freshwater habitats (Beklemishev, 1957; Cushing, 1959b; Digby, 1953;

* As regards the theory that the primary production should be utilized by oceanic suspension feeders in the dissolved state reference is made to the arguments put forward above for suspension feeders from coastal waters.

Fleming, 1939; Gauld, 1950; Gillbricht, 1959; McAllister *et al.*, 1960; Pennington, 1941; Riley *et al.*, 1949; Steele, 1956, 1959, 1961; Steemann Nielsen, 1958c, 1962). In oceanic environments, especially in tropical and subtropical waters characterized by a plankton standing stock and productivity that varies little throughout the year, the steady state of the phytoplankton population is assumed mainly to be maintained by the grazing activity of the zooplankton. In its absence, the phytoplankton production and standing stock might increase by 100 per cent in a few days and a phytoplankton maximum or bloom might arise (McAllister *et al.*, 1960; Steemann Nielsen, 1962).

It can be shown that if the maintenance of a constant phytoplankton population is mainly due to grazing by the zooplankton the dominating grazing organisms cannot clear the water at such low rates that their food requirements can only be covered by detritus. In typical oceanic environments the organic detritus is ultimately largely derived from the phytoplankton. Its turnover time is therefore determined by the fraction of phytoplankton converted into detritus and the concentration to which the detritus builds up. It was argued above (p. 236) that the turnover time of detritus may be some 25 times that of the phytoplankton; or in other words, only about $\frac{1}{25}$ of the detritus is exchanged in the time in which an amount of phytoplankton corresponding to the standing stock is exchanged. Either, therefore, phytoplankton must be selectively taken in by the suspension feeders, leaving the detritus particles behind, or, if no selection takes place in the feeding organs, only a small fraction of the detritus gathered can be digested, the remaining passing the intestinal tract to be repeatedly recycled. In neither case can organic detritus be the main food for the zooplankton, under conditions where this controls the phytoplankton population by its grazing activity.

In view of the weight of the evidence supporting the theory that grazing is the main factor in controlling the phytoplankton productivity, the detritus should probably be ascribed only a minor role as a food for dominating zooplankton organisms within the photosynthetic zone of most natural waters (Jørgensen, 1962). The dominating oceanic zooplankton organisms, as various copepods, euphausids, and the various forms of pelagic tunicates must consequently be able to clear the surrounding water of phytoplankton at the rates needed to cover their energy requirements on an algal diet. Of these zooplankton organisms at least the appendicularians and the Thaliacea are exclusively suspension feeders. It cannot be excluded, however, that suspension feeding forms of minor quantitative importance may transport water at low rates and be adapted to a diet mainly consisting of organic detritus and bacteria. A preliminary to this possibility is of course that the concentrations of the detritus–bacteria fraction are many times higher than the concentrations of phytoplankton in the oceanic environment. Presumably, determinations of the rates at which suspension feeders from the euphotic zone of the oceans clear the water of particulate matter are needed before the problem of the type of food of the various forms can be finally settled.

3.3. DEEP SEA

3.31. VERTICAL DISTRIBUTION OF ANIMALS IN THE OCEANS

As we have practically no direct knowledge about the food of the deep sea suspension feeders, we must content ourselves with conclusions and suggestions that can be obtained from circumstantial evidence of various kinds. Ultimately, the greater part of the food to the deep sea fauna must be derived from the euphotic zone of the oceans. Generally, therefore, the food supply gets sparser with increasing depth, and the problem arises whether lack of food may in places prevent the establishment of populations of suspension feeders, or other higher animals. We may therefore get some information about food conditions in the deep sea by studying the distribution and abundance of the fauna, especially with reference to the relative importance of the suspension feeders.

During recent years measurements have become available of the biomass of animal populations down to the greatest depths. Some of the data, which mainly originate from Russian deep sea expeditions, are presented in Tables 3.2 and 3.3. It appears that both the benthic and pelagic faunas decrease regularly in mass with increasing depth, and that the decrease continues even down to the greatest abyssal and ultra-abyssal, or hadal, depths. For example, in the region of the Kurile-Kamchatka Trench the

TABLE 3.2. RELATIONSHIPS BETWEEN BIOMASS OF ZOOPLANKTON AND DEPTH IN THE OCEAN

Region of ocean	Depth (m)	Biomass (mg/m^3)	References
N.W. Pacific Ocean, boreal region	Surface	*ca.* 300	Bogorov, 1958
	3,000	*ca.* 10	Bogorov, 1958
N.W. Pacific Ocean, tropical region	Surface	*ca.* 50	Bogorov, 1958
	2,000–4,000	*ca.* 3	Bogorov, 1958
N.W. Pacific Ocean, boreal region	Surface	100–1,000	Vinogradov, 1962b
	3,000	10	Vinogradov, 1962b
W. Pacific Ocean, tropical region	Surface	8–70	Vinogradov, 1962b
	3,000	0·2	Vinogradov, 1962b
Kurile-Kamchatka Trench	6,000–9,000	0·48	Birshstein, 1959*
Kermadec Trench	6,000–9,000	0·085	Birshstein, 1959*
Bougainville Trench	6,000–9,000	0·01	Birshstein, 1959*
Kurile-Kamchatka Trench	Surface	100–1,000	Vinogradov, 1959a
	1,000	30	Vinogradov, 1959a
	4,000	5	Vinogradov, 1959a
	6,000	2	Vinogradov, 1959a
	7,000–8,000	0·4	Vinogradov, 1959a
Atlantic Ocean, Sargasso Sea	0–50	65	Jashnov, 1961
	1,000–2,000	1	Jashnov, 1961
	2,000–4,000	0·5	Jashnov, 1961
	4,000–5,000	0·1	Jashnov, 1961
	Coastal areas	a few grams	Zenkevitch, 1961
	Abyssal	0·17	Zenkevitch, 1961

* Quantity of plankton is primarily determined by the abundance of the plankton biomass in the surface layer.

TABLE 3.3. RELATIONSHIPS BETWEEN BIOMASS OF BENTHOS AND DEPTH IN THE OCEAN

Region of ocean	Depth (m)	Biomass (g/m²)	References
Atlantic Ocean	83	49	Kusnetzov, 1960
	350–1,800	0·1–3,804, av. 1,135	
	2,000–3,300	0·2–23, av. 5·4	
	3,300–5,400	0–0·05 (18), av. 0·02 (3)*	
N.E. Pacific Ocean	100–1,000	45	Filatova and Levenstein, 1961
	1,000–2,000	36	
	2,000–4,000	0·08–2·9, av. 2·2	
	4,000–6,000	0·03–3·5, av. 0·8†	
Antarctic waters	100– 200	1,347	Belyaev and Ushakov, 1957
	1,000–3,200	1·3	
Okhotsk Sea	Littoral zone	*ca.* 1,200	Savilov, 1957
	Sublittoral zone	77–359	
	Slope, 250–500	13	
	Slope, 500–1,000	36	
Coastal regions	—	Up to a few kg/m²	Zenkevitch, 1961
Abyssal regions (below 2,000m):			
Between Antarctic and Africa–Australia	—	0·1–>1	
Northern and western Pacific Ocean	—	0·01–0·05	
Indian Ocean; Atlantic Ocean off Brazil	—	*ca.* 0–<0·01	
Kermadec Trench	9,000–11,000	0·26–0·93	Birshstein, 1959‡
Bougainville Trench	9,000–11,000	0·007	
Tonga Trench	9,000–11,000	0·001	
Various trenches	6,000–11,000	10–40, 0·3	Wolff, 1960

* Only one sample out of six above 0·05. † Only one sample above 0·26. ‡ Benthos biomass depends upon latitudinal position and distance of the trenches from land.

biomass of the plankton at depths of 1,000m was only about $\frac{1}{10}$ of the biomass in the surface layer; at 5,000m the proportion had fallen to $\frac{1}{100}$ and in the trench proper at 7,000–8,000m to $\frac{1}{1000}$ (Vinogradov, 1959a). In other regions of the oceans similar results were obtained. The decrease in biomass of the benthos with increasing depth is not less pronounced than of the plankton, with biomasses at abyssal depths ranging from less than 10mg to one g or more per m², or from $\frac{1}{100}$ to $\frac{1}{10000}$ the amounts typical of shallow coastal waters.

It has long been suspected that the size of the deep sea populations is related to the productivity in the surface layer immediately above, and the quantitative studies of the later years have been confirmatory. Thus, in eutrophic areas with a surface population of zooplankton amounting to several hundreds of mg per m^3 the biomass at 4,000m may amount to 5mg/m^3, whereas below an oligotrophic area with less than 100mg/m^3 only 0·5mg/m^3 may be found at 4,000m (Jashnov, 1961). The relationship between surface productivity and size of deep sea populations seems to hold good for the benthos too. Relatively high deep sea populations may, however, also be found in areas not directly beneath highly productive regions, but which receive water currents from such regions, e.g. currents of deep water from boreal regions towards the equator (Beklemishev, 1960; Vinogradov, 1959b, 1962b), or turbidity currents transporting shallow water sediments relatively rich in particulate food to great depths of the sea (Heezen, 1959; Heezen *et al.,* 1959). These exceptions from the rule of a direct relationship between the size of the deep sea population and the productivity in the surface layer immediately above are therefore still consistent with the notion of a rather direct transfer of food from its place of origin to the consuming deep sea populations (see also p. 280).

It is of special interest to us that the relative importance of the suspension feeders within the animal assemblages seem generally to decline with increasing depth, both within the benthos (Sokolova, 1959) and the plankton (Bogorov, 1958). For example, in the Okhotsk Sea the proportion of suspension feeders was lower on the slope down to 1,000m than in the littoral and sublittoral zones (Savilov, 1957). In antarctic waters the biomass of sponges decreased from 74·2 per cent of the total biomass at depths of 100–200m to 15·4 per cent at 1,000–3,200m; and of ascidians from 7 per cent to 0 (Belyaev and Ushakov, 1957). In the hadal depths of the oceanic trenches, ranging from 6,000–7,000m to the maximum depths of 11,000m, the dominant groups are deposit feeders or are living on larger prey: coelenterates (actinians), polychaetes, crustaceans (isopods and amphipods), molluscs (gastropods and bivalves), and echinoderms (holothurians); whereas most suspension feeding groups either seem to be completely missing, e.g. brachiopods, or to be weakly represented: tunicates, cirripeds, bryozoans and sponges (Wolff, 1960). The bivalves seem to be the only group of typical suspension feeders that constitutes an important part of the abyssal and hadal faunas, but, by closer inspection, this group too bears out the general rule of the relative paucity of suspension feeders at great depths, because the deep-sea bivalves are dominated by the deposit eating forms, especially protobranchs (Clarke, 1962; Filatova, 1959; Ockelmann, 1958). Also within the echinoderms the forms that are most abundant and reach farthest down in the deep sea are deposit feeders, as the holothurians. Of the suspension feeding crinoids only one species has been found below 6,000m, three below 5,000m and 11 below 4,000m (Madsen, 1954). In the plankton, the proportion of raptorial forms amongst copepods and the other crustaceans, which dominate the plankton, increases with depths (Bogorov, 1958; Vinogradov, 1959a). Moreover, the planktonic suspension-feeding stages of the bottom invertebrates, typical of shallow water communities outside arctic regions, are missing at great depths. The abandoning of a long pelagic stage in the larval development of the invertebrates in the deep sea has been attributed to the shortage in suspended food (Thorson, 1950, 1951).

It is thus indicated that conditions for life deteriorates more rapidly with depth for suspension feeders than for other feeding types. However, even when the deep sea is very sparsely populated, the fact remains that suspension feeders are able to live at great depths. Thus several suspension feeding bivalves are characterized by a very wide vertical range of distribution, e.g. *Anomia squamata, Crenella decussata, Cyclopecten* spp., *Pycnodonta cochlear* and *Modiola phaseolina* which are all distributed from shallow water throughout the bathyal region, down to about

2,000m (Clarke, 1962; Madsen, 1949; Ockelmann, 1958). Abyssal forms include *Thyasira* and the pectinid *Propeamussium*; and suspension feeding bivalves have been recorded to the greatest depths of about 10,000m in some of the trenches, e.g. *Lima* and *Thyasira* (Filatova, 1959; Wolff, 1960). Also abyssal and hadal forms of sponges, crinoids and ascidians are known. Contrary to their relatives from shallower waters, which generally belong to the epifauna, the deep sea forms of sponges (*Hyalonema*), crinoids and ascidians are often adapted for life on the soft bottom (Fage, 1954). The stalked Bathycrinoidae, which are amongst the crinoids that have reached the greatest known depths (*ca.*8, 100m), anchor themselves in the soft bottom (Gislén, 1954). Also some deep sea ascidians are stalked, e.g. *Culeolus* spp. However, in the ascidians the stalk is probably not rigid enough to hold the body off the bottom, but together with filamentous processes on the test may prevent the animal from sinking into the soft substratum (Millar, 1959). Many of the benthic deep sea suspension feeders thus seem to take their feeding currents from the free water masses and not from the transition zone between the mud and the water.

In the plankton, the suspension feeding forms seem to decrease in importance with increasing depth more rapidly than in the benthos. It is even doubtful whether suspension feeding occurs in the copepods and other planktonic crustaceans in the deep sea (Wickstead, 1962). And important suspension feeders in oceanic plankton as appendicularians and Thaliacea may not live at true abyssal depths, but only in the bathyal down to perhaps 2,000–3,000m (Harant, 1948; Marshall, 1954).

It thus seems that benthic, but not pelagic suspension feeders can penetrate to the greatest oceanic depths. This indicates that feeding conditions for suspension feeders at abyssal depths are generally poorer in the mid water than close to the bottom. A striking parallel to this situation has been described for benthic and pelagic fish by Marshall (1960). About a hundred species of pelagic fish are known to live at depths from 1,000 to 4,000m. They have reduced tissues compared with the oceanic fish living above 1,000m. Especially skeleton and musculature are reduced, and a swimbladder is absent; the fish are likely to be neutrally buoyant. However, about 250 species of benthic fish that have been taken below 2,000m, many down to 5,000m, are highly organized compared with the deeper bathypelagic fish; and more than half of them have well-developed swim-bladders. This striking difference in development of benthic and pelagic deep sea fish Marshall ascribes to a difference in food supply, the strong reduction of the tissues of the deep mid-water fish representing an adaptation to the very limited food supply reaching this region.

As mentioned, the concentration of particulate food may perhaps decrease in the deep sea to levels too low to support a pelagic suspension feeding population. However, the particulate matter is not regularly distributed in the deep sea, but zones of higher concentrations may arise at various levels, whereas intervening regions may be more or less depleted of particulate matter (see p. 281). Obviously processes which result in concentration of particulate matter at various levels in the deep sea may influence the food conditions and thus the distribution of pelagic suspension feeders, but the importance of the phenomenon is unknown.

Close to the bottom of the deep sea the particle concentration may rise again. The Swedish Deep-Sea Expedition, 1947–48, often found increased amounts of suspended matter in the first tens of metres above the bottom. The highly stratified distribution indicated that the turbulence was low, yet still sufficient to counteract the sinking of particles due to gravity. In other localities of the deep sea the bottom suspension was absent and the water extremely clear in close proximity to the bottom. The presence or absence of suspended matter to a great extent depends upon the relief of the bottom and the strength of the bottom currents (Jerlov, 1953).

As amounts of suspended food are probably generally correlated with total amounts of suspended matter, a correlation is to be expected between the density of the bottom suspensions and the size of the population of benthic suspension feeders. Such a correlation was demonstrated by Sokolova (1959) who found in the Kurile-Kamchatka Trench down to a depth of 10,000m that in zones where the conditions favoured high concentrations of suspended matter over the bottom the suspension feeders formed a much greater proportion of the total biomass of the benthos than in zones with presumably little suspended matter.

3.32. TRANSFER OF FOOD TO THE DEEP SEA

We know of several mechanisms by which organic matter can be transferred from the euphotic zone to the deep sea, but our knowledge is still incomplete about the relative importance of the various mechanisms, and about the transformations which the organic matter may undergo before it is presented as food to the suspension feeders in the deep sea.

3.321. SINKING OF PHYTOPLANKTON AND DEAD ORGANIC REMAINS

The simplest way of imagining how food from the photosynthetic zone may become available to the suspension feeders below is to assume that this food sinks out of the surface layers and down into the deep sea. This rain of phytoplankton algae and other particulate organic matter has long been a cliché used in explaining the transfer of food to the deep sea animals; but it has often been emphasized that the passive sinking of living or dead phytoplankton cells seems unable to transport significant amounts of food very far down due to the low rate of sinking and the rapid break-down of the organic matter. Living phytoplankton cells mostly sink only slowly, or may not sink at all, because of specific buoyancy mechanisms (Gross and Zeuthen, 1948; Denton, 1963); but part of the production cannot escape being lost from the euphotic zone by sinking.

The rates at which particles are sinking through the water mainly depend upon the specific gravity and diameter of the particles and may for phytoplankton cells amount to some metres a day (Andersen, in Steemann Nielsen and Jensen, 1957; Grim, 1939; Krey, 1958a; Steele, 1959). At the high sinking rate of 10m a day an intact phytoplankton cell would take 500 days to reach a depth of 5,000m. Actual sinking rates of particles about one or some few μ in diameter, which seem to dominate particulate matter in the oceans (Jerlov, 1953) are, however, much lower, perhaps as low as 20m/year (Kullenberg, 1953), in agreement with a turnover time of detritus of perhaps some hundred years (see p. 237).

In fact, no well defined sinking rates apparently exist for the suspended particulate matter in the oceans, because the particles are not sinking at constant rates, but various factors, e.g. changes in size and density of particles due to bacterial decomposition, increase in viscosity of the water with depth, establishment of discontinuity layers between water masses, tend to cause an accumulation of the matter at certain levels in the sea (Jerlov, 1953; Cooper, 1961; Vinogradov, 1961). It is of interest that many zooplankton, for instance copepods, cladocerans, and larvae of cirripeds, have been found in laboratory experiments actively to concentrate at the interface between water masses of different densities (Harder, 1957). Such behaviour would tend towards bringing food and consumers together. It would therefore be of interest to know whether the accumulation of particulate matter at discontinuity layers and the preference of many zooplankton organisms for such layers influence the distribution of suspension feeders in the deep sea.

On the way down, the properties of the passively sinking phytoplankton, and other organic particulate matter, are profoundly changing in composition, fast in the upper layers where the temperature and the concentrations of bacteria and zooplankton are high, and slowly in the deep sea below about 1,000m.

The sinking, living phytoplankton cells continue to respire and will thus reduce their biomass before death. The depths a sinking autotrophic phytoplankton cell may reach alive probably seldom exceed 1,000m. Thus in the equatorial Pacific Ocean, Hasle (1959) found that at a depth of some hundred metres the phytoplankton cells were generally empty, with the dinoflagellates as an exception. In the Sargasso Sea, Yentsch and Ryther (1959a) found that at a depth of 1,000m chlorophyll was absent from the particulate matter, but some was still present at 300–500m.

After death, the decay of the plankton organisms rapidly sets in. The sequence of processes going on during the decomposition has been studied in laboratory experiments by von Brand *et al.* (1937–42). Nitrogen oxidation was found to take place in three steps, namely liberation of ammonia, mainly due to deamination of proteins, followed by oxidation of the ammonia to nitrite, which is finally oxidized to nitrate. In nature, ammonia liberation may largely take place within the upper few hundred metres, whereas the final conversion into nitrate has a maximum at, e.g. 600–800m, together with a maximum in phosphate regeneration (Wattenberg, 1937). The bacterial decay of the organic matter consumes oxygen; oxygen minima are therefore often met with in the layers of rapid decomposition, and especially pronounced in the highly productive parts of the oceans; in some areas practically oxygen free layers are established (Brandhorst, 1959–60; Plunkett and Rakestraw, 1955; Riley, 1951).

Thus, at a depth of about 1,000m the greater part of the easily decomposable constituents of the sinking phytoplankton and organic detritus seems to have been re-mineralized; and the detritus has presumably been enriched with respect to organic matter resistant to bacterial attack. Most likely the detritus has also increased in resistance towards attack of the digestive enzymes of most animals that may happen to ingest the particles: during the sinking from the photic layer to some 1,000m the nutritional value of the particulate organic matter is therefore probably strongly reduced.

As the sinking rate increases with the diameter of the body, larger particles or organisms belonging to the zooplankton should have a better chance than phytoplankton of reaching the deep sea before disintegrating. Dead zooplankton organisms and faecal materials have therefore also been considered an important source in the "rain" of food particles from the surface layers to the deep sea. However, Seiwell and Seiwell (1938) found in laboratory experiments that sinking rates of zooplankton decreased from about 320m/day to 2·6m/day within 363 hr after the death of the organisms; and in nature the rate of sinking of the zooplankton seems to be slower than observed in the laboratory. Thus shell-bearing pteropods were observed to fall at a rate corresponding to 1–2,000m/day, and should accordingly be able to reach the greatest depths in a few days without disintegrating, but actually only 10–20 per cent of pteropod shells sampled below 500–1,000m contained organic remains; below 2,000m practically all shells were empty, and below 3,000m no shells containing organic remains have been obtained (Vinogradov, 1961). Even zooplankton organisms as large as 1–2mm therefore seem to dissolve before reaching abyssal depths by sinking. However, large dead animals, e.g. fish and whales, and their excreta presumably sink fast enough to carry their organic matter practically untaxed to the greatest depths. In fact, in the open ocean such large bodies would seem to be the only significant source contributing organic matter to the deep sea under the force of gravity (Krogh, 1934b).*

3.322. TRANSPORT BY CURRENTS

Fast transport of phytoplankton to the deep sea may occur in certain areas with deep currents. However, apparently this method of food transfer from the productive surface layers to the deep sea can by its nature mainly be of local importance in the oceans.

3.323. FOOD TRANSPORT BY VERTICALLY MIGRATING ANIMALS

It is well known that many zooplankton organisms perform vertical migrations. Often these have the character of food migrations: the organisms may rise into the surface layers during the night to graze on the phytoplankton, and descend again at daytime, transferring organic matter to the lower level by their excreta or their own bodies. The vertical extension of the daily migrations varies much from species to species, and may extend to some hundreds of metres. But even more extensive migrations may take place. Thus, important copepods in the Norwegian Sea, as *Calanus finmarchicus* and *Pseudocalanus minutus,* have been found to descend to and stay at depths between 600 and 2,000m (the bottom) during the autumn and winter and to ascend to the surface layers in the early spring for spawning and growth of the new generation (Østvedt, 1955). This type of vertical migration will also result in a net transfer of organic matter to deeper regions in the sea. Apparently vertically migrating forms are not

* Surprisingly, many large debris of plants have been found at abyssal and hadal depths, and even far from land. Trees and their parts, e.g. carried out by rivers, or large sea weeds must contribute relatively significantly amongst the energy sources to the deep sea because they can be carried to the greatest depths with little loss *en route* (Bruun and Wolff, 1961). These large remains of plants are utilized by wood eating species, amongst which the wood-boring bivalves Xylophagidae are important. They are commonly found in larger or smaller pieces of wood or other plant material at abyssal and hadal depths. It appears that the wood constitutes their main food, and it remains uncertain whether suspended matter contributes significantly to the food (Knudsen, 1961, see also p. 89).

restricted to the bathyal region of the oceans, but occur also amongst abyssal populations, and some species are stated to perform short-timed migrations of magnitudes up to 2–4,000m. These vertically migrating species have been ascribed an important role in carrying organic matter down to the deep sea (Vinogradov, 1955; Wickstead, 1962). Animals sinking from the surface zone meet the inhabitants of the deep waters, which feed on them and on their remains, and which, descending to still greater depths become, in their turn, the prey of animals living in the deepest layers. Thus the organic matter produced in the surface zone is actively carried to the greatest oceanic depths along such a "ladder of migrations".

It is as yet not quite clear whether species living in oceanic trenches, in depths greater than 6,000m, are included in this system of migrations (Vinogradov, 1962a).

This system of active transport of organic matter to the deep sea is mainly built up as prey-predator relationships, but some finely particulate organic matter of nutritional value to suspension feeders may of course also arise as by-products, either directly from disintegrating organisms or indirectly as heterotrophic unicellular organisms feeding on the remains and dissolved substances from the predators or their prey, but the quantitative importance of this source of suspended matter is difficult to ascertain.

3.324. BACTERIA AND OTHER HETEROTROPHIC MICRO-ORGANISMS IN THE DEEP SEA

The deep sea contains a population of heterotrophic organisms which have quite naturally been considered of importance in the food chains of the deep sea. Besides bacteria, yeasts and other better known groups of heterotrophs, various other groups of unicellular organisms have been described which presumably nourish on the dissolved and dead particulate organic matter present in the water. Thus, Hentschel (1928) described olive green cells which were present in decreasing numbers, down to the greatest depths in the Atlantic Ocean. According to Hentschel the olive green cells constitute the characteristic nannoplankton organisms in the deep sea. Similar cells were also described in the deep layer of the Mediterranean (Schiller, 1931). The cells vary in size from 1 to 15 μ in diameter, and less than 50 cells are present per litre of water at depths below 4,000m. If we assume the very high average diameter of the cells of 10 μ (1 μ is probably more correct), and 100 cells to be present per litre, the biomass amounts to *ca.* 50,000 μ^3/l. or *ca.* 0·01 μg organic matter/l. This is of the same order of magnitude as the bacteria organic matter in the deep sea found by Kriss (1959). Exceptionally high amounts of heterotrophic deep sea algae are recorded by Bernard (1958) from the Mediterranean and Atlantic, especially *Coccolithus*. The presence of algal cells in the deep sea is confirmed by stomach analyses of the filter feeding ascidian *Culeolus suhmi* from a depth of 5,000m. Some of the algae were identified as Coccolithophoridaceae-cells, fairly frequent; brown-walled cells, 3–10 μ, similar to non-motile stages of the genus Chrysochromulina, very frequent; various plant cells, apparently naked, possibly cells which have lost their coccoliths; other Chrysophyceae, and occasional dinoflagellates (Millar, 1959). The source of this material Millar assumed to be the slow rain of cells from the upper layers of water, but as mentioned above the phytoplankton cells of the photic zone are not likely to reach the deep sea by sinking alone. It is not clear to what extent the algal cells found in the ascidians may have originated locally in the deep sea.

If the available analyses of bacterial populations are correct (Table 2.21), the average concentration of bacterial food of about 0·03 μg/l. in the deep sea amounts to only about 1/10,000 of the amounts needed to cover food requirements in a suspension feeder clearing about 15l. of water for each ml. of oxygen consumed (p. 251). Even when we do not know to how low concentrations of particulate food the various types of suspension feeders are capable of adapting (Jørgensen, 1955), the concentrations of heterotrophs recorded in deep sea water would seem to be below the limiting level for suspension feeders. It is therefore indicated that where suspension feeders are present in the deep sea their food must either be derived from other sources, that is, organic detritus, or the concentrations of heterotrophic microorganisms must be much higher than indicated by the available figures. It is not known to what extent bacterial populations may be locally concentrated in the midwater, but close to the bottom the concentrations presumably often increase, especially in areas where currents may resuspend bottom deposits, which may contain 10^6 to 10^7 bacteria per ml or about 0·1–1mg organic matter per litre (ZoBell, 1954; ZoBell and Morita, 1957–58) (see p. 218). In such localities the concentration of bacteria may be expected to be high enough to support a population of benthic suspension feeders.

3.325. DISSOLVED ORGANIC MATTER

Also the organic matter dissolved in the deep sea water mostly originates from the euphotic zone of the sea, but whereas the amounts of particulate matter that are passively and actively carried down into the deep sea to a large extent depend upon the productivity at the surface layer above, the dissolved organic matter is more uniformly distributed in the water of the oceans. This is a consequence of its slow rate of turnover, which seems to be of the same order of magnitude as the rate of turnover of the oceans themselves (see p. 239). As mentioned above the abundance of the deep sea fauna varies with the productivity in the euphotic layer above. If dissolved organic matter did contribute significantly as food to the deep sea animals no such relationships would be expected. Therefore the distribution and biomass of the deep sea faunas, together with previous arguments (pp. 264–70), speaks against an important direct utilization of the dissolved organic substances as food of deep sea animals, suspension feeders included.

However, dissolved organic matter together with generally non-digestible components of the detritus, as cellulose, lignin, chitin, keratin, etc., may be brought back into a utilizable form by the heterotrophic micro-organisms in the deep sea. It may therefore be appropriate to end this chapter by discussing again the nature of the dissolved organic matter and the role played by the heterotrophs in bringing it back into circulation in a form that may be utilized by suspension feeders.

As mentioned, the organic matter dissolved in sea water must range from substances given off from living or autolyzing cells, and presumably most easily utilizable by aquatic animals to which the substances are accessible, to "humic" and other substances that are not digestible by animals and even attacked only with difficulty by bacteria. However, much remains to be known about the chemistry of the substances; only a minor fraction of the total amount of dissolved organic matter in deep sea water is chemically identified (see Table 2.19). We do not know, therefore, what proportion of the organic matter is utilizable by aquatic animals; but a little more is known about the accessibility of the dissolved organic matter as food for the bacteria.

It is well known that in sea water brought into the laboratory, oxygen is consumed simultaneously with a vigorous growth of bacteria. The oxygen consumption has been shown to be due to bacterial decomposition of the organic matter in the sea water filtered through filters fine enough to remove all particulate or even colloidal matter and afterwards inoculated with bacteria (Keys *et al.,* 1935; Waksman and Carey, 1935b). Dissolved organic matter utilizable by bacteria is therefore present in sea water. It has further been found that oxygen consumption declines or stops before all of the dissolved organic matter is presumably decomposed. Measurements of oxygen consumption in stored sea water have therefore been used to characterize the dissolved organic matter in sea water with respect to the ease with which it is utilized by bacteria. By means of this technique it should be possible to estimate how much of the dissolved organic matter is easily decomposed by bacteria; and, if the total content of dissolved organic matter is known, also how much is refractory. However, the results of measurements of oxygen consumption are not easily interpreted in terms of degree of refractoriness of the organic matter, because other factors, such as the temperature and the size of surface in relation to volume of the water sample, may strongly influence the oxygen consumption in incubated sea water. The explosive increase in bacterial growth and consequently decomposition of organic matter that generally takes place in a sample of sea water isolated from the surrounding water, e.g. by being transferred into a bottle, is thus most likely mainly due to the increase in the size of the surface available for attachment and adsorption of bacteria and organic matter. The inner surface of a one litre bottle is about 100,000 times as large as the total surface of particulate material in one litre of oceanic surface waters (Steemann Nielsen, 1955). As in nature the turnover time of the dissolved organic matter may be about 1,000 years, whereas in the laboratory under favourable conditions half of the organic matter in an incubated sample of water may be decomposed in about a week (Waksman and Carey, 1935b; ZoBell and Anderson, 1936), the rate of turnover may be increased many thousand times.

As the amounts of oxygen consumed in stored sea water vary with other factors than the refractoriness of the organic matter, the measurements of oxygen consumption allow only rough estimates, which tend to provide minimum values, to be made of the proportions of dissolved organic matter that is easily utilized by marine bacteria. In Table 3.4 are collated measurements of oxygen consumption in samples of sea water from various parts and depths of the sea. The values of oxygen consumed may be converted into the approximate amounts of organic matter combusted by assuming that 1ml of oxygen is utilized in oxidizing 1mg of organic matter. It may be seen from the Table 3.4 that from 0·1 to 2·5ml oxygen, or some 0·1 to 2·5mg organic matter, is utilized per litre during the periods of incubation, varying from less than 2–44 weeks. In the experiments of long duration it was found that oxygen consumption did not proceed at a measurable rate for more than some weeks or at most months, even at low temperatures of 4–5°C. The proportion of organic matter finally oxidized apparently depended more upon the temperature of incubation than upon the oceanic depth from which the water originated. Thus, in the experiments of Keys *et al.* (1935) and Rakestraw (1947) the organic matter oxidized by bacteria did not significantly vary with sampling depths from 0 to 1,800 or 1,200 metres. Seiwell (1937), though, in experiments of only 10 days' duration, states that the higher values of oxygen consumption characterized the water samples, which were unfiltered, from below a depth of 1,000m.

From the available data on oxygen consumption in samples of sea water we can conclude that under favourable conditions up to *ca.* 2mg/l. of dissolved organic matter may be rapidly oxidized by bacteria (Waksman and Carey, 1935b; ZoBell and Anderson, 1936). As the amounts of organic matter simultaneously incorporated in the bacterial biomass may be up about half the amount oxidized (Rahn, 1940; Waksman and Carey, 1935b; ZoBell and Grant, 1943), the amounts of dissolved

TABLE 3.4. BACTERIAL OXIDATION OF DISSOLVED ORGANIC MATTER IN SEA WATER

Source of water sample	Depth (m)	Temp. (°C)	Oxygen consumed (ml/l.)	Period of incubation (days)	References
Oslo Fjord, Norway	—	—	2	—	Gran and Ruud, 1926
Buzzard Bay, Vineyard Sound, Mass., U.S.A.	Surface	—	0·5	20	ZoBell, 1940
Woods Hole Harbor, Mass., U.S.A.	—	—	0·8	14	Waksman and Renn, 1936
	—	—	1·4	63	
Woods Hole Harbor, Mass., U.S.A.	—	21–23	0·35	20	Keys *et al.*, 1935
Oceanic waters	50	21–23	0·7	17	
	1,800	21–23	0·45	17	
Atlantic Ocean, 32°N, 66°W	0–4,600	24	0·6–1·7, av. 1·2	10	Seiwell, 1937
		11	0·2–0·7, av. 0·5	10	
Atlantic Ocean, 39°N, 71°W and 36°N, 68°W	Surface	25	*ca.* 0·6	82–172	Rakestraw, 1947
	400	8–9	*ca.* 0·1	20–172	
	400	25	*ca.* 0·6	212	
	1,200	4–5	*ca.* 0·1	82–305	

organic matter that may easily be utilized by bacteria increases to about 3mg/l.

Further information about the biological properties of the dissolved organic matter has been obtained from experiments in which the rates of oxygen consumption have been measured in samples of sea water without and with addition of known amounts of various organic substances. Thus Waksman and Carey (1935b) found that the amount of dissolved organic matter that was utilized at the same rate as added glucose corresponded to an oxygen consumption of 1·28ml/l., or about 1·3mg/l. of organic matter. This, of course, does not mean that significant amounts of glucose were present in the sample of sea water; a great many water soluble organic substances are oxidized by bacteria at about the same rate as glucose. It can be concluded, though, that the water, probably surface water, contained significant amounts of organic matter about as easily oxidizable as glucose.*

The total content of dissolved organic matter found in unpolluted sea water varies from 1 to 5mg/l. (Tables 2.17, 2.18). Thus in the water from the North Atlantic Ocean, 30°N, 69°W, sampled at depths from 0 to 5,400m, Krogh (1934a, b) found 4·7mg/l. In water sampled from 0 to 4,600m some degrees farther north, 36–39°N, 68–71°W, and incubated for 10 days at 24°C, Seiwell (1937) measured an oxygen consumption of 0·6–1·7ml/l., corresponding to an oxidation of some 0·6–1·7mg/l. of organic matter. Considering that the amounts of organic matter incorporated in the bacteria may amount to about half the amounts oxidized, the total amounts of organic matter utilized in 10 days

* According to ZoBell (1943) the bacterial decomposition of dilute solutions of easily soluble substances as glucose and aminoacids is not enhanced by an increase in surface relative to volume of the culture medium, whereas the effect of surfaces for attachment and adsorption of bacteria and organic substances is especially conspicuous for organic substances in colloidal solution. It should accordingly be possible to differentiate between organic substances in natural sea water that are easily decomposed in solution and such that are decomposed mainly after adsorption on surfaces; the latter fraction should be dominant in oligotrophic sea water (Steemann Nielsen, 1955). However, Heukelekian and Heller (1940) found that also glucose in dilute solutions (0·5 and 2·5mg/l.) was rapidly utilized only if sufficient surface was supplied to the culture medium, e.g. in form of inert glass beads.

increase to some 0·9–2·5mg/l. It is thus indicated that 20–50 per cent of the total dissolved organic matter in Atlantic oceanic water are easily utilized by bacteria, the higher values prevailing in the upper 1,000 metres. But farther north, at 50–60°N, Atlantic water from a depth of 0 to 3,000m was found to contain about 1mg/l. of dissolved organic matter (Duursma, 1961), or from about the same to less than the amounts of easily oxidizable matter estimated from the measurements of Seiwell.

Obviously, our present knowledge does not allow accurate estimates to be made of the proportions of the total dissolved organic matter in the various regions of the sea that can be rapidly oxidized by bacteria and those that are only broken down with difficulty or are not broken down at all at a measurable speed even under favourable conditions in the laboratory; but it is suggested that the non-refractory and refractory parts may be of the same order of magnitude.

The refractory fraction, which has sometimes been assumed to comprise by far the greater part of the organic matter in the oceans at least in the deeper parts, has often been considered as a pool of organic matter more or less out of the organic circulation (Fraga, 1960; Keys *et al.,* 1935; Kriss, 1959a; Kriss *et al.,* 1960; Plunkett and Rakestraw, 1955; Steemann Nielsen, 1955). It is unlikely, though, that any significant fraction of the dissolved organic matter is completely out of circulation because this would imply that the concentration of the substances constituting this fraction had been constantly increasing during as long a period of time as the present conditions of life in the oceans have prevailed, probably millions of years. More likely the organic matter dissolved in the oceans represents a steady state (Krey, 1961c; Vallentyne, 1962) so that the rates at which the individual substances comprising the dissolved organic matter are produced equal the rates at which these substances are removed from the solution in the sea, for instance by being directly metabolized by living organisms, or by being adsorbed to surfaces of particles suspended in the water and eventually sedimented.

If the refractory and non-refractory parts of the dissolved organic matter are of the same order of magnitude, it follows that both have about the same turnover times. It is therefore indicated that in nature, the degree of refractoriness towards bacterial decomposition is of minor importance in determining the rate of turnover of the dissolved organic substances. One main factor responsible for the very slow turnover is probably the lack of surface without which the dilute solutions of organic substances are not accessible to the bacteria, and without which the refractory matter cannot be extracted from the solution by adsorption. It would thus seem that the particles, organic as inorganic, with their attached bacteria may play a major role in keeping the organic matter circulating (Armstrong, 1958), and thus enabling the bacteria to convert the dissolved organic matter in a form more easily accessible to suspension feeders and other organisms (Butterfield and Purdy, 1931; Butkevich and Butkevich, 1936; Kriss, 1959; Waksman and Carey, 1935a). Any increase in supply of particulate matter to the surface water, whether inorganic of terrigenic origin or of organic origin from the primary production, may therefore also increase the amounts of the dissolved organic matter that can be made available to the animal populations in the deep sea immediately below.

We have outlined the various mechanisms by which organic matter may be transferred from the euphotic zone to the deep sea, as well as the transformations it may undergo *en route,* and we should finally comment upon the possible importance of the various mechanisms and processes in providing food for the suspension feeders of the deep sea.

The finding that the biomass of the fauna in the deep sea seems to be correlated with the productivity in the euphotic zone above indicates that the deep sea fauna is mainly maintained on food that has been rather directly transferred from the surface layer down to the consumers. If the dependency of the deep sea fauna upon the productivity in the surface layer also applies to the suspension feeders it is further indicated that the amounts of particulate food transferred to the deep sea depends directly upon the primary production in the euphotic zone. However, it is unlikely that this particulate food is to any extent the dead remains of the plankton sinking towards the bottom of the oceans under the force of gravity. There is good evidence that the turnover time of the detritus in the oceans, or the time taken for detritus particles passively to sink to abyssal depths, amounts to scores of years, during which time any original nutritional value to a metazoan of the detritus particles probably has disappeared. However, a nutritional value of the sinking particles may be maintained by the bacteria growing on the particles. The growth of the bacteria is probably not only based on the organic matter originally present in the detritus particles, but the bacteria may utilize the dissolved organic matter in the surrounding water. Therefore, the passive transport by sinking of particulate matter of nutritional value for suspension feeders may presumably also occur by means of inorganic particles that can serve as substrate for bacteria.

Particulate organic matter is present throughout the deep sea. Measurements are scarce and the amounts found vary, but values of the same order of magnitude as those found for phytoplankton in oceanic surface waters have been recorded even down to great depths (Table 2.16). However, if only the fraction consisting of heterotroph micro-organisms is of nutritional value, the concentration of food particles to suspension feeders may be low in the deep parts of the oceans; according to present knowledge the heterotrophs constitute only a small fraction of total particulate matter in the deep sea.

The standing stock of pelagic suspension feeders decreases rapidly with increasing depths, perhaps to disappear completely at about 3,000m; or even less, since records of maximum depth of pelagic animals that also live at lower depths may be misleading. The vertical distribution of pelagic suspension feeders is thus closer correlated to the variations with depth in the standing stock of heterotroph micro-organisms than to the variation in total particulate organic matter, relations that support the suggestion of a low value as food to the pelagic suspension feeders of deep sea detritus.

The apparent absence of a pronounced overall decrease in concentration of particulate organic matter with depth in the deep sea is also in agreement with the lack of pelagic suspension feeders in these regions of the oceans.

A rapid transfer of organic matter to the deep sea with less loss of easily digestible materials due to bacterial decomposition results from the vertical migration of animals and from rapidly sinking larger bodies. Presumably most of the material brought down in these ways is too large to be directly utilized by suspension feeders, but it represents a potential food source that becomes utilizable by suspension feeding when it disintegrates. The quantitative significance of the two sources in providing food for the suspension feeders in the deep sea is unknown, but it would seem that they are insufficient to maintain an abyssal population of planktonic suspension feeders. Their main importance may therefore be the provision of food, directly or indirectly via the heterotrophic micro-organisms, to the benthic suspension feeders.

REFERENCES

AACH, H. G. (1952) Über Wachstum und Zusammensetzung von *Chlorella pyrenoidosa* bei unterschiedlichen Lichtstärken und Nitratmengen. *Arch. Mikrobiol.* **17,** 213–46.

AFZELIUS, B. A. (1961) Flimmer-flagellum of the sponge. *Nature, Lond.* **191,** 1318–19.

AIELLO, E. L. (1960) Factors affecting ciliary activity on the gill of the mussel *Mytilus edulis*. *Physiol. Zoöl.* **33,** 120–35.

ALFIMOV, N. M. (1954) Comparative evaluation of methods for the determination of bacterial counts in sea water. *Microbiology, Moscow,* **23,** 693 (in Russian).

ALLEN, E. J. (1919) A contribution to the quantitative study of plankton. *J. Mar. biol. Ass. U.K.* **12,** 1–8.

ALLEN, G. D. (1919) Quantitative studies on the rate of respiratory metabolism in *Planaria*. *Amer. J. Physiol.* **49,** 420–73.

ALLEN, J. A. (1954) On the structure and adaptations of *Pandora inaequivalvis* and *P. pinna*. *Quart. J. micr. Sci.* **95,** 473–82.

ALLEN, J. A. (1958a) On the basic form and adaptations to habitat in the Lucinacea (Eulamellibranchia). *Phil. Trans. roy. Soc. B.* **241,** 421–84.

ALLEN, J. A. (1958b) Observations on *Cochlodesma praetenne* (Pulteney) (Eulamellibranchia). *J. Mar. biol. Ass. U.K.* **37,** 97–112.

ALLEN, J. A. (1962) Preliminary experiments on the feeding and excretion of bivalves using *Phaeodactylum* labelled with ^{32}P. *J. Mar. biol. Ass. U.K.* **42,** 609–23.

ALLEN, W. R. (1921) Studies on the biology of fresh-water mussels. Experimental studies on the food relations of certain Unionidae. *Biol. Bull.* **40,** 210–41.

ANDERSON, W. W., GEHRINGER, J. W., and COHEN, E. (1956a, b) Physical oceanographic, biological, and chemical data, South Atlantic Coast of the United States. M/V Theodore N. Gill, Cruise 1 and 2. *U.S. Fish and Wildlife Serv., Spec. Sci. Rep. Fish.* **178** and **198.**

ANKEL, W. E. (1948) Über Fragen der Formbildung und der Zelldetermination bei Süsswasserschwämmen. *Verh. dtsch. Zool. Kiel.* 1948, 59–66.

ANKEL, W. E. (1950) Blick in den lebenden Schwamm. *Umschau.* 1950, 87–90.

ANSELL, A. D. (1961) The functional morphology of the British species of Veneracea (Eulamellibranchia). *J. Mar. biol. Ass. U.K.* **41,** 489–515.

ANSELL, A. D. (1962) The functional morphology of the larva, and the post-larval development of *Venus striatula* (Da Costa). *J. Mar. biol. Ass. U.K.* **42,** 419–44.

ARMSTRONG, F. A. J. (1958) Inorganic suspended matter in sea water. *J. Mar. Res.* **17,** 23–34.

ARMSTRONG, F. A. J. and ATKINS, W. R. G. (1950) The suspended matter of sea water. *J. Mar. biol. Ass. U.K.* **29,** 139–44.

ATKINS, D. (1936) On the ciliary mechanisms and interrelationships of lamellibranchs. Part I. Some new observations on sorting mechanisms in certain lamellibranchs. *Quart. J. micr. Sci.* **79,** 181–308.

ATKINS, D. (1937a) On the ciliary mechanisms and interrelationships of lamellibranchs. Part II. Sorting devices on the gills. *Quart. J. micr. Sci.* **79,** 339–73.

ATKINS, D. (1937b) On the ciliary mechanisms and interrelationships of lamellibranchs. Part III. Types of lamellibranch gills and their food currents. *Quart. J. micr. Sci.* **79,** 375–421.

ATKINS, D. (1938) On the ciliary mechanisms and interrelationships of lamellibranchs. Part VII. Latero-frontal cilia of the gill filaments and their phylogenetic value. *Quart. J. micr. Sci.* **80,** 345–436.

ATKINS, D. (1943) On the ciliary mechanisms and interrelationships of lamellibranchs. Part VIII. Notes on gill musculature in Microciliobranchia. *Quart. J. micr. Sci.* **84,** 187–256.

ATKINS, D. (1956) Ciliary feeding mechanisms of brachiopods. *Nature, Lond.* **177,** 706–7.

ATKINS, D. (1960) The ciliary feeding mechanism of the Megathyridae (Brachiopoda), and the growth stages of the lophophore. *J. Mar. biol. Ass. U.K.* **39,** 459–79.

ATKINS, D. and RUDWICK, M. J. S. (1962) The lophophore and ciliary feeding mechanisms of the brachiopod *Crania anomala* (Müller). *J. Mar. biol. Ass. U.K.* **42,** 469–80.

ATKINS, W. R. G. (1922) The respirable organic matter of sea water. *J. Mar. biol. Ass. U.K.* **12,** 772–80.

ATKINS, W. R. G. (1923) Note on the oxidisable organic matter of sea water. *J. Mar. biol. Ass. U.K.* **13,** 160–3.

ATKINS, W. R. G. and PARKE, M. (1951) Seasonal changes in the phytoplankton as indicated by chlorophyll estimations. *J. Mar. biol. Ass. U.K.* **29,** 609–18.

ATKINS, W. R. G. JENKINS, P. G. and WARREN, F. J. (1954) The suspended matter in sea water and its seasonal changes as affecting the visual range of the Secchi discs *J. Mar. biol. Ass. U.K.* **33,** 497–509.

BACQ, Z. M. (1935) Observations physiologiques sur le cœur, les muscles et le système nerveux d'une ascidie, *Ciona intestinalis. Arch. int. Physiol.* **40,** 357–73.

BALLE, P. (1959a) Análisis cuantitativo del fitoplancton del puerto de Palma de Mallorca (desde noviembre de 1953 a marzo de 1956). *Bol. Inst. esp. Oceanogr.* **95,** 1–13.

BALLE, P. (1959b) Nuevos datos sobre el desarrollo del fitoplancton del puerto de Palma de Mallorca (desde abril de 1956 a diciembre de 1957). *Bol. Inst. esp. Oceanogr.* **94,** 1–8.

BALLENTINE, D. (1953) Comparison of the different methods of estimating nannoplankton. *J. Mar. biol. Ass. U.K.* **32,** 129–47.

BALLENTINE, D. and ABBOTT, B. C. (1957) Toxic marine flagellates; their occurrence and physiological effects on animals. *J. gen. Microbiol.* **16,** 274–81.

BALLENTINE, D. and MORTON, J. E. (1956) Filtering, feeding, and digestion in the lamellibranch *Lasaea rubra. J. Mar. biol. Ass. U.K.* **35,** 241–74.

BANG, B. G. (1961) The surface pattern of the nasal mucosa and its relation to mucous flow. A study of chicken and herring gull nasal mucosae. *J. Morph.* **109,** 57–72.

BANSE, K. (1956) Produktionsbiologische Serien-Bestimmungen im südlichen Teil der Nordsee im März 1955. *Kieler Meeresforsch.* **12,** 166–79.

BARKLEY, E. (1940) Nahrung und Filterapparat des Walkrebses *Euphausia superba* Dana. *Z. Fisch. u. Hilfswissensch. Beih.* 1, *Walforsch.* **1,** 65–156.

BARNES, G. E. (1955) The behaviour of *Anodonta cygnea* L., and its neurophysiological basis. *J. exp. Biol.* **32,** 158–74.

BARNES, H. (1959) Stomach contents and microfeeding of some common cirripedes. *Canad. J. Zool.* **37,** 231–6.

BARNES, H. and REESE, E. S. (1959) Feeding in the pedunculate cirripede *Pollicipes polymerus* J. B. Sowerby. *Proc. Zool. Soc. Lond.* **132,** 569–85.

BARRINGTON, E. J. W. (1958) The localization of organically bound iodine in the endostyle of Amphioxus. *J. Mar. biol. Ass. U.K.* **37,** 117–26.

BATHAM, E. J. (1945) *Pollicipes spinosus* Quoy and Gaimard. I. Notes on biology and anatomy of adult barnacle. *Trans. roy. Soc. N.Z.* **74,** 359–74.

BAYLOR, E. R. and SUTCLIFFE, W. H., Jr. (1963) Dissolved organic matter in sea water as a source of particulate food. *Limnol. Oceanogr.* **8,** 369–71.

BEKLEMISHEV, K. V. (1954) Feeding of some common planktonic copepods in far-east seas. *Zool. J. U.S.S.R.* **33,** 1210–30 (in Russian).

BEKLEMISHEV, K. V. (1957) The spatial interrelationships of marine zoo- and phytoplankton. *Trans. Inst. Okeanol.* **20,** 206–26. English translation. 1959: Marine Biology. Ed. B. N. Nikitin.

BEKLEMISHEV, K. V. (1959) Anatomy of mouth parts of copepods. II. Masticatory surfaces of the mandibles in some calanids and eucalanids. *Trudy Inst. Okeanol.* **30,** 148–55 (in Russian).

BEKLEMISHEV, K. V. (1960) The role of deep waters of polar and subpolar origin in enriching the deep sea plankton. *Bull. Soc. Nat. Moscow, Ser. Biol.* **65,** 46–52 (in Russian).

BEKLEMISHEV, K. V. (1962) Superfluous feeding of marine herbivorous zooplankton. *Rapp. Cons. Explor. Mer,* **153,** 108–13.

BELIATSKAIA, I. S. (1958) Seasonal changes of the total number and biomass of bacteria in waters of three lakes of different types. *Microbiology, Moscow* **27,** 112–17 (in Russian).

BELSER, W. L. (1959) Bioassay of organic micronutrients in the sea. *Proc. nat. Acad. Sci.* **45,** 1533–42.

BELYAEV, G. M. and USHAKOV, P. V. (1957) Some regularities in the quantitative distribution of the benthic fauna in antarctic waters. *Dokl. biol. Sci. Sect.* **112,** 116–19.

BERE, R. (1933) Numbers of bacteria in inland lakes of Wisconsin as shown by the direct microscopic method. *Int. Rev. ges. Hydrobiol. & Hydrogr.* **29,** 248–63.

BERG, K. and OCKELMANN, K. W. (1959) The respiration of freshwater snails. *J. exp. Biol.* **36,** 690–708.

BERGE, G. (1958) The primary production in the Norwegian Sea in June, 1954, measured by an adapted ^{14}C technique. *Rapp. Cons. Explor. Mer,* **144,** 85–91.

BERNARD, F. (1958) Données récentes sur la fertilité élémentaire en Méditerranée. *Rapp. Cons. Explor. Mer,* **144,** 103–108.

BERNER, Å. (1962) Feeding and respiration in the copepod *Temora longicornis* (Müller). *J. Mar. biol. Ass. U.K.* **42,** 625–40.

BERRILL, N. J. (1950) *The Tunicata with an Account of the British Species.* London, Ray Society.

BIDDER, G. P. (1920) *Syncrypta spongiarum, nova. J. Linn. Soc.* **34,** 305–13.

BIDDER, G. P. (1923) The relation of the form of a sponge to its currents. *Quart. J. micr. Sci.* **67,** 293–323.

BIDDER, G. P. (1937) The perfection of sponges. *Proc. Linn. Soc. Lond.* **149,** 119–46.

BIRGE, E. A. and JUDAY, C. (1934) Particulate and dissolved organic matter in inland lakes. *Ecol. Monogr.* **4,** 440–74.

BIRSHSTEIN, J. A. (1959) The ultra-abyssal fauna of the Pacific Ocean. *Int. Oceanogr. Congr.* 31 Aug.–12 Sept., 1959. A.A.A.S. Wash. D.C. Preprints pp. 370–71.

BISHOP, C. T., ADAMS, G. A. and HUGHES, E. O. (1954) A polysaccharide from the blue-green alga, *Anabaena cylindrica. Canad. J. Chem.* **32,** 999–1004.

BLEGVAD, H. (1914) Undersøgelser over Næring og Ernæringsforhold hos Havbundens invertebrate Dyresamfund i danske Farvande. *Ber. danske biol. Sta.* **22.**

BODINE, J. H. (1921) Factors influencing the water content and the rate of metabolism of certain Orthoptera. *J. exp. Zool.* **32,** 137–64.

BOETTGER, C. R. (1930) Studien zur Physiologie der Nahrungsaufnahme festgewachsener Schnecken. Die Ernährung der Wurmschnecke *Vermetus. Biol. Zbl.* **50,** 581–97.

BOGOROV, B. G. (1958) Biogeographical regions of the plankton of the northwestern Pacific Ocean and their influence on the deep sea. *Deep-Sea Res.* **5,** 149–61.

BOND, R. M. (1933) A contribution to the study of the natural food-cycle in aquatic environments. *Bull. Bingham oceanogr. Coll.* **4**(4), 1–89.

BONE, Q. (1958a) Nervous control of cilia in Amphioxus (Branchiostoma). *Nature, Lond.* **181,** 193–94.

BONE, Q. (1958b) Observations upon the living larva of Amphioxus. *Pubbl. Staz. Zool. Napoli,* **30,** 458–71.

BONE, Q. (1958c) The asymmetry of the larval Amphioxus. *Proc. Zool. Soc. Lond.* **130,** 289–93.

BONE, Q. (1961) The organization of the atrial nervous system of Amphioxus [*Branchiostoma lanceolatum* (Pallas)]. *Phil. Trans. roy. Soc. B.* **243,** 241–69.

BRAARUD, T. (1958) Counting methods for determination of the standing crop of phytoplankton. *Rapp. Cons. Explor. Mer,* **144,** 17–19.

VON BRAND, T. (1935) Methods for the determination of nitrogen and carbon in small amounts of plankton. *Biol. Bull.* **59,** 221–32.

VON BRAND, T. (1937) Observations upon the nitrogen of the particulate matter in the sea. *Biol. Bull.* **72,** 1–6.

VON BRAND, T. (1938) Quantitative determination of the nitrogen in the particulate matter of the sea. *J. Cons. int. Explor. Mer,* **13,** 187–96.

VON BRAND, T. and RAKESTRAW, N. W. (1940) Decomposition and regeneration of nitrogenous organic matter in sea water. III. Influence of temperature and source and condition of water. *Biol. Bull.* **79,** 231–6.

VON BRAND, T. and RAKESTRAW, N. W. (1941a) Decomposition and regeneration of nitrogenous organic matter in sea water. IV. Interrelationship of various stages; influence of concentration and nature of particulate matter. *Biol. Bull.* **81,** 63–9.

VON BRAND, T. and RAKESTRAW, N. W. (1941b) The determination of dissolved organic nitrogen in sea water. *J. Mar. Res.* **4,** 76–80.

VON BRAND, T., NOLAN, M. O. and MANN, E. R. (1948) Observations on the respiration of *Australorbis glabratus* and some other aquatic snails. *Biol. Bull.* **95,** 199–213.

VON BRAND, T., RAKESTRAW, N. W. and RENN, C. E. (1937) The experimental decomposition and regeneration of nitrogenous organic matter in sea water. *Biol. Bull.* **72,** 165–75.

VON BRAND, T., RAKESTRAW, N. W. and RENN, C. E. (1939) Further experiments on the decomposition and regeneration of nitrogenous organic matter in sea water. *Biol. Bull.* **77,** 285–96.

VON BRAND, T., RAKESTRAW, N. W. and ZABOR, I. W. (1942) Decomposition and regeneration of nitrogenous organic matter in sea water. V. Factors influencing the length of the cycle; observations upon the gaseous and dissolved organic nitrogen. *Biol. Bull.* **83,** 273–82.

BRANDHORST, W. (1960) Nitrification and denitrification in the eastern tropical North Pacific. *J. Cons. int. Explor. Mer,* **25,** 3–20.

BRANDT, K. and RABEN, E. (1920) Zur Kenntnis der chemischen Zusammensetzung des Planktons und einiger Bodenorganismen. *Wiss. Meeresunters. Abt. Kiel,* **19,** 175–210.

BRIEN, P. (1948) Embranchement des Tuniciers. Morphologie et reproduction. *Traité de Zoologie* (Ed. P. P. Grassé), **11,** 554–894.

BROECKER, W. S., GERARD, R. D., EWING, M. and HEEZEN, B. C. (1961) *Geochemistry and physics of oceanic circulation. Oceanography,* Ed. M. Sears. Washington D.C., U.S.A. pp. 301–22.

BROUARDEL, J. and RINCK, E. (1956) Détermination de la production de matière organique en Méditerranée à l'aide du ^{14}C. *C. R. Acad. Sci.* **243,** 1797–1800.

Brown, A. H. (1953) The effects of light on respiration, using isotopically enriched oxygen. *Amer. J. Bot.* **40,** 719–29.

Brown, F. A., Bennett, M. F., Webb, H. M. and Ralph, C. L. (1956) Persistent daily, monthly and 27-day cycles of activity in the oyster and quahog. *J. exp. Zool.* **131,** 235–62.

Bruce, J. R., Knight, M. and Parke, M. W. (1940) The rearing of oyster larvae on an algal diet. *J. Mar. biol. Ass. U.K.* **24,** 337–74.

Bruun, A. F. and Wolff, T. (1961) *Abyssal benthic organisms: nature, origin, distribution, and influence on sedimentation. Oceanography,* Ed. M. Sears. Washington D.C., U.S.A. pp. 391–7.

Bsharah, L. (1957) Plankton of the Florida Current. V. Environmental conditions, standing crop, seasonal and diurnal changes at a station 40 miles east of Miami. *Bull. Mar. Sci. Gulf and Caribbean,* **7,** 201–51.

von Buddenbrock, W. (1939) *Grundriss der vergleichenden Physiologie,* vol. 2. Berlin.

von Buddenbrock, W. (1956) *Vergleichende Physiologie. III. Ernährung, Wasserhaushalt und Mineralhaushalt der Tiere.* Birkhäuser, Basel.

Bülbring, E., Burn, J. H. and Shelley, H. J. (1953) Acetylcholine and ciliary movement in the gill plates of *Mytilus edulis. Proc. roy. Soc. Lond. B.* **141,** 445–66.

Buley, H. M. (1936) Consumption of diatoms and dinoflagellates by the mussel. *Bull. Scripps Instn. Oceanogr. Tech. Ser.* **4,** 19–27.

Burkholder, P. R. and Sieburth, J. M. (1961) Phytoplankton and chlorophyll in the Gerlache and Bransfield Straits of Antarctica. *Limnol. Oceanogr.* **6,** 45–52.

Burlew, J. S., Ed. (1953) *Algal Culture. From Laboratory to Pilot Plant.* Carnegie Institute.

Bursche, E. M. (1959) Untersuchungen über den Detritusgehalt in Seen und Flüssen. *Int. Rev. Hydrobiol.* **44,** 439–48.

Bursche, E. M., Kühl, H. and Mann, H. (1958) Hydrochemie und Phytoplankton in der Unterelbe. *Veröff. Inst. Meeresforsch. Bremerhaven,* **5,** 165–92.

Bursche, E. M., Kühl, H. and Mann, H. (1959) Beziehungen zwischen dem Chemismus und der Phytoplanktonentwicklung auf der unteren Weser. *Int. Rev. Hydrobiol.* **44,** 277–98.

Burtt, E. T. (1940) A filter-feeding mechanism in a larva of the Chironomidae (Diptera: Nematocera). *Proc.* R. *ent. Soc. Lond.* (*A*). **15,** 113–21.

Butkevich, N. V. and Butkevich, V. S. (1936) Multiplication of sea bacteria depending on the composition of the medium and on temperature. *Microbiology, Moscow* **5,** 322–43 (in Russian, English summary).

Butkevich, V. S. (1938) On the bacterial population of the Caspian and Azov seas. *Microbiology, Moscow,* **7,** 1005–1021 (in Russian, English summary).

BUTTERFIELD, C. T. and PURDY, W. C. (1931) Some interrelationships of the plankton and bacteria in natural purification of polluted water. *Ind. Eng. Chem.* **23,** 213–18.

CAMPS, J. M., SELGA, J. and ARIAS, E. (1960a) Determinatión del carbono orgánico en el aqua de mar. *Invest. Pesq.* **16,** 139–50.

CAMPS, J. M., SELGA J. and ARIAS, E. (1960b) *Determinatión del carbono orgánico en el aqua de mar, datos sobre las costa de Castellón.* Inst. Invest. Pesq. IV Reunion, pp. 26–28.

CANNON, H. G. (1929) On the feeding mechanism of the copepods, *Calanus finmarchicus* and *Diaptomus gracilis. Brit. J. exp. Biol.* **6,** 131–44.

CANNON, H. G. (1933a) On the feeding mechanisms of the Branchiopoda. *Phil. Trans. roy. Soc. B.* **222,** 267–352.

CANNON, H. G. (1933b) On the feeding mechanisms of certain marine ostracods. *Trans. roy. Soc. Edinb.* **57,** 739–64.

CANNON, H. G. (1935) A further account of the feeding mechanism of *Chirocephalus diaphanus. Proc. roy. Soc. B.* **117,** 455–70.

CANNON, H. G. and MANTON, S. M. (1927) On the feeding mechanism of a mysid crustacean *Hemimysis lamornæ. Trans. roy. Soc. Edinb.* **55,** 219–53.

CANNON, H. G. and MANTON, S. M. (1928) On the feeding mechanism of the syncarid Crustacea. *Trans. roy. Soc. Edinb.* **56,** 175–89.

CARLISLE, D. B. (1950) Alcune osservazioni sulla meccanica dell'alimentazione della Salpa. *Pubbl. Staz. zool. Napoli,* **22,** 146–54.

CARRIKER, M. R. (1961) Interrelation of functional morphology, behaviour, and autecology in early stages of the bivalve *Mercenaria mercenaria. J. Elisha Mitchell sci. Soc.* **77,** 168–241.

CASPERS, H. (1940) Über Nahrungserwerb und Darmverlauf bei *Nucula. Zool. Anz.* **129,** 48–55.

CASSIE, M. and CASSIE, V. (1960) Primary production in a New Zealand West Coast phytoplankton bloom. *N. Z. J. Sci.* **3,** 173–99.

CHESTNUT, A. F. (1946) *Some observations on the feeding of oysters with special reference to the tide.* Conv. nat. Shellfish Ass. New York, June 5–7, 1946, pp. 22–27.

CHIBA, K. and OSHIMA, Y. (1957) Effects of suspending particles on the pumping of marine bivalves, especially of Japanese neck-clam. *Bull. Jap. Soc. sci. Fish.* **23,** 348–53.

CHIPMAN, W. A. and HOPKINS, J. G. (1954) Water filtration by the bay scallop, *Pecten irradians,* as observed with the use of radioactive plankton. *Biol. Bull.* **107,** 80–91.

CHOLODNY, N. (1929) Zur Methodik der quantitativen Erforschung des bakteriellen Planktons. *Centralbl. Bakt. Abt.* II, **77,** 179–93.

CHUANG, S. H. (1956) The ciliary feeding mechanisms of *Lingula unguis* (L.) (Brachiopoda). *Proc. zool. Soc. Lond.* **127,** 167–89.

CHURCHILL, E. P. and LEWIS, S. T. (1924) Food and feeding in freshwater mussels. *Bull. U.S. Bur. Fish.* **39,** 437–69.

CLARKE, A. H., Jr. (1962) Annotated list and bibliography of the abyssal marine molluscs of the world. *Nat. Mus. Canada Bull. No. 181, Biol. Ser.* No. 67, **114** pp.

Clarke, G. L. and Gellis, S. S. (1935) The nutrition of copepods in relation to the food-cyle of the sea. *Biol. Bull.* **68,** 231–46.

Cleland, D. M. (1954) A study of the habits of *Valvata piscinalis* (Müller) and the structure and function of the alimentary canal and reproductive system. *Proc. malac. Soc. Lond.* **30,** 167–203.

Cloudsley-Thompson, J. L. (1960) Eaters of mud. *New Scientist,* **7,** 1487–89.

Coe, W. R. (1948) Nutrition, environmental conditions, and growth of marine bivalve molluscs. *J. Mar. Res.* **7,** 586–601.

Coe, W. R. and Fox, D. L. (1944) Biology of the California sea mussel (Mytilus californianus). III. Environmental conditions and rate of growth. *Biol. Bull.* **87,** 59–72.

Cole, H. A. (1937) Experiments in the breeding of oysters (*Ostrea edulis*) in tanks, with special reference to the food of the larva and spat. *Fish. Invest. Lond. Ser. II,* **15** (4).

Cole, H. A. (1938) The fate of the larval organs in the metamorphosis of *Ostrea edulis. J. Mar. biol. Ass. U.K.* **22,** 469–84.

Collander, R. (1930) Permeabilitätsstudien an *Chara ceratophylla.* I. Die normale Zusammensetzung des Zellsaftes. *Acta bot. fenn.* **6,** 1–20.

Collier, A. (1958a) Some biochemical aspects of red tides and related oceanographic problems. *Limnol. Oceanogr.* **3,** 33–9.

Collier, A. (1958b) Gulf of Mexico physical and chemical data from *Alaska* cruises. *U.S. Fish and Wildlife Serv., Spec. Sci. Rep. Fish.* **249, 1.**

Collier, A. (1959) Some observations on the respiration of the American oyster *Crassostrea virginica* (Gmelin). *Publ. Inst. Mar. Sci.* **6,** 92–108.

Collier, A. and Ray, S. M. (1948) An automatic proportioning apparatus for experimental study of the effect of chemical solutions on aquatic animals. *Science,* **107,** 576.

Collier, A., Ray, S. M. and Magnitzky, W. (1950) A preliminary note on naturally occurring organic substances in sea water affecting the feeding of oysters. *Science,* **111,** 151–2.

Collier, A., Ray, S. M., Magnitzky, A. W. and Bell, J. O. (1953) Effect of dissolved organic substances on oysters. *Fish. Bull., U.S. Fish and Wildlife Serv.* **54,** 167–85.

Collyer, D. M. (1962) Method for determination of fat percentage in unicellular algae. *J. Mar. biol. Ass. U.K.* **42,** 485–92.

Conover, R. J. (1956) Oceanography of Long Island Sound, 1952–54. VI. Biology of the copepod genus *Acartia. Bull. Bingham oceanogr. Coll.* **15,** 156–233.

Conover, R. J. (1960) The feeding behaviour and respiration of some marine planktonic Crustacea. *Biol. Bull.* **119,** 399–415.

Conover, R. J. (1962) Metabolism and growth in *Calanus hyperboreus* in relation to its life cycle. *Rapp. Cons. Explor. Mer,* **153,** 190–7.

Cook, J. R. (1961) *Euglena gracilis* in synchronous division. II. Biosynthetic rates over the life cycle. *Biol. Bull.* **121,** 277–89.

Cook, P. M. (1949) A ciliary feeding mechanism in *Viviparus viviparus* L. *Proc. malac. Soc. Lond.* **27,** 265–71.

COOPER, L. H. N. (1958) Consumption of nutrient salts in the English Channel as a means of measuring production. *Rapp. Cons. Explor. Mer,* **144,** 35–37.

COOPER, L. H. N. (1961) *Vertical and horizontal movements in the ocean.* Oceanography, Ed. M. Sears, Washington D.C., U.S.A. pp. 599–621.

CORLETT, J. (1958) Measurements of primary production in the Western Barents Sea. *Rapp. Cons. Explor. Mer,* **144,** 76–78.

CORNER, E. D. S. (1961) On the nutrition and metabolism of zooplankton. I. Preliminary observations on the feeding of the marine copepod, *Calanus helgolandicus* (Claus). *J. Mar. biol. Ass. U.K.* **41,** 5–16.

COSTLOW, J. D. and BOOKHOUT, C. G. (1953) Moulting and growth in *Balanus improvisus. Biol. Bull.* **105,** 420–33.

COSTLOW, J. D. and BOOKHOUT C. G. (1957) Larval development of *Balanus eburneus* in the laboratory. *Biol. Bull.* **112,** 313–24.

COWEY, C. B. (1956) A preliminary investigation of the variation of vitamin B_{12} in oceanic and coastal waters. *J. Mar. biol. Ass. U.K.* **35,** 609–20.

COWEY, C. B. and CORNER, E. D. S. (1963) On the nutrition and metabolism of zooplankton. II. The relationship between the marine copepod *Calanus helgolandicus* and particulate material in Plymouth sea water, in terms of amino acid composition. *J. Mar. biol. Ass. U.K.* **43,** 495–511.

CREAC'H, P. (1955) Sur la présence des acides citrique et malique dans les eaux marines littorales. *C. R. Acad. Sci. Paris,* **240,** 2551–53.

CREEK, G. A. (1960) The development of the lamellibranch *Cardium edule* L. *Proc. zool. Soc. Lond.* **135,** 243–60.

CREITZ, G. J. and RICHARDS, F. A. (1955) The estimation and characterization of plankton populations by pigment analysis. III. A note on the use of 'millipore' membrane filters in the estimation of plankton pigments. *J. Mar. Res.* **14,** 211–16.

CRISP, D. J. and SOUTHWARD, A. J. (1961) Different types of cirral activity of barnacles. *Phil. Trans. roy. Soc. B.* **243,** 271–307.

CURRIE, R. I. (1958) Some observations on organic production in the North-East Atlantic. *Rapp. Cons. Explor. Mer,* **144,** 96–102.

CURRIE, R. I. (1962) Pigments in zooplankton fæces. *Nature, Lond.* **193,** 956–7.

CUSHING, D. H. (1955) Production and a pelagic fishery. *Fish. Invest. Lond. Ser. II,* **18,** No. 7.

CUSHING, D. H. (1957) Production of carbon in the sea. *Nature, Lond.* **179,** 876.

CUSHING, D. H. (1958a) The effect of grazing in reducing the primary production: A Review. *Rapp. Cons. Explor. Mer,* **144,** 149–54.

CUSHING, D. H. (1958b) The estimation of carbon in phytoplankton. *Rapp. Cons. Explor. Mer,* **144,** 32–3.

CUSHING, D. H. (1959a) On the nature of production in the sea. *Fish. Invest. Lond. Ser. II,* **22,** No. 6, 1–40.

CUSHING, D. H. (1959b) The seasonal variation in oceanic production as a problem in population dynamics. *J. Cons. int. Explor. Mer,* **24,** 455–64.

CUSHING, D. H. and NICHOLSON, H. F. (1958) The measurement of the carbon content of diatoms, using the ^{14}C technique: a preliminary note. *Rapp. Cons. Explor. Mer,* **144,** 34.

CUSHING, D. H. and VUCETIC, T. (1963) Studies on a *Calanus* patch. III. The quantity of food eaten by *Calanus finmarchicus*. *J. Mar. biol. Ass. U.K.* **43,** 349–71.

CZECZUGA, B. (1958) Researches on the chlorophyll content in the phytoplankton of the lakes of Rajgród. *Acta Soc. Bot. Polon.* **27,** 541–61 (in Polish, English summary).

DAISLEY, K. W. and FISHER, L. R. (1958) Vertical distribution of vitamin B_{12} in the sea. *J. Mar. biol. Ass. U.K.* **37,** 683–6.

DAKIN, W. J. and DAKIN, C. M. G. (1925) The oxygen requirements of certain aquatic animals and its bearing upon the source of food supply. *Brit. J. exp. Biol.* **2,** 293–322.

DALES, R. P. (1957a) The feeding mechanism and structure of the gut of *Owenia fusiformis* delle Chiaje. *J. Mar. biol. Ass. U.K.* **36,** 81–90.

DALES, R. P. (1957b) Some quantitative aspects of feeding in sabellid and serpulid fan worms. *J. Mar. biol. Ass. U.K.* **36,** 309–16.

DALES, R. P. (1961) Observations on the respiration of the sabellid polychaete *Schizobranchia insignis*. *Biol. Bull.* **121,** 82–91.

DALHAMN, T. (1956) Mucous flow and ciliary activity in the trachea of healthy rats exposed to respiratory irritant gases. *Acta physiol. scand.* **36,** Suppl. 123.

VAN DAM, L. (1957) On the respiration in scallops (Lamellibranchiata). *Biol. Bull.* **107,** 192–202.

DATZKO, V. G. (1939) Organic matter in water of certain seas. *C. R. Acad. Sci. U.R.S.S.* **24,** 294 (in Russian).

DATZKO, V. G. (1951) The vertical distribution of organic matter in the Black Sea. *C. R. Acad. Sci. U.R.S.S.* **77,** 1059–62., (in Russian).

DATZKO, V. G. and DATZKO, V. E. (1950) Method for the determination of organic carbon in natural waters. *Dokl. Akad. Nauk SSSR,* **73,** 337–9 (in Russian).

DAVIS, H. C. (1953) On food and feeding of larvae of the American oyster, *C. virginica*. *Biol. Bull.* **104,** 334–50.

DAVIS, H. C. and GUILLARD, R. R. (1958) Relative value of ten genera of micro-organisms as food for oyster and clam larvae. *Fish. Bull., Fish and Wildlife Serv.* **58,** 293–304.

DAVIS, P. S. (1959) Some aspects of the ecology of Lake Macquarie N.S.W., with regard to an alleged depletion of fish v. Chlorophyll distribution. *Aust. J. Mar. Freshw. Res.* **10,** 316–21.

DEAN, D. (1957) *The experimental feeding of oysters.* Ph.D. Thesis. Rutgers Univ. New Brunswick, N.J.

VON DENFFER, D. (1948) Die planktische Massenkultur pennater Grunddiatomeen. *Arch. Mikrobiol.* **14,** 159–202.

DENNELL, R. (1950) Note on the feeding of Amphioxus (*Branchiostoma bermudae*). *Proc. roy. Soc. Edinb. B.* **64,** 229–34.

DENTON, E. J. (1963) Buoyancy mechanisms of sea creatures. *Endeavour,* **22,** 3–8.

DESCHAMPS, P. (1957) La nutrition chez les mollusques Térédinides: possibilité de rétraction et fermeture de l'étui calcaire. Chromatographie quantitative des sucres. *Bull. Soc. zool. Fr.* **82,** 72–81.

DIGBY, P. S. B. (1953) Plankton production in Scoresby Sound, East Greenland. *J. Anim. Ecol.* **22,** 289–322.

DODD, J. M. (1950) Ciliary feeding mechanisms in anuran larvae. *Nature, Lond.* **165,** 283.

DODGSON, R. W. (1928) Report on mussel purification. *Fish. Invest. Lond. Ser. II,* **10.**

DOTY, M. S. and OGURI, M. (1956) The island mass effect. *J. Cons. int. Explor. Mer,* **22,** 33–7.

DRAGOVICH, A. (1961) Relative abundance of plankton off Naples, Florida, and associated hydrographic data, 1956–57. *U.S. Fish and Wildlife Serv., Spec. Sci. Rep. Fish.* **372.**

DRINKWAARD, A. C. (1959) The quality of oysters in relation to environmental conditions in the Oosterschelde in 1957. *Ann. Biol.* **14,** 217–23.

DRINKWAARD, A. C. (1960) The quality of oysters in relation to environmental conditions in the Oosterschelde in 1958. *Ann. Biol.* **15,** 224–33.

DRISKO, R. W. and HOCHMAN, H. (1957) Amino acid content of marine borers. *Biol. Bull.* **112,** 325–9.

DROOP, M. R. (1957) Vitamin B_{12} in marine ecology. *Nature, Lond.* **180,** 1041–2.

DROOP, M. R. (1958) Requirement for thiamine among some marine and sub-littoral protista. *J. Mar. biol. Ass. U.K.* **37,** 323–9.

DUURSMA, E. K. (1961) Dissolved organic carbon, nitrogen and phosphorus in the sea. *Nethl. J. Sea Res.* **1,** 1–148.

DUURSMA, E. K. (1963) The production of dissolved organic matter in the sea, as related to the primary gross production of organic matter. *Nethl. J. Sea Res.* **2,** 85–94.

EDMONDSON, W. T. and EDMONDSON, Y. H. (1947) Measurements of production in fertilized saltwater. *J. Mar. Res.* **6,** 228–46.

EINARSSON, H. (1945) Euphausiacea. I. Northern Atlantic species. *Dana-Rep.* **27,** 1–191.

ELSEY, C. R. (1935) On the structure and function of the mantle and gill of *Ostrea gigas* (Thunberg) and *Ostrea lurida* Carpenter. *Trans. roy. Soc. Canada,* **79,** Sect. V, 131–60.

ENDERS, H. E. (1909) A study of the life history and habits of *Chaetopterus variopedatus,* Renier et Claparède. *J. Morph.* **20,** 479–532.

ERIKSSON, S. (1934) Studien über die Fangapparate der Branchiopoden. *Zool. Bidr. Uppsala,* **15,** 23–287.

ERMAN, L. A. (1956) The quantitative aspect of feeding of Rotatoria. *Zool. Zhurn.* **35,** 965–71 (in Russian).

FAGE, L. (1954) Remarques sur les conditions de vie de la fauna benthique abyssale. *Int. Union biol. Sci. Ser. B.* **16,** 12–18.

FAULKNER, G. H. (1931) Notes on the feeding mechanism and on intestinal respiration in *Chaetopterus variopedatus*. *Biol. Bull.* **61,** 472–7.

FEDELE, M. (1923) Le attività dinamiche ed i rapporti nervosi nella vita dei Dolioli. *Pubbl. Staz. zool. Napoli,* **4,** 129–240.

FEDELE, M. (1933) Sul complesso delle funzioni che intervengono nel meccanismo ingestivo dei Salpidae. *Ren. Acc. Lincei,* **17,** 241–5.

FERGUSSON, G. J. (1959) Radiocarbon ages of ocean water and the exchange time of carbon dioxide from the atmosphere to the ocean. *Int. Oceanogr. Congr. A.A.A.S. Washington D.C.,* p. 431.

FILATOVA, Z. A. (1957) General review of the bivalve mollusks of the northern seas of the U.S.S.R. *Trans. Inst. Oceanol.* **20,** 1–44. English translation 1959: Marine Biology. Ed. B. N. Nikitin.

FILATOVA, Z. A. (1959) Bivalve molluscs of the abyssal zone of the northwestern Pacific. *Proc. XV. Int. Congr. Zool. Lond.* 1958, 221–2.

FILATOVA, Z. A. and LEVENSTEIN, R. J. (1961) The quantitative distribution of the deep-sea bottom fauna in north-eastern Pacific. *Trudy Inst. Okeanol. Akad. Nauk SSSR,* **45,** 190–213 (in Russian).

FISHER, W. K. and MACGINITIE, G. E. (1928) The natural history of an echiuroid worm. *Ann. Mag. nat. Hist.* **10,** 204–13.

FJERDINGSTAD. E. J. (1961a) The ultrastructure of choanocyte collars in *Spongilla lacustris* (L.). *Z. Zellforsch.* **53,** 645–57.

FJERDINGSTAD, E. J. (1961b) Ultrastructure of the collar of the choanoflagellate *Codonosiga botrytis* (Ehrenb.). *Z. Zellforsch.* **54,** 499–510.

FLEMING, R. H. (1939) The control of diatom populations by grazing. *J. Cons. int. Explor. Mer,* **14,** 210–27.

FOGG, G. E. (1952) The production of extracellular nitrogenous substances by a blue-green alga. *Proc. roy. Soc. B.* **139,** 372–97.

FOGG, G. E. (1953) *The Metabolism of Algae.* Methuen and Co., Lond.

FOGG, G. E. (1958) Extracellular products of phytoplankton and the estimation of primary production. *Rapp. Cons. Explor. Mer,* **144,** 56–60.

FOGG, G. E. (1959) Dissolved organic matter in oceans and lakes. *New Biology,* **29,** 31–48.

FOGG, G. E. (1962) *Extracellular products. The Physiology and Biochemistry of Algae.* Ed. R. A. Lewin. Academic Press, New York.

FOGG, G. E. and NALEWAJKO, C. (1963) *Proc. roy. Soc. B.* **157,** 381–2. Abstract.

FOL, A. (1872) Etudes sur les Appendiculaires de détroit de Messine. *Mém. Soc. Phys. Hist. Nat. Genève,* **21,** 445–99.

FOL, H. (1876) Ueber die Schleimdrüse oder den Endostyl der Tunicaten. *Morph. Jb.* **1,** 222–42.

FORSBLAD, J. (1955) Eine Mikromethode zur Bestimmung von organisch gebundenen Kohlenstoff in Seewasser. *Mikrochim. Acta,* 1955, 176.

FOX, D. L., Editor (1936) The habitat and food of the California sea-mussel. *Bull. Scripps Instn. Oceanogr. Tech. Ser.* **4,** 1–64.

FOX, D. L. (1950) Comparative metabolism of organic detritus by inshore animals. *Ecology,* **31,** 100–8.

Fox, D. L. (1957) Particulate organic detritus. *Mem. geol. Soc. Amer.* **67,** I, 383–90.

Fox, D. L. and Coe, W. R. (1943) Biology of the California sea-mussel (*Mytilus californianus*). II. Nutrition, metabolism, growth and calcium deposition. *J. exp. Zool.* **93,** 205–49.

Fox, D. L., Isaacs, J. D. and Corcoran, E. F. (1952) Marine leptopel, its recovery, measurement and distribution. *J. Mar. Res.* **11,** 29–46.

Fox, D. L., Oppenheimer, C. H. and Kittredge, J. S. (1953) Microfiltration in oceanographic research. II. Retention of colloidal micelles by adsorptive filters and by filter-feeding invertebrates; proportions of dispersed organic to dispersed inorganic matter and to dissolved solutes. *J. Mar. Res.* **12,** 233–43.

Fox, D. L., Sverdrup, H. N. and Cunningham, J. P. (1937) The rate of water propulsion by the California mussel. *Biol. Bull.* **72,** 417–38.

Fraga, F. (1957) Nitrógeno orgánico suspendido y disuelto en la Ría de Vigo. *Reunión Product. y Pesq.* **3,** 26–28.

Fraga, F. (1959) Determinación de nitrógeno orgánico suspendido y disuelto en el agua de mar. *Invest. Pesq.* **14,** 121–7.

Fraga, F. (1960) *La materia organica nitrogenada inerte disuelta en el agua de mar.* Inst. Invest. Pesq. IV Reunion, 29–31.

Freeden, F. J. H. (1960) Bacteria as a source of food for black-fly larvae. *Nature, Lond.* **187,** 963.

Fretter, V. (1953) Experiments with radioactive strontium (^{90}Sr) on certain molluscs and polychaetes. *J. Mar. biol. Ass. U.K.* **32,** 367–84.

Fretter, V. and Graham, A. (1962) *British Prosobranch Molluscs.* Ray Society Lond.

Fryer, G. (1957) The feeding mechanism of some freshwater cyclopoid copepods. *Proc. zool. Soc. Lond.* **129,** 1–25.

Fryer, G. (1962) Secretions of the labral and trunk limb glands in the cladoceran *Eurycercus lamellatus. Nature, Lond.* **195,** 97.

Fuller, J. L. (1937) Feeding rate of *Calanus finmarchicus* in relation to environmental conditions. *Biol. Bull.* **72,** 233–46.

Gaarder, T. and Eliassen E. (1954) The energy-metabolism of *Ostrea edulis. Univ. Bergen Årb. naturv.* R. No. 3, 1–6.

Gaarder, T. and Gran, H. H. (1927) Investigations of the production of plankton in the Oslo Fjord. *Rapp. Cons. Explor. Mer,* **42,** 1.

Gaevskaya, N. S. (1949) On the selectivity of feeding in filter feeding organisms. *Trudy Vsesojuzn. gidrobiol. obshch.* **1,** 159–74 (in Russian).

Galtsoff, P. S. (1926) New methods to measure the rate of flow produced by the gills of oyster and other Mollusca. *Science,* **63,** 233–4.

Galtsoff, P. S. (1928a) Experimental study of the function of the oyster gills and its bearing on the problems of the oyster culture and sanitary control of the oyster industry. *Bull. U.S. Bur. Fish.* **44,** 1–39.

Galtsoff, P. S. (1928b) The effect of temperature on the mechanical activity of the gills of the oyster (*Ostrea virginica* Gm). *J. gen. Physiol.* **11,** 415–31.

Galtsoff, P. S. (1946) Reaction of oysters to chlorination. *U.S. Fish and Wildlife Serv. Res. Rep.* **11.**

GALTSOFF, P. S. and WHIPPLE, D. V. (1931) Oxygen consumption of normal and green oysters. *Bull. U.S. Bur. Fish.* **46,** 489–508.

GALTSOFF, P. S., CHIPMAN Jr., W. A., ENGLE, J. B. and CALDERWOOD, H. N. (1947) Ecological and physiological studies of the effect of sulphate pulp mill wastes on oysters in the York River, Virginia. *Fish. Bull., U.S.,* **51,** 58–186.

VAN GANSEN, P. (1960) Adaptations structurelles des animaux filtrants. *Ann. Soc. Roy. Zool. Belg.* **90,** 161–231.

GARDINER, A. C. (1943) Measurement of phytoplankton population by the pigment extraction methods. *J. Mar. biol. Ass. U.K.* **25,** 739–44.

GAULD, D. T. (1950) A fish cultivation experiment in an arm of a sea-lock. III. The plankton of Kyle Scotnish. *Proc. roy. Soc. Edinb. B.* **64,** 36–64.

GAULD, D. T. (1951) The grazing rate of planktonic copepods. *J. Mar. biol. Ass. U.K.* **29,** 695–706.

GAULD, D. T. (1953). Diurnal variations in the grazing of planktonic copepods. *J. Mar. biol. Ass. U.K.* **31,** 461–74.

GAULD, D. T. (1959) Swimming and feeding in crustacean larvae: the nauplius larva. *Proc. zool. Soc. Lond.* **132,** 31–50.

GAVARD (1927) De quoi se nourrissent les huîtres? Leur nourriture envisagée au point de vue 'Ostréiculture'. *Bull Trav. Stat. Aquic. Pêche Castiglione Alger,* **2,** 237–54.

GELLIS, S. S. and CLARKE, G. L. (1935) Organic matter in dissolved and colloidal form as food for *Daphnia magna*. *Physiol. Zool.* **8,** 127–37.

GIBOR, A. (1956) Some ecological relationships between phyto- and zooplankton. *Biol. Bull.* **111,** 230–4.

GIBOR, A. (1957) Conversion of phytoplankton to zooplankton. *Nature, Lond.* **179,** 1304.

GILLBRICHT, M. (1952) Untersuchungen zur Produktionsbiologie in der Kieler Bucht. *Kieler Meeresforsch.* **8,** 173–91.

GILLBRICHT, M. (1959) Die Planktonverteilung in der Irminger-See im Juni 1955. *Ber. dtsch. Wiss. Komm. Meeresforsch.* **15,** 260–75.

GILLBRICHT, M. (1961) Über den Zusammenhang zwischen verschiedenen Messungen im Meer. *Helgoländ. wiss. Meeresunters.* **7,** 238–51.

GISLÉN, T. (1924) Echinoderm studies. *Zool. Bidr. Uppsala,* **9,** 1–316.

GISLÉN, T. (1954) On the bathymetric distribution of the deep sea crinoids. *Int. Union biol. Sci. Ser. B. No.* **16,** 38–9.

GOLDBERG, E. D., BAKER, M. and FOX, D. L. (1952) Microfiltration in oceanographic research. I. Marine sampling with the molecular filter. *J. Mar. Res.* **11,** 194–204.

GOLDBERG, E. D., MCBLAIR, W. and TAYLOR, K. M. (1951) The uptake of vanadium by tunicates. *Biol. Bull.* **101,** 84–94.

GOODBODY, I. (1960). The feeding mechanism in the sand dollar *Mellita sexiesperforata* (Leske). *Biol. Bull.* **119,** 80–6.

GOSSELIN, R. E. (1961) The cilio-excitatory activity of serotonin. *J. cell. comp. Physiol.* **58,** 17–26.

GOSSLER, O. (1950) Funktionsanalysen am Räderorgan von Rotatorien durch optische Verlangsamung. *Öst. zool. Z.* **2,** 568–84.

GRAHAM, A. (1931) The morphology, feeding mechanisms and digestion of *Ensis siliqua* (Schumacher). *Trans. roy. Soc. Edinb.* **56,** 725–51.

GRAHAM, A. (1938) On a ciliary process of food-collecting in the gastropod *Turritella communis* Risso. *Proc. zool. Soc. Lond.* A, **108,** 453–63.

GRAHAM, H. W. (1943) Chlorophyll-content of marine plankton. *J. Mar. Res.* **5,** 153–60.

GRAN, H. H. and RUUD, B. (1926) Untersuchungen über die im Meerwasser gelösten organischen Stoffe und ihr Verhältnis zur Planktonproduction. *Avh. norske vidensk. Akad.* **6.**

GRAVE, C. (1916) The process of feeding in the oyster. *Science,* **44,** 178–81.

GREENFIELD, L. J. and LANE, C. E. (1953) Cellulose digestion in *Teredo. J. biol. Chem.* **204,** 669–72.

GRIM, J. (1939) Beobachtungen am Phytoplankton des Bodensees (Obersee) sowie deren rechnerische Auswertung. *Int. Rev. Hydrobiol.* **39,** 193–315.

GRØNTVED, J. (1949) Investigations on the phytoplankton in the Danish Wadden Sea in July, 1941. *Medd. Komm. Danmarks Fisk.- og Havunders. Ser. Plankton,* **5,** No. 2, 1–55.

GRØNTVED, J. (1958) Planktological Contributions. III. Investigations on the phytoplankton and the primary production in an Oyster culture in the Limfjord. *Medd. Danmarks Fisk.- og Havunders. N.S.* **2.** No. 17, 1–15.

GRØNTVED, J. (1960a) Planktological contributions IV. Taxonomical and productional investigations in shallow coastal waters. *Medd. Danmarks Fisk.- og Havunders. N.S.* **3,** No. 1, 1–18.

GRØNTVED, J. (1960b) On the productivity of microbenthos and phytoplankton in some Danish fjords. *Medd. Danmarks Fisk.- og Havunders. N.S.* **3,** No. 3, 55–92.

GRØNTVED, J. (1962) Preliminary report on the productivity of microbenthos and phytoplankton in the Danish Wadden Sea. *Medd. Danmarks Fisk.- og Havunders. N.S.* **3,** No. 12, 347–78.

GRØNTVED, J. and STEEMANN NIELSEN, E. (1957) Investigations on the phytoplankton in sheltered Danish marine localities. *Medd. Komm. Danmarks Fisk.- og Havunders. Ser. Plankton,* **5,** No. 6, 1–52.

GROSS, F. (1940) The osmotic relations of the plankton diatom *Ditylum brightwelli* (West.). *J. Mar. biol. Ass. U.K.* **24,** 381–415.

GROSS, F. and ZEUTHEN, E. (1948) The buoyancy of plankton diatoms: a problem of cell physiology. *Proc. roy. Soc. B.* **135,** 382.

GRY, H. (1942) Das Wattenmeer bei Skallingen. No. 1. Quantitative Untersuchungen über den Sinkstofftransport durch Gezeitenströmungen. *Folia geogr. dan.* **11,** No. 1, 1–138.

GUILLARD, R. R. L. (1959) Further evidence of the destruction of bivalve larvae by bacteria. *Biol. Bull.* **117,** 258–66.

GUILLARD, R. R. L. and WANGERSKY, P. J. (1958) The production of extracellular carbohydrates by some marine flagellates. *Limnol. Oceanogr.* **3,** 449–54.

HALL, V. E. (1931) The muscular activity and oxygen consumption of *Urechis caupo. Biol. Bull.* **61,** 400–16.

HANAOKA, T. (1958) Suspended matter as an index of the productivity of the sea. *Rapp. Cons. Explor. Mer,* **144,** 28–31.

HANSEN, V. K. (1959) Danish investigations on the primary production and the distribution of chlorophyll a at the surface of the North Atlantic during summer. *Int. Cons. Explor. Mer, I.G.Y. Meeting,* No. 5.

HARANT, H. (1948) Systématique et éthologie des Tuniciers. *Traité de Zoologie* (Ed. P. P. Grassé), **11,** 894–918.

HARDER, W. (1957) Verhalten von Organismen gegenüber Sprungschichten. *Ann. Biol.* **33,** 227–32.

HARLEY, M. B. (1950) Occurrence of a filter-feeding mechanism in the polychaete *Nereis diversicolor. Nature, Lond.* **165,** 734–5.

HARRIS, E. and RILEY, G. A. (1956) Oceanography of Long Island Sound 1952–54. VIII. Chemical composition of the plankton. *Bull. Bingham oceanogr. Coll.* **15,** 315–23.

HARTLAND-ROWE, R. (1953). Feeding mechanism of an Ephemeropteran nymph. *Nature, Lond.* **172,** 1109–10.

HARTMAN, R. T. (1958) Studies of plankton centrifuge efficiency. *Ecology,* **39,** 374–6.

HARVEY, H. W. (1934) Measurements of phytoplankton population. *J. Mar. biol. Ass. U.K.* **19,** 761–73.

HARVEY, H. W. (1937) Note on selective feeding by *Calanus. J. Mar. biol. Ass. U.K.* **22,** 97–100.

HARVEY, H. W. (1950) On the production of living matter in the sea off Plymouth. *J. Mar. biol. Ass. U.K.* **29,** 97.

HARVEY, H. W. (1953) Synthesis of organic nitrogen and chlorophyll by *Nitzschia closterium. J. Mar. biol. Ass. U.K.* **31,** 477–87.

HASLE, G. R. (1959) A quantitative study of phytoplankton from the equatorial Pacific. *Deep-Sea Res.* **6,** 38–59.

HAVINGA, B. (1954) *Hydrografie van het IJsselmeer. Veranderingen in de Flora en Fauna van de Zuiderzee.* Ed. L. F. de Beaufort. Den Helder.

HECHT, S. (1916) The water current produced by *Ascidia atra* Lesueur. *J. exp. Zool.* **20,** 429–34.

HECHT, S. (1918a) The physiology of *Ascidia atra* Lesueur. I. General physiology. *J. exp. Zool.* **25,** 229–60.

HECHT, S. (1918b) The physiology of *Ascidia atra* Lesueur. II. Sensory physiology. *J. exp. Zool.* **25,** 261–99.

HEEZEN, B. C. (1959) Modern turbidity currents. *Int. Oceanogr. Congr. A.A.A.S. Washington D.C.,* pp. 458–60.

HEEZEN, B. C., MENZIES, R. J. and EWING, M. (1959) Influence of modern turbidity currents on abyssal productivity. Preprints *Int. Oceanogr. Congr. A.A.A.S. Washington D.C.,* p. 375.

HENSEN, V. (1887) *Ueber die Bestimmung des Planktons.* 5ter Bericht Komm. Wiss. Untersuch. dtsch. Meere, Kiel 1882–6, pp. 1–109.

HENTSCHEL, E. (1928) Die Grundzüge der Planktonverteilung im Südatlantischen Ozean. *Int. Rev. Hydrobiol.* **21,** 1–16.

HENZE, M. (1908) Bemerkungen zu der Anschauungen Pütters über den Gehalt des Meeres an gelösten organischen Kohlenstoffverbindungen und deren Bedeutung für den Stoffhaushalt des Meeres. *Arch. ges. Physiol.* **123,** 487–90.

HERSH, G. L. (1960) A method for the study of the water currents of invertebrate ciliary filter feeders. *The Veliger* **2,** 77–83.

HEUKELEKIAN, H. and HELLER, A. (1940). Relation between food concentration and surface for bacterial growth. *J. Bact.* **40,** 547–58.

VAN HEYNINGEN, H. E. (1954) *A Study of the Food of Some Daphniidae*. Københavns Univ. Ferskvandsbiol. Lab. Hillerød.

HILL, A. V. (1911–12) The total energy exchanges of intact coldblooded animals at rest. *J. Physiol.* **43,** 379–94.

HOLMES, R. W. (1958a) Surface chlorophyll 'A', surface primary production, and zooplankton volumes in the eastern Pacific Ocean. *Rapp. Cons. Explor. Mer,* **144,** 109–16.

HOLMES, R. W. (1958b) Size fractionation of photosynthesizing phytoplankton. *U.S. Fish and Wildlife Serv., Spec. Sci. Rep. Fish.* **279,** 69–71.

HOPKINS, A. E. (1931) Temperature and the shell movement of oysters. *Bull. U.S. Bur. Fish.* **47,** 1–14.

HOPKINS, A. E. (1933) Experiments on the feeding behavior of the oyster *Ostrea gigas*. *J. exp. Zool.* **64,** 469–94.

HOPKINS, A. E. (1935) Temperature optima in the feeding mechanism of the oyster, *O. gigas*. *J. exp. Zool.* **71,** 195–208.

HOPKINS, A. E. (1936a) Activity of the adductor muscle in oysters. *Physiol. Zoöl.* **9,** 498–507.

HOPKINS, A. E. (1936b) Adaptation of the feeding mechanism of the oyster (*O. gigas*) to changes in salinity. *Bull. U.S. Bur. Fish.* **48,** 345–63.

HOYLE, G. (1953) Spontaneous squirting of an ascidian, *Phallusia mammillata* Cuvier. *J. Mar. biol. Ass. U.K.* **31,** 541–62.

HULBURT, E. M., RYTHER, J. H. and GUILLARD, R. R. L. (1960). The phytoplankton of the Sargasso Sea off Bermuda. *J. Cons. int. Explor. Mer,* **25,** 115–28.

HUMPHREY, G. F. (1960) The concentration of plankton pigments in Australian waters. *Div. Fish. and Oceanogr. Tech. Paper,* No. 9, 1–27.

HUNT, O. D. (1925) The food of the bottom fauna of the Plymouth fishing grounds. *J. Mar. biol. Ass. U.K.* **13,** 560–99.

HUTCHINSON, G. E. (1957) *A Treatise on Limnology*. Vol. I. New York, Wiley.

HYMAN, L. H. (1919) Physiological studies on *Planaria*. I. Oxygen consumption in relation to feeding and starvation. *Amer. J. Physiol.* **49,** 377–402.

HYMAN, L. H. (1925) Respiratory differences along the axis of the sponge *Grantia*. *Biol. Bull.* **48,** 379–89.

ICHIMURA, S. and SAIJŌ, Y. (1959) Chlorophyll content and primary production of the Kuroshio off the Southern Midcoast of Japan. *Bot. Mag. Tokyo,* **72,** 193–202.

IMAI, T. and HATANAKA, M. (1950) Studies on marine non-colored flagellates, *Monas* sp., favorite food of larvae of various marine animals. I. Preliminary research on cultural requirements. *Sci. Rep. Tôhoku Univ.* 4 *Ser. Biol.* **18,** 304–15.

JANNASCH, H. W. (1953) Zur Methode der quantitativen Untersuchung von Bakterienkulturen in flüssigen Medien. *Arch. Mikrobiol.* **18,** 425–30.

JANNASCH, H. W. (1958) Studies on planktonic bacteria by means of a direct membrane filter method. *J. gen. Microbiol.* **18,** 609–620.

JANNASCH, H. W. and JONES, G. E. (1959) Bacterial populations in sea water as determined by different methods of enumeration. *Limnol. Oceanogr.* **4,** 128–39.

JASHNOV, V. A. (1961) Vertical distribution of the mass of zooplankton throughout the tropical zone of the Atlantic. *Dokl. Akad. Nauk SSSR,* **136,** 705–8 (in Russian).

JEFFREY, S. W. (1962) Purification of chlorophyll c from *Sargassum flavicans. Nature, Lond.* **194,** 600.

JENKIN, P. M. (1957) The filter-feeding and food of flamingoes (Phoenicopteri). *Phil. Trans. roy. Soc. Lond. B.* **240,** 401–93.

JENKINS, P. G. (1955) Seasonal changes in the phytoplankton as indicated by spectrophotometric chlorophyll estimations 1952–53. *Deep-Sea Res.* **3,** Suppl. 58–67.

JENSEN, P. B. (1915) Studies concerning the organic matter of the sea-bottom. *Rep. Danish biol. Sta.* **22,** 3–39.

JENSEN, P. B. and PETERSEN, C. G. J. (1911) Valuation of the sea. I. Animal life of the sea-bottom, its food and quantity. *Rep. Danish biol. Sta.* **20,** 1–81.

JERLOV, N. G. (1951) Optical measurement of particle distribution in the sea. *Tellus,* **3,** 122–8.

JERLOV, N. G. (1953) Particle distribution in the ocean. *Swedish Deep-Sea Exp.* 1947–48. *Rep. 3, Physics and Chem.* **3,** 73–97.

JERLOV, N. G. (1955) The particulate matter in the sea as determined by means of the Tyndall meter. *Tellus,* **7,** 218–25.

JERLOV, N. G. (1959) Maxima in the vertical distribution of particles in the sea. *Deep-Sea Res.* **5,** 173–84.

JERLOV, N. G. and KULLENBERG, B. (1953) The Tyndall effect of uniform minerogenic suspensions. *Tellus,* **5,** 306–7.

JØRGENSEN, C. B. (1943) On the water transport through the gills of bivalves. *Acta physiol. scand.* **5,** 297–304.

JØRGENSEN, C. B. (1949a) The rate of feeding by *Mytilus* in different kinds of suspension. *J. Mar. biol. Ass. U.K.* **28,** 333–44.

JØRGENSEN, C. B. (1949b) Feeding-rates of sponges, lamellibranchs and ascidians. *Nature, Lond.* **163,** 912.

JØRGENSEN, C. B. (1952a) On the relation between water transport and food requirements in some marine filter feeding invertebrates. *Biol. Bull.* **103,** 356–63.

JØRGENSEN, C. B. (1952b) Efficiency of growth in *Mytilus edulis* and two gastropod veligers. *Nature, Lond.* **170,** 714.

JØRGENSEN, C. B. (1955) Quantitative aspects of filter feeding in invertebrates. *Biol. Rev.* **30,** 391–454.

JØRGENSEN, C. B. (1960) Efficiency of particle retention and rate of water transport in undisturbed lamellibranchs. *J. Cons. int. Explor. Mer,* **26,** 94–116.

JØRGENSEN, C. B. (1962) The food of filter feeding organisms. *Rapp. Cons. Explor. Mer,* **153,** 99–106.

JØRGENSEN, C. B. and GOLDBERG, E. D. (1953) Particle filtration in some ascidians and lamellibranchs. *Biol. Bull.* **105,** 477–89.

JOHANSSON, K. E. (1927) Beiträge zur Kenntniss der Polychaeten-Familie Hermellidae, Sabellidae and Serpulidae. *Zool. Bidr. Uppsala,* **11,** 1–184.

JOHNSTON, R. (1963) Sea water, the natural medium of phytoplankton. I. General features. *J. Mar. biol. Ass. U.K.* **43,** 427–56.

JONES, D. and WILLS, M. S. (1956) The attenuation of light in sea and estuarine waters in relation to the concentration of suspended solid matter. *J. Mar. biol. Ass. U.K.* **35,** 431–44.

JONES, G. E. and JANNASCH, H. W. (1959) Aggregates of bacteria in sea water as determined by treatment with surface active agents. *Limnol. Oceanogr.* **4,** 269–76.

JORDAN, H. J. (1913) *Vergleichende Physiologie wirbelloser Tiere.* I. Bd. Die Ernährung. Jena.

JORDAN, H. J. and HIRSCH, G. C. (1927) Einige vergleichend–physiologische Probleme der Verdauung bei Metazoen. *Handb. norm. u. pathol. Physiol.* **3,** 24–101.

JOSEPH, J. (1954) Die Trübungsverhältnisse in der südwestlichen Nordsee während der 'Gauss'-Fahrt im Februar/März 1952. *Ber. dtsch. Wiss. Komm. Meeresforsch.* **13,** 93–103.

JOSEPH, J. (1959) Die Trübungsverhältnisse in der Irminger See im Juni 1955 ihre hydrographischen Ursachen. *Ber. dtsch. Wiss. Komm. Meeresforsch.* **15,** 238–59.

JUDAY, C., BLAIR, J. M. and WILDA, E. F. (1943) The photosynthetic activities of the aquatic plants of Little John Lake, Vilas County, Wisconsin. *Amer. Midl. Nat.* **30,** 426–46.

KAIN, J. M. and FOGG, G. E. (1958) Studies on the growth of marine phytoplankton. I. *Asterionella japonica* Gran. *J. Mar. biol. Ass. U.K.* **37,** 397–413.

KAISER, P. (1962) 'Misst' die Köcherfliegenlarve *Hydropsyche angustipennis* Curt. beim Bau des Netzes die Maschenweite? *Naturwissensch.* **49,** 116.

KALLINOWSKY, H. (1955) Über den Einfluss exogener Faktoren (Ernährung, Temperatur, Lebensraum) auf Wachstum, Körpergrösse und Lebensdauer von *Chirocephalus grubei* Dyb. *Z. Morph. Ökol. Tiere,* **44,** 196–221.

KAY, H. (1954a) Eine Mikromethode zur chemischen Bestimmung des organisch gebundenen Kohlenstoffs im Meerwasser. *Kieler Meeresforsch.* **10,** 26–36.

KAY, H. (1954b) Untersuchungen zur Menge und Verteilung der organischen Substanz im Meerwasser. *Kieler Meeresforsch.* **10,** 202–14.

KEISER-NIELSEN, H. (1953) '*Mucin*'. Diss. Copenhagen.

KELLOGG, J. L. (1915) Ciliary mechanisms of lamellibranchs. *J. Morph.* **26,** 625–701.

KENT. P. W. (1962). The chemistry of mucoproteins: An introduction to gastrointestinal mucus. *Gastroenterology* **43,** 292–303.

KETCHUM, B. H. and REDFIELD, A. C. (1949) Some physical and chemical characteristics of algae grown in mass culture. *J. cell. comp. Physiol* **33,** 281–300.

KEYS, A., CHRISTENSEN, E. H. and KROGH, A. (1935) The organic metabolism of sea-water with special reference to the ultimate food cycle in the sea. *J. Mar. biol. Ass. U.K.* **20,** 181–96.

KILIAN, E. F. (1952) Wasserströmung und Nahrungsaufnahme beim Süsswasserschwamm *Ephydatia fluviatilis*. *Z. vergl. Physiol.* **34,** 407–47.

KNUDSEN, J. (1961) The bathyal and abyssal *Xylophaga* (Pholadidae, Bivalvia). *Galathea Rep.* **5,** 163–209.

KOBLENZ-MISHKE, O. J. (1960) On the study of primary production in the sea by Soviet scientists. *Int. Rev. Hydrobiol.* **45,** 319–26.

KOCH, H. J. and HERS, M. J. (1943) Influence de facteurs respiratoires sur les interruptions de la ventilation par la siphon exhalant chez *Anodonta cygnea* L. *Ann. Soc. Roy. Zool. Belg.* **74,** 32–44.

KÖRNER, W. F. (1952) Untersuchungen über die Gehäusebildung bei Appendicularien. *Z. Morph. Ökol. Tiere,* **41,** 1–53.

KOFOID, C. A. (1897) On some important sources of error in the plankton method. *Science N.S.* **6,** 829–32.

KORRINGA, P. (1949) More light upon the problem of the oyster's nutrition? *Bijdr. Dierkunde,* **28,** 237–48.

KORRINGA, P. (1952) Recent advances in oyster biology. *Quart. Rev. Biol.* **27,** 266–308, 339–65.

KORRINGA, P. (1957) The quality of marketable oysters from the Zealand waters in 1955. *Ann. Biol.* **12,** 225–8.

KORRINGA, P. (1958) The quality of marketable oysters from the Zealand waters in 1956. *Ann. Biol.* **13,** 238–41.

KORRINGA, P. and POSTMA, H. (1957) Investigations into the fertility of the Gulf of Naples and adjacent salt water lakes, with special reference to shellfish cultivation. *Pubbl. Staz. zool. Napoli,* **29,** 229–84.

KOSHTOYANTS, C. S. and SALÁNKI, J. (1958) On the physiological principles underlying the periodical activity of *Anodonta*. *Acta biol. Hung.* **8,** 362–6.

KOYAMA, T. and THOMPSON, T. G. (1959) Organic acids of sea water. *Int. Oceanogr. Congr. A.A.A.S. Washington D.C.* pp. 925–6.

KOZHOVA, O. M. (1953) Feeding of *Epischura baicalensis* Sars (Copepoda, Calanoidea) in Lake Baikal. *Dokl. Akad. Nauk SSSR,* **90,** 299 (in Russian).

KOZHOVA, O. M. (1956) On the biology of *Epischura baicalensis* Sars in the Baikal Lake. *Isv. biol. geogr. Nautj.- isled. Inst.* **16,** 92–120 (in Russian).

KREPS, E. and VERJBINSKAYA, N. (1930) Seasonal changes in the phosphate and nitrate content and in hydrogen ion concentrations in the Barents Sea. *J. Cons. int. Explor Mer,* **5,** 327–46.

KREY, J. (1939) Die Bestimmung des Chlorophylls in Meerwasser-Schöpfproben. *J. Cons. int. Explor. Mer,* **14,** 201–9.

KREY, J. (1949) Über Art und Menge des Seston im Meere. *Verh. dtsch. zool. Mainz* 1949, p. 295.

KREY, J. (1950) Eine neue Methode zur quantitativen Bestimmung des Planktons. *Kieler Meeresforsch.* **7,** 58–75.

KREY, J. (1952) Untersuchungen zum Sestongehalt des Meereswassers. I. Der Sestongehalt in der westlichen Ostsee und unter Helgoland. *Ber. dtsch. Wiss. Komm. Meeresforsch.* **12,** 431–56.

KREY, J. (1953) Plankton- und Sestonuntersuchungen in der südwestlichen Nordsee auf der Fahrt der 'Gauss' Februar-März 1952. *Ber. dtsch. Wiss. Komm. Meeresforsch.* **13,** 136–53.

KREY, J. (1956) Die Trophie küstennaher Meeresgebiete. *Kieler Meeresforsch.* **12,** 46–71.

KREY, J. (1957) Ergebnisse eines hydrographisch-produktions-biologischen Längsschnitts durch die Ostsee im Sommer 1956. III. Die Verteilung des Gesamtphosphors. *Kieler Meeresforsch.* **13,** 202–11.

KREY, J. (1958a) Die produktionbiologischen Untersuchungen auf den Fahrten der deutschen Schiffe im internationalen geophysikalischen Jahr 1957/8. *Geogr. Rundschau,* **10,** 374–81.

KREY, J. (1958b) Chemical methods of estimating standing crop of phytoplankton. *Rapp. Cons. Explor. Mer,* **144,** 20–7.

KREY, J. (1961a) The vertical distribution of seston in the northern Atlantic, as observed during the cruises of 'Gauss' and 'Anton Dohrn' in late winter and late summer of 1958. *Rapp. Cons. Explor. Mer,* **149,** 194–9.

KREY, J. (1961b) Beobachtungen über den Gehalt an Mikrobiomasse und Detritus in der Kieler Bucht 1958–1960. *Kieler Meeresforsch.* **17,** 163–75.

KREY, J. (1961c) *The balance between living and dead matter in the oceans.* Oceanography, Ed. M. Sears, Washington D.C., U.S.A. pp. 539–48.

KREY, J. (1961d) Der Detritus im Meere. *J. Cons. int. Explor. Mer,* **26,** 263–80.

KREY, J., BANSE, K. and HAGMEIER, E. (1957) Über die Bestimmung von Eiweiss im Plankton mittels der Biuretreaktion. *Kieler Meeresforsch,* **13,** 35–40.

KREY, J., HANTSCHMANN, D. and WELLERSHAUS, S. (1959) Der Sestongehalt entlang eines Schnittes von Kap Farvel bis zur Flämischen Kappe im April und September, 1958. *Dtsch. hydrogr. Z. Ergänzungsheft Reihe B,* No. **3,** 73–81.

KRISS, A. E. (1959a) The role of micro-organisms in the primary production of the Black Sea. *J. Cons. int. Explor. Mer,* **24,** 221–30.

KRISS, A. E. (1959b) Microbiology and the chief problems in the Black Sea. *Deep-Sea Res.* **5,** 193–200.

KRISS, A. E. (1961) *Meeresmikrobiologie.* Gustav Fischer Verlag Jena.

KRISS, A. E., ABYZOV, S. S. and MITZKEVICH, I. N. (1960) Micro-organisms as indicators of hydrological phenomena in seas and oceans. III. Distribution of water masses in the central part of the Pacific Ocean (according to microbiological data). *Deep-Sea Res.* **6,** 335–45.

KROGH, A. (1930) Über die Bedeutung von gelösten organischen Substanzen bei der Ernährung von Wassertieren. *Z. vergl. Physiol.* **12,** 668–81.

KROGH, A. (1931) Dissolved substances as food of aquatic organisms. *Rapp. Cons. Explor. Mer,* **75,** 1–36.

KROGH, A. (1934a, b) (1) Conditions of life in the ocean. (2) Conditions of life at great depths in the ocean. *Ecol. Monogr.* **4,** 421–9, 430–9.

KROGH, A. and KEYS, A. (1934) Methods for the determination of dissolved organic carbon and nitrogen in sea water. *Biol. Bull.* **67,** 132–44.

KROGH, A. and LANGE, E. (1931) Quantitative Untersuchungen über Plankton, Kolloide und gelöste organische und anorganische Substanzen in dem Furesee. *Int. Rev. Hydrobiol.* **26,** 20–53.

KROGH, A., LANGE, E. and SMITH, W. (1930) On the organic matter given off by algae. *Biochem. J.* **24,** 1666–71.

KRÜGER, F. (1959) Zur Ernährungsphysiologie von *Arenicola marina* (L). *Zool. Anz.* **22,** Suppl. 115–20.

KUFFERATH, J. (1957) Tentative d'estimation gravimétrique de l'ultraplanc. ton dans l'eau de la Méditerranée. *Bull. Inst. Oceanogr. Monaco,* **1088, 1**–8.

KULLENBERG, B. (1953) Absolute chronology of deep sea sediments and the deposition of clay on the ocean floor. *Tellus,* **5,** 302–5.

KUSNETZOV, A. (1960) Quantitative distribution of the sea-floor fauna of the Atlantic Ocean. *Dokl. Akad. Nauk SSSR,* **130,** 1345–8 (in Russian).

KUSNEZOW, S. L. (1959) *Die Rolle der Mikroorganismen im Stoffkreislauf der Seen.* VEB. Deutscher Verlag der Wissenschaften Berlin.

KWART, H. and SHACHOUA, V. E. (1957) The structure and constitution of mucus. *Trans. N.Y. Acad. Sci.* **19,** 595–612.

LAEVASTU, T. (1957) *Review of the methods used in plankton research and conversion tables for recording the data and recommendations for standardization.* FAO/57/7/4472 (mimeo).

LAEVASTU, T. (1958) *Zooplankton: factors affecting its abundance and distribution with special consideration to the relation to phytoplankton and fisheries.* FAO Fish. Div., F.B./58/T 5.

LANE, C. E. and TIERNEY, J. Q. (1951) Hydrodynamics and respiration in *Teredo. Bull. Mar. Sci.* **1,** 104–10.

LASKER, R. (1960) Utilization of organic carbon by a marine crustacean: Analysis with Carbon-14. *Science,* **131,** 1098–1100.

LASKER, R. and HOLMES, R. W. (1957) Variability in retention of marine phytoplankton by membrane filters. *Nature, Lond.* **180,** 1295.

LASKER, R. and LANE, C. E. (1953) The origin and distribution of nitrogen in *Teredo Bartschi* Clapp. *Biol. Bull.* **105,** 316–19.

LEATHERS, A. L. (1922) Ecological study of aquatic midges and some related insects with special reference to feeding habits. *Bull. U.S. Bur. Fish.* **38,** 1–62.

LEBOUR, M. V. (1931) The larval stages of *Nassarius reticulatus* and *Nassarius incrassatus. J. Mar. biol. Ass. U.K.* **17,** 797–807.

LEBOUR, M. V. (1933) The importance of larval Mollusca in the plankton. *J. Cons. int. Explor. Mer,* **8,** 335–43.

LEFÈVRE, M. (1942) L'utilisation des algues d'eau douce par les Cladocères. *Bull. biol. France-Belg.* **76,** 250–76.

LEWIN, R. A. (1956) Extracellular polysaccharides of green algae. *Canad. J. Microbiol.* **2,** 665–72.

LEWIS, Jr., G. J. and RAKESTRAW, N. W. (1955) Carbohydrate in sea water. *J. Mar. Res.* **14,** 253–8.

LISITZIN, A. P. (1959) Suspended matter in the ocean. *Int. Oceanogr. Congr. A.A.A.S. Washington D.C.* pp. 470–1.

LISITZIN, A. P. and ZHIVAGO, A. V. (1960) Marine geological work of the Soviet Antarctic Expedition, 1955, 1957. *Deep-Sea Res.* **6,** 77–87.

LOCHHEAD, Y. H. (1936) On the feeding mechanism of a ctenopod cladoceran *Penilia ovirostris* Dana. *Proc. zool. Soc. Lond.* 1936, Part 1, 335–55.

LOHMANN, H. (1899a) Das Gehäuse der Appendicularien, sein Bau, seine Funktion und seine Entstehung. *Schr. naturw. Ver. Schl.-Holst.* **11,** 347–407.

LOHMANN, H. (1899b) Das Gehäuse der Appendicularien nach seiner Bildungsweise, seinem Bau und seiner Funktion. *Zool. Anz.* **22,** 206–14.

LOHMANN, H. (1902) Die Coccolithophoridae, eine Monographie der Coccolithen bildenden Flagellaten. *Arch. Protistenk.* **1,** 89–165.

LOHMANN, H. (1903) Neue Untersuchungen über den Reichtum des Meeres an Plankton und über die Brauchbarkeit der verschiedenen Fangmethoden. *Wiss. Meeresunters. Abt. Kiel,* **7,** 1–86.

LOHMANN, H. (1908) Untersuchungen zur Feststelling des vollständigen Gehaltes des Meeres an Plankton. *Wiss. Meeresunters. Abt. Kiel,* **10,** 129–370.

LOHMANN, H. (1909) Die Gehäuse und Gallertblasen der Appendicularien und ihre Bedeutung für die Erforschung des Lebens im Meer. *Verh. dtsch. zool. Ges. Vers.* **19,** 200–39.

LOHMANN, H. (1933) Appendiculariae, in Kükenthal and Krumbach. *Handb. Zool. Berl.* **5,** 2nd half 15–202.

LOOSANOFF, V. L. (1939) Effect of temperature upon shell movements of clams, *Venus mercenaria* (L.). *Biol. Bull.* **76,** 171–82.

LOOSANOFF, V. L. (1942) Shell movements of the edible mussel, *Mytilus edulis* (L.) in relation to temperature. *Ecology,* **23,** 231–4.

LOOSANOFF, V. L. (1949) On the food selectivity of oysters. *Science,* **110,** 122.

LOOSANOFF, V. L. (1958a) Some aspects of behavior of oysters at different temperatures. *Biol. Bull.* **114,** 57–70.

LOOSANOFF, V. L. (1958b) Challenging problems in shellfish biology. Perspectives in Marine Biology. Ed. A. A. Buzzati-Traverso. *U.I.S.B. Sér. B.* **27,** 483–96.

LOOSANOFF, V. L. (1961) Effects of turbidity on some larval and adult bivalves. *Proc. Gulf and Caribbean Fish. Inst. 14. ann. ses.* 80–95.

LOOSANOFF, V. L. and DAVIS, H. C. (1963) Rearing of bivalve mollusks. *Adv. Mar. Biol.* **1,** 1–16.

LOOSANOFF, V. L. and ENGLE, J. B. (1947) Effect of different concentrations of micro-organisms on the feeding of oysters (*O. virginica*). *Fish. Bull., U.S.* **51,** 31–57.

LOOSANOFF, V. L. and NOMEJKO, C. A. (1946) Feeding of oysters in relation to tidal stages and to periods of light and darkness. *Biol. Bull.* **90,** 244–64.

LOOSANOFF, V. L. and TOMMERS, F. D. (1947) Effect of low pH upon rate of water pumping of oysters, *Ostrea virginica*. *Anat. Rec.* **99,** 112–13.

LOOSANOFF, V. L. and TOMMERS, F. D. (1948) Effect of suspended silt and other substances on rate of feeding of oysters. *Science,* **107,** 69–70.

LOOSANOFF, V. L., DAVIS, H. C. and CHAULEY, P. E. (1955) Food requirements of some bivalve larvae. *Proc. nat. Shellfish. Ass.* **45,** 66–83.

LOOSANOFF, V. L., MILLER, W. S. and SMITH, P. B. (1951) Growth and setting of larvae of *Venus mercenaria* in relation to temperature. *J. Mar. Res.* **10,** 59–81.

LOWNDES, A. G. (1933) The feeding mechanism of *Chirocephalus diaphanus. Proc. zool. Soc. Lond.* 1093–1118.

LOWNDES, A. G. (1935) The swimming and feeding of certain calanoid copepods. *Proc. zool. Soc. Lond.* Pt. 3, 687–715.

LUCAS, A. M. (1931a) An investigation of the nervous system as a possible factor in regulation of ciliary activity of the lamellibranch gill. *J. Morph.* **51,** 147–94.

LUCAS, A. M. (1931b) The distribution of the branchial nerve in *Mytilus edulis* and its relation to the problem of nervous control of ciliary activity. *J. Morph.* **51,** 195–205.

LUCAS, A. M. and DOUGLAS, L. C. (1934) Principles underlying ciliary activity in the respiratory tract. II. A comparison of nasal clearance in man, monkey and other mammals. *Arch. Otolaryn.* **20,** 518–41.

LUCAS, C. E. (1947) The ecological effects of external metabolites. *Biol. Rev.* **22,** 270–95.

LUCAS, C. E. (1958) External metabolites and productivity. *Rap. Cons. Explor. Mer,* **144,** 155–8.

LUCAS, C. E. (1961) Interrelationships between aquatic organisms mediated by external metabolites. Oceanography, Ed. M. Sears. Washington D.C., U.S.A. pp. 499–517.

LÜNEBURG, H. (1952) Beiträge zur Hydrografie der Wesermünding. *Veröff. Inst. Meeresforsch. Bremerhaven,* **1,** 91–114.

LUND, E. J. (1957) A quantitative study of clearance of a turbid medium and feeding by the oyster. *Inst. Mar. Sci.* **4,** 296–312.

MACGINITIE, G. E. (1932) The role of bacteria as food for bottom animals. *Science,* **76,** 490.

MACGINITIE, G. E. (1939a) The method of feeding of *Chaetopterus. Biol. Bull.* **77,** 115–18.

MACGINITIE, G. E. (1939b) The method of feeding of tunicates. *Biol. Bull.* **77,** 443–47.

MACGINITIE, G. E. (1941) On the method of feeding of four Pelecypods. *Biol. Bull.* **80,** 18–25.

MACGINITIE, G. E. (1945) The size of the mesh openings in mucous feeding nets of marine animals. *Biol. Bull.* **88,** 107–11.

MACGINITIE, G. E. and MACGINITIE, N. (1949) *Natural History of Marine Animals.* New York.

MACKIN, J. G. (1962) Canal dredging and silting in Louisiana Bays. *Publ. Inst. Mar. Sci. Texas,* **7,** 262–314.

MADSEN, F. J. (1947) *Octacnemus ingolfi* n. sp., an atlantic representative of the peculiar tunicate-family Octacnemidae. *Vidensk. Medd. dansk naturh. Foren. Kbh.* **110,** 31–44.

MADSEN, F. J. (1949) Marine Bivalvia. *The Zoology of Iceland,* **4,** (63) 1–116.

MADSEN, F. J. (1954) Some general remarks on the distribution of the echinoderm fauna of the deep-sea. *Union int. Sci. biol. Sér. B,* **16,** 30–7.

MAGNUS, D. B. E. (1963) Der Federstern *Heterometra savignyi* in Roten Meer. *Natur u. Mus.* **93,** 355–68.

MALOVITSKAYA, L. M. and SOROKIN, YU. I. (1961a) On the feeding of some species of diaptomids (Copepoda, Calanoidea) on bacteria. *Dokl. biol. Sci. sect.* **136,** 94–5.

MALOVITSKAYA, L. M. and SOROKIN, YU. I. (1961b) Experimental investigation of nutrition in *Diaptomus* (Crustacea, Copepoda) with carbon-14. *Trudy Inst. Biol. Vodoxranilisch,* **4,** 262–72 (in Russian).

MANNING, W. M. and JUDAY, R. E. (1941) The chlorophyll content and productivity of some lakes in north-eastern Wisconsin. *Trans. Wis. Acad. Sci. Arts. Lett.* **33,** 363–93.

MANUILOVA, E. F. (1958) The question of the role of bacterial numbers in the development of Cladocera in natural conditions. *Dokl. biol. Sci. Sect.* **120,** 438–41.

MARGALEF, R. (1954) Consideraciónes sobre la determinación cuantitativa del fitoplancton por la valoración de pigmentos solubles y los factores que afectan a la relación entre cantidad de pigmento y peso seco. *Publ. Inst. Biol. apl. Barcelona,* **16,** 71–84.

MARGALEF, R., DURAN, M. and SAIZ, F. (1955) El fitoplancton de la ría de Vigo de enero de 1953 a marzo de 1954. *Invest. Pesq.* **2,** 85–129.

MARGALEF, R., HERRERA, J. and ARIAS, E. (1959) Hidrografía y fitoplancton de las costás de Castellón, de julio de 1957 a junio de 1958. *Invest. Pesq.* **15,** 3–38.

MARGALEF, R., MUÑOZ, F. and HERRERA, J. (1957) Fitoplancton de las costás de Castellón de enero de 1955 a junio de 1956. *Invest. Pesq.* **7,** 3–31.

MARSHALL, N. (1956) Chlorophyll a in the phytoplankton in coastal waters of the eastern Gulf of Mexico. *J. Mar. Res.* **15,** 14–32.

MARSHALL, N. B. (1954) *Aspects of deep sea biology*. Hutchinson, London.

MARSHALL, N. B. (1960) Swimbladder structure of deep-sea fishes in relation to their systematics and biology. *Discovery Rep.* **31,** 1–122.

MARSHALL, S. M. and ORR, A. P. (1955a) *The biology of a marine copepod Calanus finmarchicus* (*Gunnerus*). Edinburgh.

MARSHALL, S. M. and ORR, A. P. (1955b) On the biology of *Calanus finmarchicus.* VIII. Food uptake, assimilation and excretion in adult and stage V *Calanus. J. Mar. biol. Ass. U.K.* **34,** 495–529.

MARSHALL, S. M. and ORR, A. P. (1955c) Experimental feeding of the copepod *Calanus finmarchicus* (Gunnerus) on phytoplankton cultures labelled with radio-active carbon (^{14}C). *Deep-Sea Res.* suppl. **3,** 110–14.

MARSHALL, S. M. and ORR, A. P. (1956) On the biology of *Calanus finmarchicus* IX. Feeding and digestion in the young stages. *J. Mar. biol. Ass. U.K.* **35,** 587–604.

MARSHALL, S. M. and ORR, A. P. (1961) On the biology of *Calanus finmarchicus* XII. The phosphorus cycle: excretion, egg production, autolysis. *J. Mar. biol. Ass. U.K.* **41,** 463–88.

MARSHALL, S. M. and ORR, A. P. (1962) Food and feeding in copepods. *Rapp. Cons. Explor. Mer,* **153,** 92–8.

MARSHALL, S. M., NICHOLLS, A. G. and ORR, A. P. (1935) On the biology of *Calanus finmarchicus* VI. Oxygen consumption in relation to environmental conditions. *J. Mar. biol. Ass. U.K.* **20,** 1–27.

MARTIN, G. W. (1928) Experimental feeding of oysters. *Ecology,* **9,** 49–55.

MATTHEWS, L. H. and PARKER, H. W. (1950) Notes on the anatomy and biology of the basking shark. *Proc. zool. Soc. Lond.* **120,** 555–76.

MAUCHLINE, J. (1959) The biology of the euphausiid crustacean, *Meganyctiphanes norvegica* (M. Sars). *Proc. roy. Soc. Edinb.* **67,** 141–79.

MCALLISTER, C. D., PARSONS, T. R. and STRICKLAND, J. D. H. (1960) Primary productivity at station 'P' in the North-East Pacific Ocean. *J. Cons. int. Explor. Mer,* **25,** 240–59.

MCALLISTER, C. D., PARSONS, T. R., STEPHENS, K. and STRICKLAND, J. D. H. (1961) Measurements of primary production in coastal sea water, using a large-volume plastic sphere. *Limnol. Oceanogr.* **6,** 237–58.

MENZEL, D. W. and RYTHER, J. H. (1960) The annual cycle of primary production in the Sargasso Sea off Bermuda. *Deep-Sea Res.* **6,** 351–67.

MENZEL, R. W. (1955) Some phases of the biology of *Ostrea equestris* Say and comparison with *Crassostrea virginica* (Gmelin). *Publ. Inst. Mar. Sci. Texas,* **4,** 69–153.

MILLAR, R. H. (1959) Ascidiacea. *Galathea Rep.* **1,** 189–209.

MILLER, S. M. and MOORE, H. B. (1953) Significance of nannoplankton. *Nature, Lond.* **171,** 1121–2.

MILLER, S. M., MOORE, H. B. and KVAMMEN, K. R. (1953) Plankton of the Florida Current. I. General conditions. *Bull. Mar. Sci. Gulf and Caribbean,* **2,** 465–85.

MITCHELL, P. H. (1914) The oxygen requirements of shellfish. *Bull. U.S. Bur. Fish.* **32,** 207–22.

MONAKOV, A. V. and SOROKIN, YU. I. (1960) Application of isotop method in studying nutrition of cyclops and their nauplial stages on bacterial food. *Izvestia Akad. Nauk SSSR Ser. Biol.* 917–22 (in Russian).

MONAKOV, A. V. and SOROKIN, YU. I. (1961a) An experimental investigation of *Daphnia* nutrition using C^{14}. *Dokl. biol. Sci. Sect.* **135,** 925–6.

MONAKOV, A. V. and SOROKIN, YU. I. (1961b) Quantitative data on feeding in daphnians. *Trudy Inst. Biol. Vodoxranilisch,* **4,** 251–61 (in Russian).

MOORE, B., EDIE, E. S., WHITLEY, E. and DAKIN, W. J. (1912) The nutrition and metabolism of marine animals in relation to (*a*) dissolved organic matter and (*b*) particulate organic matter of sea water. *Biochem. J.* **6,** 255–96.

MOORE, H. F. (1910) Volumetric studies of the food and feeding of oysters. *Bull. U.S. Bur. Fish.* **28,** 1295–1308.

MOORE, W. G. (1957) Studies on the laboratory culture of *Anostraca. Trans. Amer. micr. Soc.* **76,** 159–73.

MORRIS, R. W. (1954) Some considerations regarding the nutrition of marine fish larvae. *J. Cons. int. Explor. Mer,* **20,** 255–65.

MORRIS, R. W. (1956) Some aspects of the problem of rearing marine fishes. *Bull. Inst. Océanogr. Monaco,* **1082,** 1–61.

MORTON, J. E. (1951a) The structure and adaptations of the New Zealand Vermetidae. Part I. The Genus *Serpulorbis. Trans. Proc. roy. Soc. N.Z.* **79,** 1–19.

MORTON, J. E. (1951b) The structure and adaptations of the New Zealand Vermetidae. Part II. The Genera *Stephopoma* and *Pyxipoma. Trans. Proc. roy. Soc. N.Z.* **79,** 20–42.

MORTON, J. E. (1951c) The ecology and digestive system of the Struthiolariidae (Gastropoda). *Quart. J. micr. Sci.* **92,** 1–26.

MORTON, J. E. (1954) The biology of *Limacina retroversa. J. Mar. biol. Ass. U.K.* **33,** 297–312.

MORTON, J. E. (1958) *Molluscs.* London 1958.

MORTON, J. E., BONEY, A. D. and CORNER, E. D. S. (1957) The adaptation of *Lasaea rubra* (Montagu), a small intertidal lamellibranch. *J. Mar. biol. Ass. U.K.* **36,** 383–406.

MOUL, E. T. and MASON D. (1957) Study of diatom populations on sand and mud flats in the Wood's Hole area. *Biol. Bull.* **113,** 351.

MUÑOS, F. (1960) *Fitoplancton en el Golfo de Cadiz.* Inst. Invest. Pesq. IV Reunion, 35–7.

MYERS, F. J. (1941) *Lecane curvucornis* var. *miamiensis,* new variety of rotatoria, with observations on the feeding habits of rotifers. *Not. nat.* **75.**

MYERS, J. and JOHNSTON, J. A. (1949) Carbon and nitrogen balance of *Chlorella* during growth. *Plant Physiol.* **24,** 111–19.

NALEWAJKO, C., CHOWDHURI, N. and FOGG, G. E. (1963) Excretion of glycollic acid and the growth of a planktonic *Chlorella. Microalgae & Photosynth. Bact.* 171–83.

NAUMANN, E. (1924) Notizen zur Ernährungsbiologie der limnischen Fauna. *Ark. Zool.* **16,** (12) 1–14.

NAUWERCK, A. (1959) Zur Bestimmung der Filtrierrate limnischer Planktontiere. *Arch. Hydrobiol. Suppl.* **25,** 83–101.

NAUWERCK, A. (1962) Nicht-algische Ernährung bei *Eudiaptomus gracilis* (Sars). *Arch. Hydrobiol. Suppl.* **25,** 393–400.

NAUWERCK, A. (1963) Die Beziehungen zwischen Zooplankton und Phytoplankton im See Erken. *Symb. bot. upsaliens.* **17** (5), 1–163.

NEEDHAM, J. G., TRAVER, J. R. and HSU, YIN-CHI (1935) *The Biology of Mayflies.* Ithaca, New York.

NELSON, D. J. (1960) Improved chlorophyll extraction method. *Science,* **132,** 351.

NELSON, T. C. (1923) The mechanism of feeding in the oyster. *Proc. Soc. exp. Biol. N.Y.* **21,** 166–8.

NELSON, T. C. (1935) Water filtration by the oyster and a new hormone effect thereon. *Anat. Rec.* **64,** Suppl. 1, 68.

NELSON, T. C. (1936) Water filtration by the oyster and a new hormone effect upon rate of flow. *Proc. Soc. exp. Biol. N.Y.* **34,** 189–90.

NELSON, T. C. (1938) The feeding mechanism of the oyster I. *J. Morph.* **63,** 1–61.

NELSON, T. C. (1941a) On the locus of action of diantlin upon the oyster's gills, as revealed by the effects of acetylcholine, eserine and adrenalin. *Anat. Rec.* **81,** 88.

NELSON, T. C. (1941b) Stimulation by the sperm of *O. cristata* of water pumpage in *O.* (*Gryphaea*) *virginica. Anat. Rec.* **81,** 88.

NELSON, T. C. (1960) The feeding mechanism of the oyster II. On the gills and palps of *Ostrea edulis, Crassostrea virginica* and *C. angulata. J. Morph.* **107,** 163–91.

NELSON, T. C. and ALLISON, J. A. (1940) On the nature and action of diantlin; a new hormone-like substance carried by the spermatozoa of the oyster. *J. exp. Zool.* **85,** 299–338.

NEUMANN, G. (1935) Cyclomyaria, in Kükenthal and Krumbach. *Handb. Zool. Berl.* V, 2nd half, 324–400.

NICHOLS, D. (1960) The histology and activities of the tube–feet of *Antedon bifida. Quart. J. micr. Sci.* **101,** 105–17.

NICOL, E. A. T. (1931) The feeding mechanism, formation of the tube, and physiology of digestion in *Sabella pavonina. Trans. roy. Soc. Edinb.* **56,** 537–99.

NICOL, E. A. T. (1932) The feeding habits of the Galatheidea. *J. Mar. biol. Ass. U.K.* **18,** 87–106.

NICOL, J. A. C. (1960) *The Biology of Marine Animals.* Sir Isaac Pitman & Sons Ltd., London.

NIELSEN, A. (1943) Postembryonale Entwicklung und Biologie der rheophilen Köcherfliege *Oligoplectrum maculatum* Fourcroy. *Biol. Medd. Vidensk. Selsk. Kbh.* **19** (2), 1–87.

NIELSEN, E. T. (1935) Über den Stofwechsel der von Grabwespen paralysierten Tiere. *Vidensk. Medd. dansk naturh. Foren. Kbh.* **99,** 149–231.

NOZAWA, A. (1929) The normal and abnormal respiration of the oyster (*Ostrea circumpicta* Pils.). *Sci. Rep. Tôhoku Imp. Univ.* 4. *Ser. Biol.* **4,** 315–25.

OCKELMANN, W. K. (1958) Marine Lamellibranchiata. *Medd. Grønland,* **122,** No. 4, 1–256.

ØSTVEDT, O. T. (1955) Zooplankton investigations from weather ship 'M' in the Norwegian Sea 1948–49. *Hvalråd. Skr.* **40,** 1–93.

OLDFIELD, E. (1955) Observations on the anatomy and mode of life of *Lasaea rubra* (Montagu) and *Turtonia minuta* (Fabricius). *Proc. malac. Soc. Lond.* **31,** 226–49.

OLDFIELD, E. (1961) The functional morphology of *Kellia suborbicularis* (Montagu), *Montacuta ferruginosa* (Montagu) and *M. substriata* (Montagu). (Mollusca, Lamellibranchiata). *Proc. malac. Soc. Lond.* **34,** 255–95.

OPPENHEIMER, C. H. and JANNASCH, H. W. (1962) Some bacterial populations in turbid and clear sea water near Port Aransas, Texas. *Publ. Inst. Mar. Sci. Texas,* **8,** 56–60.

OPPENHEIMER, C. H. and VANCE, M. H. (1960) Attachment of bacteria to micro-organisms. *Z. allg. Mikrobiol.* **1,** 47–62.

ORTON, J. H. (1913) The ciliary mechanisms on the gill and the mode of feeding in amphioxus, ascidians and *Solenomya togata. J. Mar. biol. Ass. U.K.* **10,** 19–49.

ORTON, J. H. (1914) On ciliary mechanisms in brachiopods and some polychaetes, with a comparison of the ciliary mechanisms on the gills of molluscs, protochordata, brachiopods and cryptocephalous polychaetes, and an account of the endostyle of *Crepidula* and its allies. *J. Mar. biol. Ass. U.K.* **10,** 283–311.

ORTON, J. H. (1948) Notes on the feeding habit of *Capulus ungaricus*. *Rep. Mar. biol. Sta. Port Erin,* **61,** 29–30.

OVERBACK, J. (1959) Die Menge des Nannoplanktons und die Bedeutung des Eisen-Phosphor-Verhältnisses für die Planktonproduktion im Brackwassergebiet der Rügenschen Bodden. *Naturwissensch.* **46,** 499.

OWEN, G. (1953) On the biology of *Glossus humanus* (L.) (*Isocardia cor* Lam.). *J. Mar. biol. Ass. U.K.* **32,** 85–106.

OWEN, G. (1955) Observations on the stomach and digestive diverticula of the Lamellibranchia. I. The Anisomyaria and Eulamellibranchia. *Quart. J. micr. Sci.* **96,** 517–37.

OWEN, G. (1961) A note on the habits and nutrition of *Solenomya parkinsoni*. *Quart. J. micr. Sci.* **102,** 15–21.

PACE, D. M. and BELDA, W. H. (1944) Effect of food content and temperature on respiration in *Pelomyxa carolinensis* Wilson. *Biol. Bull.* **86,** 146–53.

PACE, N. J. (1941) Pigments of the marine diatom *Nitzschia closterium*. *J. biol. Chem.* **140,** 483–9.

PARKER, G. H. (1914) On the strength and the volume of the water currents produced by sponges. *J. exp. Zool.* **16, 443**–46.

PARKER, H. W. and BOESEMAN, M. (1955) The basking shark, *Cetorhinus maximus,* in winter. *Proc. zool. Soc. Lond.* **124,** 185.

PARSONS, T. R. (1961) On the pigment composition of eleven species of marine phytoplankton. *J. Fish. Res. Bd. Canada,* **18,** 1017–25.

PARSONS, T. R. and STRICKLAND, J. D. H. (1959) Proximate analysis of marine standing crops. *Nature, Lond.* **184,** 2038–9.

PARSONS, T. R. and STRICKLAND, J. D. H. (1962a) Oceanic detritus. *Science,* **136,** 313–14.

PARSONS, T. R. and STRICKLAND, J. D. H. (1962b) On the production of particulate organic carbon by heterotrophic processes in sea water. *Deep-Sea Res.* **8,** 211–22.

PARSONS, T. R., STEPHENS, K. and STRICKLAND, J. D. H. (1961) On the chemical composition of eleven species of marine phytoplankton. *J. Fish. Res. Bd. Canada,* **18,** 1001–16.

PAVLOVA, E. V. (1959) On food of *Penilia avirostris* Dana. *Trudy Sevastopol. biol. Sta.* **11,** 63–71 (in Russian).

PEDERSEN, E. (1947) Østersens respiration. *Fiskeridir. Skr. Havundersøk.* **8,** No. 10.

PENNINGTON, W. (1941) The control of the number of freshwater phytoplankton by small invertebrate animals. *J. Ecol.* **29,** No. 2.

PETIPA, T. S. (1959) Feeding of the copepod *Acartia clausi* Giesbr. *Trudy Sevastopol. biol. Sta.* **11,** 72–100 (in Russian).

PETIPA, T. S. (1960) Role of *Noctiluca milians* Sur. in the feeding of *Calanus helgolandicus* Claus. (*Dokl. Akad. Nauk SSSR,* **132,** 961–963.) *Dokl. biol. Sci. Sect.* **132,** 488–90.

PETTERSSON, H. (1934) Scattering and extinction of light in sea water. *Göteborgs Vetensk. Samh. Handl.,* Ser. 5, **4,** No. 4.

Plankton Symposium: Measurements of Primary Production in the Sea (1958). *Rapp. Cons. Explor. Mer,* **144,** 1–158.

PLOUGH, H. H. and JONES, N. (1939) *Ecteinascidia tortugensis,* Species Nova. With a review of the Perophoridae (Ascidiacea) of the Tortugas. *Carneg. Inst. Wash. Publ.* No. 517, 47–60.

PLUNKETT, H. A. and RAKESTRAW, N. W. (1955) Dissolved organic matter in the sea. *Deep-Sea Res.* **3,** Suppl. 12–14.

POPHAM, M. L. (1940) The mantle cavity of some of the Erycinidae, Montacutidae and Galeommatidae with special reference to the ciliary mechanisms. *J. Mar. biol. Ass. U.K.* **24,** 549–87.

POSTMA, H. (1954) Hydrography of the Dutch Wadden Sea. (A study of the relations between water movement, the transport of suspended materials and the production of organic matter.) *Arch. neérl. Zool.* **10,** 405–511.

POTTS, F. A. (1923) The structure and function of the liver of *Teredo*, the shipworm. *Proc. Camb. Phil. Soc.* **1,** 1–17.

POURBAIX, N. (1931) Contribution a l'étude de la nutrition chez les spongiaires (Eponges Siliceuses). *Bull. Sta. océanogr. Salammbô,* **23,** 1–19.

POURBAIX, N. (1932) Note sur la nutrition bactérienne des Eponges. *Ann. Soc. Roy. Zool. Belg.* **63,** 11–15.

POURBAIX, N. (1933a) Mecanisme de la nutrition chez les Spongillidae. *Ann. Soc. Roy. Zool. Belg.* **64,** 11–20.

POURBAIX, N. (1933b) Recherches sur la nutrition des spongiaires. *Inst. Español Oceanogr. Notas y Resúmenes Ser.* II, No. **69,** 1–42.

POURRIOT, R. (1957) Sur la nutrition des Rotifères à partir des algues d'eau douce. *Hydrobiologica.* **9,** 50–9.

PRATT, D. M. and BERKSON, H. (1959) Two sources of error in the oxygen light and dark bottle method. *Limmol. Oceanogr.* **4,** 328–34.

PRINGSHEIM, E. G. (1963) Die Ernährung der Algen. *Naturwissensch.* **50,** 146–50.

PRITCHARD, G. G., GRIFFIN, W. J. and WHITTINGHAM, C. P. (1962). The effect of carbon-dioxide concentration, light intensity and isonicotinyl hydrazide on the photosynthetic production of glycollic acid by *Chlorella. J. exp. Bot.* **13,** 176–84.

PROETZ, A. W. (1943) *Applied Physiology of the Nose.* Saint Louis.

PROVASOLI, L., SHIRAISHI, K. and LANCE, J. R. (1959) Nutritional idiosyncrasies of *Artemia* and *Tigriopus* in monoxenic culture. *Ann. NY. Acad. Sci.* **77,** art. **2,** 250–61.

PÜTTER, A. (1909) *Die Ernährung der Wassertiere und der Stoffhaushalt der Gewässer.* Fischer, Jena.

PÜTTER, A. (1914) Der Stoffwechsel der Kieselschwämme. *Z. allg. Physiol.* **16,** 65–114.

PURCHON, R. D. (1941) On the biology and relationships of the lamellibranch *Xylophaga dorsalis* (Turton). *J. Mar. biol. Ass. U.K.* **25,** 1–41.

PURCHON, R. D. (1954) A note on the biology of the lamellibranch *Rocellaria* (*Gastrochaena*) *cuneformis* Spengler. *Proc. zool. Soc. Lond.* **124,** 17–33.

PURCHON, R. D. (1955a) The functional morphology of the rock-boring lamellibranch *Petricola pholadiformis* Lamarck. *J. Mar. biol. Ass. U.K.* **34,** 257–78.

PURCHON, R. D. (1955b) A note on the biology of *Tridacna crocea* Lam. *Proc. malac. Soc. Lond.* **31,** 95–112.

PURCHON, R. D. (1955c) The structure and function of the British Pholadidae (Rock-boring Lamellibranchia). *Proc. zool. Soc. Lond.* **124,** 859–911.

PURCHON, R. D. (1956) A note on the biology of *Martesia striata* L. (Lamellibranchia) *Proc. zool. Soc. Lond.* **126,** 245–58.

PURCHON, R. D. (1958a) Phylogeny in the lamellibranchia. Proc. Cent. and Bicent. Congr. of Biol. Singapore 1958, 69–82.

PURCHON, R. D. (1958b) The stomach in the Eulamellibranchia; stomach type IV. *Proc. zool. Soc. Lond.* **131,** 487–525.

PURCHON, R. D. (1959) Phylogenetic classification of the Lamellibranchia, with special reference to the Prosobranchia. *Proc. malac. Soc. Lond.* **33,** 224–30.

QUAYLE, D. B. (1948) *The rate of water propulsion in Venerupis pullastra (Montagu)* D. Thesis. Univ. Glasgow.

RABEN, E. (1910) Ist organisch gebundener Kohlenstoff in nennenswerter Menge im Meerwasser gelöst vorhanden *Wiss. Meeresunters. Abt. Kiel,* **11,** 109–17.

RABINOWITCH, E. I. (1945–1951) *Photosynthesis and related processes.* I–II. New York.

RAGOTZKIE, R. A. (1959) Plankton productivity in estuarine waters of Georgia. *Publ. Inst. Mar. Sci. Texas,* **6,** 146–58.

RAHN, O. (1940) Efficiency of energy utilization in the growth of bacteria. *Growth,* **4,** 77–80.

RAKESTRAW, N. W. (1947) Oxygen consumption in sea water over long periods. *J. Mar. Res.* **6,** 259–63.

RAKESTRAW, N. W. (1958) Particulate matter in the oxygen-minimum layer. *J. Mar. Res.* **17,** 429–31.

RAO, K. P. (1953) Rate of water propulsion in *Mytilus californianus* as a function of latitude. *Biol. Bull.* **104,** 171–81.

RAO, K. P. (1954) Tidal rhythmicity of rate of water propulsion in *Mytilus,* and its modifiability by transplantation. *Biol. Bull.* **106,** 353–9.

RAO, K. P. and GOLDBERG, E. D. (1954) Utilization of dissolved calcium by a pelecypod. *J. cell. comp. Physiol.* **43,** 283–92.

RAPSON, A. M. (1954) Feeding and control of toheroa (*Amphidesma ventricosum* Gray) (Eulamellibranchiata) populations in New Zealand. *Aust. J. Mar. Freshw. Res.* **5,** 486–512.

RASMONT, R. (1959) L'ultrastructure des choanocytes d'éponges. *Ann. Sci. Nat. Zool. Sér.* 12, **1,** 253–62.

RASMONT, R. (1961) Une technique de culture des éponges d'eau douce en milieu controlé. *Ann. Soc. Roy. Zool. Belg.* **91,** 147–56.

RAYMONT, J. E. G. and CONOVER, R. J. (1961) Further investigations on the carbohydrate content of marine zooplankton. *Limnol. Oceanogr.* **6,** 154–64.

RAYMONT, J. E. G. and GAULD, D. T. (1951) The respiration of some planktonic copepods. *J. Mar. biol. Ass. U.K.* **29,** 681–93.

RAYMONT, J. E. G. and GROSS, F. (1942) On the feeding and breeding of *Calanus finmarchicus* under laboratory conditions. *Proc. roy. Soc. Edinb.* B, **61,** 267–87.

REMANE, A. (1932) Netzfilter und Strudelapparate bei Rädertiere. *Zool. Anz.* **100,** 326–32.

REMANE, A. (1952) Die Besiedelung des Sandbodens im Meer und die Bedeutung der Lebenstypen für die Ökologie. *Verh. dtsch. zool. Ges.* 1951, 327–59.

RICCI, E. (1956) Teneur en matières organiques dissoutes dans le golfe de Tunis. *Bull. Sta. océanogr. Salammbô,* **53,** 69–74.

RICE, T. R. and SMITH, R. J. (1958) Filtering rates of the hard clam (*Venus mercenaria*) determined with radioactive phytoplankton. *Fish. Bull.,* U.S. **58,** 73–82.

RICHARDS, F. A. with THOMPSON, T. G. (1952) The estimation and characterization of plankton populations by pigment analyses. II. A spectrophotometric method for the estimation of plankton pigments. *J. Mar. Res.* **11,** 156–72.

RICHMAN, S. (1958) The transformation of energy by *Daphnia pulex. Ecol. Monogr.* **28,** 273–91.

RIDEWOOD, W. G. (1903) Structure of gills of lamellibranchs. *Phil. Trans. roy. Soc. Lond.* B, **195,** 147–284.

RILEY, G. A. (1937) The significance of the Mississippi River drainage for biological conditions in the northern Gulf of Mexico. *J. Mar. Res.* **1,** 60–74.

RILEY, G. A. (1938a) The measurement of phytoplankton. *Int. Rev. Hydrobiol.* **36,** 371–3.

RILEY, G. A. (1938b) Plankton studies. I. A preliminary investigation of the plankton of the Tortugas region. *J. Mar. Res.* **1,** 335–52.

RILEY, G. A. (1939) Plankton studies. II. The western North Atlantic, May–June, 1939. *J. Mar Res.* **2,** 145–62.

RILEY, G. A. (1940) Limnological studies in Connecticut. Part III. The plankton of Linsley Pond. *Ecol. Monogr.* **10,** 279–306.

RILEY, G. A. (1941a) Plankton studies IV. Georges Bank. *Bull. Bingham oceanogr. Coll.* **7,** Art. 4, 1–73.

RILEY, G. A. (1941b) Plankton studies III. Long Island Sound. *Bull. Bingham oceanogr. Coll.* **7,** Art. 3, 1–93.

RILEY, G. A. (1942) The relationship of the vertical turbulence and spring diatom flowerings. *J. Mar. Res.* **5,** 67–87.

RILEY, G. A. (1951) Oxygen, phosphate, and nitrate in the Atlantic Ocean. *Bull. Bingham oceanogr. Coll.* **13,** art. 1, 1–126.

Riley, G. A. (1955) Review of the oceanography of Long Island Sound. *Deep-Sea Res.* **3,** Suppl. 224–38.

Riley, G. A. (1957) Phytoplankton of the North Central Sargasso Sea, 1950–52. *Limnol. Oceanogr.* **2,** 252–70.

Riley, G. A. (1959) Note on particulate matter in Long Island Sound. *Bull. Bingham oceanogr. Coll.* **17,** 83–6.

Riley, G. A. and Conover, S. A. M. (1956) Oceanography of Long Island Sound. *Bull. Bingham oceanogr. Coll.* **15,** 47–61.

Riley, G. A. and Gorgy, S. (1948) Quantitative studies of summer plankton populations of the western North Atlantic. *J. Mar. Res.* **7,** 100–21.

Riley, G. A., Stommel, H. and Bumpus, D. F. (1949) Quantitative ecology of the plankton of the western North Atlantic. *Bull. Bingham oceanogr. Coll.* **12,** art. 3, 1–169.

Rochford, D. J. (1952) The application of studies on estuarine hydrology to certain problems in Australian oyster biology. *Rapp. Cons. Explor. Mer,* **131,** 35–7.

Rodhe, W. (1958) The primary production in lakes: some results and restrictions of the ^{14}C method. *Rapp. Cons. Explor. Mer,* **144,** 122–8.

Rodhe, W. (1963) *Marine Biology. I.* Ed. G. A. Riley. Amer. Inst. Biol. Sci. Washington, U.S.A.

Rodhe, W., Vollenweider, R. A. and Nauwerck, A. (1958) *The primary production and standing crop of phytoplankton. Perspectives in Marine Biology.* Ed. A. A. Buzzati-Traverso. Univ. Calif. Press. pp. 299–322.

Rodina, A. G. (1948a) Dissolved organic substances in the nourishment of Cladocera. *Zool. J. Moscow,* **27,** 403–10 (in Russian).

Rodina, A. G. (1948b) Bacteria as a food for fresh-water mollusks. *Microbiology, Moscow,* **17,** 232–9 (in Russian).

Rodina, A. G. (1960) Bacterioplankton of weed-bed zones in lakes—the food base of aquatic invertebrates. *Dokl. biol. Sci. Sect.* **127,** 667–9.

Roule, J. (1884) Recherches sur les ascidies simples des cotes de Provence (Phallusiadées). *Ann. Mus. Hist. nat. Marseille, Zool. II Mém.* **1,** 1–270.

Roushdy, H. M. and Hansen, V. K. (1960) Ophiuroids feeding on phytoplankton. *Nature, Lond.* **188,** 517–18.

Roushdy, H. M. and Hansen, V. K. (1961) Filtration of phytoplankton by the octocoral *Alcyonium digitatum* L. *Nature, Lond.* **190,** 649–50.

Rubentschik, L., Roisin, R. B. and Bieljansky, F. J. (1936) Adsorption of bacteria in salt lakes. *J. Bact.* **32,** 11–31.

Rylov, V. M. (1930) Einige Beobachtungen über den Sestonerwerb bei *Diaptomus coerulens* Fisher. *Trudy Leningr. o-va estestvoisp.* **60,** 149–76 (in Russian, German summary).

Ryther, J. H. (1954) Inhibitory effects of phytoplankton upon the feeding of *Daphnia magna* with reference to growth, reproduction, and survival. *Ecology,* **35,** 522–33.

Ryther, J. H. (1956) The measurement of primary production. *Limnol. Oceanogr.* **1,** 72–84.

RYTHER, J. H. and MENZEL, D. W. (1960) The seasonal and geographical range of primary production in the western Sargasso Sea. *Deep-Sea Res.* **6,** 235–8.

RYTHER, J. H. and YENTSCH, C. S. (1957) The estimation of phytoplankton production in the ocean from chlorophyll and light data. *Limnol. Oceanogr.* **2,** 281–6.

RYTHER, J. H., YENTSCH, C. S., HULBURT, E. M. and VACCARO, R. F. (1958) The dynamics of a diatom bloom. *Biol. Bull.* **115,** 257–68.

SALIMOVSKAYA-RODINA, A. G. (1940) Bacteria and yeast as food of Cladocera. *Dokl. Akad. Nauk SSSR,* **29,** No. 3 (in Russian).

SATTLER, W. (1958) Beiträge zur Kenntnis von Lebensweise und Körperbau der Larve und Puppe von *Hydropsyche* Pict (Trichoptera) mit besonderer Berücksichtigung des Netzbaues. *Z. Morph. Ökol. Tiere,* **47,** 115–92.

SATTLER, W. and KRACHT, A. (1963) Drift-Fang einer Trichopterenlarve unter Ausnutzung der Differenz von Gesamtdruck und statischem Druck des fliessenden Wassers. *Naturwissensch.* **50,** 362.

SAUNDERS, G. W. (1957) Interrelations of dissolved organic matter and phytoplankton. *Bot. Rev.* **23,** 389–409.

SAVAGE, R. E. (1925) The food of the oyster. *Fish. Invest. Lond. Ser.* II. **8,** (1).

SAVAGE, R. M. (1950) Feeding mechanisms in anuran tadpoles. *Nature, Lond.* **166,** 155.

SAVILOV, A. I. (1957) Biological aspects of the bottom fauna groupings of the north Okhotsk Sea. *Trans. Inst. Oceanol.* **20,** 67–136. English translation. 1959: Marine Biology. Ed. B. N. Nikitin.

SCHÄFER, H. (1953) Beiträge zur Ernährungsbiologie einheimischer Süsswasserprosobranchier. *Z. Morph. Ökol. Tiere,* **41,** 247–64.

SCHILLER, J. (1931) Über autochthone pflanzliche Organismen in der Tiefsee. *Biol. Zbl.* **51,** 329–34.

SCHROLL, F. (1959) Zur Ernährungsbiologie der steirischen Ammocöten *Lampetra planeri* (Bloch) und *Eudontomyzon danfordi* (Regan). *Int. Rev. Hydrobiol.* **44,** 395–429.

SEGEL, E., RAO, K. P. and JAMES, T. W. (1953) Rate of activity as a function of intertidal height within populations of some littoral molluscs. *Nature, Lond.* **172,** 1108–9.

SEIWELL, H. R. (1937) Consumption of oxygen in sea water under controlled laboratory conditions. *Nature, Lond.* **140,** 506–7.

SEIWELL, H. R. and SEIWELL, G. E. (1938) The sinking of decomposing plankton in sea water and its relationship to oxygen consumption and phosphorus liberation. *Proc. Amer. Phil. Soc.* **78,** 465–81.

SEMINA, G. I. (1958) Typing of different areas of the neritic zone of the boreal region of the Pacific Ocean. *Dokl. biol. Sci. Sect.* **118,** 87–89.

SEMINA, H. J. (1960) The influence of vertical circulation on the phytoplankton in the Bering Sea. *Int. Rev. Hydrobiol.* **45,** 1–10.

SENIOR-WHITE, R. (1928) Algae and the food of anophiline larvae. *Indian J. med. Res.* **15,** 969–88.

SETNA, S. B. (1930) Neuromusculature mechanism of the gill of *Pecten. Quart. J. micr. Sci.* **73,** 365–91.

SHARMAN, M. (1956) Note on *Capulus ungaricus* (L.). *J. Mar. biol. Ass. U.K.* **35,** 445–50.

SHIMADA, B. M. (1958) Diurnal fluctuation in photosynthetic rate and chlorophyll 'a' content of phytoplankton from Eastern Pacific Waters. *Limnol. Oceanogr.* **3,** 336–9.

SHIRAISHI, K. and PROVASOLI, L. (1959) Growth factors as supplements to inadequate algal foods for *Tigriopus japonicus. Tohoku J. agric. Res.* **10,** 89–96.

SIEBURTH, J. MCNEILL (1959) Antibacterial activity of antarctic marine phytoplankton. *Limnol. Oceanogr.* **4,** 419–24.

SINGH, K. R. P. and BROWN, A. W. A. (1957) Nutritional requirements of *Aedes aegyptii. J. Insect Physiol.* **1,** 199–220.

SIZER, I. W. and PEACOCK, A. C. (1947) The ultra-violet absorption of serum albumin and of its constituent amino acids as a function of pH. *J. biol. Chem.* **171,** 767–77.

SKOPINTZEV, B. A. (1959) Organic matter of sea water. Preprints *Int. Oceanogr. Congr. A.A.A.S. Washington D.C.* pp. 953–4.

SKOPINTZEV, B. A. and BROOCK, E. A. (1940) A study of the oxidation processes occurring in water during the aerobic decomposition of phytoplankton. *Microbiology, Moscow,* **9,** 595–607 (in Russian, English summary).

SLOWEY, J. F., JEFFREY, L. M. and HOOD, D. W. (1959) Characterization of the ethyl acetate extractable organic material of sea water. Preprints *Int. Oceanogr. Congr. A.A.A.S. Washington D.C.* p. 935.

SOKOLOVA, M. N. (1959) On the distribution of deep-water bottom animals in relation to their feeding habits and the character of sedimentation. *Deep-Sea Res.* **6,** 1–4.

SOROKIN, YU. I. (1959) Role of chemosynthesis in the production of organic matter in reservoirs. IV. Feeding aquatic invertebrates on methane and hydrogen-oxidizing autotrophic bacteria. *Microbiology, Moscow,* **28,** 856–60 (English edition).

SOROKIN, YU. I. (1962) The vertical distribution of saprophytic bacteria in the water strata of the central part of the Pacific Ocean. *Dokl. biol. Sci. Sect.* **145,** 821–3.

SOUTHWARD, A. J. (1955a) Feeding of barnacles. *Nature, Lond.* **175,** 1124–5.

SOUTHWARD, A. J. (1955b) Observations on the ciliary currents of the jelly-fish *Aurelia aurita* L. *J. Mar. biol. Ass. U.K.* **34,** 201–16.

SOUTHWARD, A. J. and SOUTHWARD, E. C. (1963) Notes on the biology of some Pogonophora. *J. Mar. biol. Ass. U.K.* **43,** 57–64.

SPOEHR, H. A. and MILNER, H. W. (1949) The chemical composition of *Chlorella*; effect of environmental conditions. *Plant Physiol.* **24,** 120–49.

STASEK, C. R. (1961) The ciliation and function of the labial palps of *Acila castrensis* (Protobranchia, Nuculidae), with an evaluation of the role of the protobranch organs of feeding in the evolution of the bivalvia. *Proc. zool. Soc. Lond.* **137,** 511–38.

STASEK, C. R. (1962) The form, growth and evolution of the Tridacnidae (giant clams). *Arch. Zool. exp. gén.* **101,** 1–40.

STEELE, J. H. (1956) Plant production on the Fladen Ground. *J. Mar. biol. Ass. U.K.* **35,** 1–33.

STEELE, J. H. (1957) A comparison of plant production estimates, using ^{14}C and phosphate data. *J. Mar. biol. Ass. U.K.* **36,** 233–41.

STEELE, J. H. (1958) Plant production in the northern North Sea. *Mar. Res.* 1958, No. 7, 1–36.

STEELE, J. H. (1959) The quantitative ecology of marine phytoplankton. *Biol. Rev.* **34,** 129–58.

STEELE, J. H. (1961) *Primary production. Oceanography,* Ed. M. Sears. Washington D.C., U.S.A. pp. 519–38.

STEELE, J. H. and BAIRD, I. E. (1961) Relations between primary production, chlorophyll and particulate carbon. *Limnol. Oceanogr.* **6,** 68–78.

STEELE, J. H. and BAIRD, I. E. (1962a) Further relations between primary production, chlorophyll, and particulate carbon. *Limnol. Oceanogr.* **7,** 42–7.

STEELE, J. H. and BAIRD, I. E. (1962b) Carbon-chlorophyll relations in cultures. *Limnol. Oceanogr.* **7,** 101–2.

STEELE, J. H. and YENTSCH, C. S. (1960) The vertical distribution of chlorophyll. *J. Mar. biol. Ass. U.K.* **39,** 217–26.

STEEMANN NIELSEN, E. (1933) Über quantitative Untersuchungen von marinem Plankton mit Utermöhls umgekehrten Mikroskop. *J. Cons. int. Explor. Mer,* **8,** 201–10.

STEEMANN NIELSEN, E. (1952) The use of radio-active carbon (C^{14}) for measuring organic production in the sea. *J. Cons. int. Explor. Mer,* **18,** 117–40.

STEEMANN NIELSEN, E. (1955) The production of antibiotics by plankton algae and its effects upon bacterial activities in the sea. *Deep-Sea Res.* **3,** Suppl. p. 281.

STEEMANN NIELSEN, E. (1958a) Experimental methods for measuring organic production in the sea. *Rapp. Cons. Explor. Mer,* **144,** 38–55.

STEEMANN NIELSEN, E. (1958b) A survey of recent Danish measurements of the organic productivity in the sea. *Rapp. Cons. Explor. Mer,* **144,** 92–5.

STEEMANN NIELSEN, E. (1958c) The balance between phytoplankton and zooplankton in the sea. *J. Cons. int. Explor. Mer,* **23,** 178–88.

STEEMANN NIELSEN, E. (1959) Untersuchungen über die Primärproduktion des Planktons in einigen Alpenseen Österreichs. *Oikos,* **10,** 24–37.

STEEMANN NIELSEN, E. (1960) Productivity of the oceans. *Ann. Rev. Plant Physiol.* **11**. 341–62.

STEEMANN NIELSEN, E. (1962) The relationship between phytoplankton and zooplankton in the sea. *Rapp. Cons. Explor. Mer,* **153,** 178–81.

STEEMAN NIELSEN, E. and HANSEN V. K. (1959a) Light adaptation in marine phytoplankton populations and its interrelation with temperature. *Physiol. Plant.* **12,** 353–70.

STEEMANN NIELSEN, E. and HANSEN, V. K. (1959b) Measurements with the carbon-14 technique of the respiration rates in natural populations of phytoplankton. *Deep-Sea Res.* **5,** 222–33.

Steemann Nielsen, E. and Jensen, E. Aa. (1957) Primary oceanic production. The autotrophic production of organic matter in the oceans. *Galathea Rep.* **1,** 49–136.

Stephens, G. C. (1960a) Uptake of glucose from solution by the solitary coral, *Fungia. Science,* **131,** 1532.

Stephens, G. C. (1960b) The mechanism of glucose uptake by the coral, *Fungia. Anat. Rec.* **137,** 395.

Stephens, G. C. (1962a) Uptake of amino acids by the bamboo worm. *Clymenella torquata. Biol. Bull.* **123,** 512.

Stephens, G. C. (1962b) Uptake of organic material by aquatic invertebrates. I. Uptake of glucose by the solitary coral, *Fungia scutaria. Biol. Bull.* **123,** 648–59.

Stephens, G. C. (1963) Uptake of organic materials by aquatic invertebrates. II. Accumulation of amino acids by the bamboo worm, *Clymenella torquata. Comp. Biochem. Physiol.* **10,** 191–202.

Stephens, G. C. and Schinske, R. A. (1957) Uptake of amino acids from sea water by ciliary-mucoid filter feeding animals. *Biol. Bull.* **113,** 356–7.

Stephens, G. C. and Schinske, R. A. (1961) Uptake of amino acids by marine invertebrates. *Limnol. Oceanogr.* **6,** 175–81.

Stommel, H. and Arons, A. B. (1960) On the abyssal circulation of the world ocean. II. An idealized model of the circulation pattern and amplitude in oceanic basins. *Deep-Sea Res.* **6,** 217–33.

Storch, O. (1926) Morphologie und Physiologie des Fangapparates der Daphniden. *Ergebn. u. Fortschr. Zool.* **6,** 125–234.

Storch, O. (1929) Die Schwimmbewegung der Copepoden, auf Grund von Mikro-Zeitlupenaufnahmen analysiert. *Zool. Anz.* **4,** Suppl. 118–29.

Storch, O. and Pfisterer, O. (1926) Der Fangapparat von *Diaptomus. Z. vergl. Physiol.* **3,** 330–76.

Strain, H. H. (1951) *The pigments of algae. In Manual of Phycology.* Waltham, Mass. U.S.A.

Strickland, J. D. H. (1960) Measuring the production of marine phytoplankton. *Bull. Fish. Res. Bd. Canada,* **122.**

Strickland, J. D. H. and Parsons, T. R. (1960) A manual of sea water analysis. *Bull. Fish. Res. Bd. Canada,* **125,** 1–185.

Stuart, C. A., McPherson, M. and Cooper, H. J. (1931) Studies on bacteriologically sterile *Moina macrocopa* and their food requirements. *Physiol. Zool.* **4,** 87–100.

Subrahmanyan, R. (1959) Studies on the phytoplankton of the west coast of India. Part I. Quantitative and qualitative fluctuation of the total phytoplankton crop, the zooplankton crop and their interrelationship, with remarks on the magnitude of the standing crop and production of matter and their relationship to fish landings. *Proc. Indian Acad. Sci. B,* **50,** 113–87.

Suess, H. E., Rakestraw, N. W. and Oeschger, H. (1959) Apparent age of deep water in the Pacific Ocean. *Int. Oceanogr. Congr. A.A.A.S. Washington D.C.* pp. 440–1.

SUSHCHENYA, L. M. (1958a) Quantitative data on the filtration feeding of planktonic Crustacea. *Dokl. Wish. Shkoli Biol. Sci.* **1,** 16–20 (in Russian).

SUSHCHENYA, L. M. (1958b) Dependence of filtration rate in planktonic Crustacea on the concentration of food particles. *Trans. biol. Sta. Lake Naroch,* **1,** 241–260. (in Russian).

SUSHCHENYA, L. M. (1959a) On the consumption of planktonic algae by zooplankton. *Proc. V. Sci. Conf. Dedicated to Research of Inland Waters of Baltic. Minsk,* pp. 32–7 (in Russian).

SUSHCHENYA, L. M. (1959b) Study of selective feeding in planktonic crustaceans. *Nauch. Dokl. Vysch. Schol., Biol. Nauk,* No. 4, 21–6 (in Russian).

SUSHCHENYA, L. M. (1963) The amount of seston in the waters of the Aegean, Ionian and Adriatic Seas. *Deep-Sea Res.* **10,** 82–6.

TAMMES, P. M. L. and DRAL, A. D. G. (1955) Observations on the straining of suspensions by mussels. *Arch. neérl. Zool.* **11,** 87–112.

THEEDE, H. (1963) Experimentelle Untersuchungen über die Filtrationsleistung der Miesmuschel *Mytilus edulis. Kieler Meeresforsch.* **19,** 20–41.

THIENEMANN, A. (1954) *Chironomus. Binnengewässer,* **20,** 1–834.

THOMPSON, T. E. (1959) Feeding in nudibranch larvae. *J. Mar. biol. Ass. U.K.* **38,** 239–48.

THORSON, G. (1946) Reproduction and larval development of Danish marine bottom invertebrates. *Medd. Komm. Danmarks Fisk.- og Havunders-Ser. Plankton,* **4.**

THORSON, G. (1950) Reproductive and larval ecology of marine bottom invertebrates. *Biol. Rev.* **25,** 1–45.

THORSON, G. (1951) Zur jetzigen Lage der marinen Bodentier-Ökologie. *Verh. dtsch. Zool. Ges. Wilhelmshaven* 1951, pp. 276–327.

TIEGS, D. W. and MANTON, S. M. (1958) The evolution of the Arthropoda. *Biol. Rev.* **33,** 255–337.

TOLBERT, N. E. and ZILL, L. P. (1956) Excretion of glycolic acid by algae during photosynthesis. *J. biol. Chem.* **222,** 895–906.

TRAGER, W. (1936) The utilization of solutes by mosquito larvae. *Biol. Bull.* **71,** 343–52.

VAN TRIGT, H. (1919) A contribution to the physiology of the fresh-water sponges (Spongillidae). *Tijdschr. ned. dierk. Ver.* **17,** 1–220.

TSIKHON-LUKANINA, E. A. (1961a) On the problem of the filtration feeding method of *Bithynia tentaculata* L. and *Valvata piscinalis* Müller. *Byull. Inst. Biol. Vodokhr.* No. 10, 28–30. English translation. Ministry of Agric., Fish & Food. Trans. N.S. No. 6, 1963.

TSIKHON-LUKANINA, E. A. (1961b) The dependence of feeding and filtration rates in some freshwater gastropod molluscs on the concentration of food particles in the environment. *Byull. Inst. Biol. Vodokhr.* No. 10, 31–4. English translation. Ministry of Agric., Fish & Food. Trans. N.S. No. 7, 1963.

USSING, H. H. (1938) The biology of some important plankton animals in the fjords of East Greenland. *Medd. Grønland,* **100.**

UTERMÖHL, H. (1931) Neue Wege in der quantitativen Erfassung des Planktons. *Verh. int. Ver. Limnol.* **5.**

VACCARO, R. F. and RYTHER, J. H. (1960) Marine phytoplankton and the distribution of nitrite in the sea. *J. Cons. int. Explor. Mer,* **25,** 260–71.

VALLENTYNE, J. R. (1957) The molecular nature of organic matter in lakes and oceans, with lesser reference to sewage and terrestrial soils. *J. Fish. Res. Bd. Canada,* **14,** 33–82.

VALLENTYNE, J. R. (1962) Solubility and decomposition of organic matter in nature. *Arch. Hydrobiol.* **58,** 423–34.

VERWEY, J. (1952) On the ecology of distribution of cockle and mussel in the Dutch Wadden Sea. Their role in sedimentation and the source of their food supply. With a short review of the feeding behaviour of bivalve molluscs. *Arch neérl. Zool.* **10,** 171–239.

VERWEY, J. (1960) Annual report of the zoological Station of the Netherlands Zoological Society for the year 1957. *Arch. neérl. Zool.* **13,** 540–55.

VIALLANES, H. (1892) Recherches sur la filtration de l'eau par les mollusques et applications à l'ostréiculture et à l'océanographie. *C.R. Acad. Sci. Paris,* **114,** 1386.

VINBERG, G. G. and KALER, V. L. (1960) Comparative study of primary plankton production by the radiocarbon and the oxygen method. *Dokl. biol. Sci. Sect.* **130,** 446–9.

VINOGRADOV, M. E. (1955) The vertical migration of zooplankton and its role as a food for bathypelagic fauna. *Trudy Inst. Okeanol. Akad. Nauk SSSR,* **13,** 71–7 (in Russian).

VINOGRADOV, M. E. (1959a) Über die quantitative Verbreitung des Tiefseeplanktons im nordwestlichen Teil des Stillen Ozeans. *Int. Rev. Hydrobiol.* **44,** 217–25.

VINOGRADOV, M. E. (1959b) Quantitative distribution of deep-sea plankton in the Pacific Ocean and its relation to deep water circulation. Preprints *Int. Oceanogr. Congr. A.A.A.S. Washington D.C.* pp. 390–1.

VINOGRADOV, M. E. (1961) Food sources of the deep sea fauna, on the rate of decomposition of dead pteropods. *Dokl. Akad. Nauk SSSR,* **138,** 1439–42 (in Russian).

VINOGRADOV, M. E. (1962a) Feeding of the deep-sea zooplankton. *Rapp. Cons. Explor. Mer,* **153,** 114–19.

VINOGRADOV, M. E. (1962b) Quantitative distribution of deep-sea plankton in the western Pacific and its relation to deep-water circulation. *Deep-Sea Res.* **8,** 251–8.

VISHNIAC, H. S. and RILEY, G. A. (1961) Cobalamin and thiamin in Long Island Sound: Patterns of distribution and ecological significance. *Limnol. Oceanogr.* **6,** 36–41.

VIVES, F. and LÓPEZ-BENITO, M. (1957) El fitoplancton de la ría de Vigo desde julio de 1955 a junio de 1956. *Invest. Pesq.* **10,** 45–146.

VIVES, F. and LÓPEZ-BENITO, M. (1958) El fitoplancton de la ría de Vigo y su relación con los factores térmicos y energéticos. *Invest. Pesq.* **13,** 87–125.

VOSKRESIENSKY, K. A. (1959) The clearance of water by Unionidae and data about the rate of water transport. *Proc. V. Sci. Conf. Dedicated to Research of Inland Waters of Baltic. Minsk,* 1959, pp. 38–43 (in Russian).

WAKSMAN, S. A. (1934) The distribution and conditions of existence of bacteria in the sea. *Ecol. Monogr.* **4,** 523–9.

WAKSMAN, S. A. and CAREY, C. L. (1935a) Decomposition of organic matter in sea water by bacteria. I. Bacterial multiplication in stored sea water. *J. Bact.* **29,** 531–43.

WAKSMAN, S. A. and CAREY, C. L. (1935b) Decomposition of organic matter in sea water by bacteria. II. Influence of addition of organic substance upon bacterial activities. *J. Bact.* **29,** 545–61.

WAKSMAN, S. A. and RENN, C. E. (1936) Decomposition of organic matter in sea water by bacteria. III. Factors influencing the rate of decomposition. *Biol. Bull.* **70,** 472–83.

WAKSMAN, S. A. and VARTIOVARA, U. (1938) The adsorption of bacteria by marine bottom. *Biol. Bull.* **74,** 56–63.

WALLENGREN, H. (1905) Zur Biologie der Muscheln. I. Wasserströmungen. *Lunds Univ. Årsskr. Afd.* 2, **1,** No. 2, 1–64.

WALLENGREN, H. (1905) Zur Biologie der Muscheln. II. Die Nahrungsaufnahme. *Lunds Univ. Årsskr. Afd.* 2, **1,** No. 3, 1–59.

WALNE, P. R. (1956) Experimental rearing of the larvae of *Ostrea edulis* L. in the laboratory. *Fish. Invest. Lond. Ser. II,* **20,** No. 9, 1–23.

WALNE, P. R. (1958) The importance of bacteria in laboratory experiments on rearing the larvae of *Ostrea edulis* (L.). *J. Mar. biol. Ass. U.K.* **37,** 415–25.

WALNE, P. R. (1959) Some observations on the feeding behaviour of oyster (*Ostrea edulis*) larvae and their relation to rearing problems. *Proc. XV. Int. Congr. Zool. Lond.* 1958, 234–6.

WALNE, P. R. (1963) Observations on the food value of seven species of algae to the larvae of *Ostea edulis*. I. Feeding experiments. *J. Mar. biol. Ass. U.K.* **43,** 767–84.

WALSHE, B. M. (1947) Feeding mechanisms of *Chironomus* larvae. *Nature, Lond.* **160,** 474.

WALSHE, B. M. (1950) Observations on the biology and behaviour of larvae of the midge *Rheotanytarsus*. *J. Quekett micr. Club,* **3,** 171–8.

WALSHE, B. M. (1951) The feeding habits of certain chironomid larvae (subfamily Tendipedinae) *Proc. zool. Soc. Lond.* **121,** 63–79.

WANGERSKY, P. J. (1952) Isolation of ascorbic acid and rhamnosides from sea water. *Science,* **115,** 685.

WANGERSKY, P. J. (1959) Dissolved carbohydrates in Long Island Sound. *Bull. Bingham oceanogr. Coll.* **17,** 87–94.

WANGERSKY, P. J. and GUILLARD, R. R. L. (1960) Low molecular weight organic base from the dinoflagellate *Amphidinium carteri. Nature, Lond.* **185,** 689–90.

WARBURTON, F. E. (1960) Influence of currents on form of sponges. *Science,* **132,** 89.

WATANABE, A. (1951) Production in cultural solution of some amino acids by the atmospheric nitrogen-fixing blue green algae. *Arch. Biochem. Biophys.* **34,** 50–55.

WATTENBERG, H. (1937) Die chemischen Arbeiten auf der 'Meteor'-Fahrt Februar bis Mai 1937. *Ann. Hydrogr. Mar. Met., Beiheft September,* pp. 17–22.

WAUGH, G. D. (1960) The ecology and mode of life of the edible molluscs. *Proc. malac. Soc. Lond.* **34,** 113–22.

VAN WEEL, P. B. (1937) Die Ernährungsbiologie von *Amphioxus lanceolatus. Pubbl. Staz. zool. Napoli,* **16,** 221–72.

VAN WEEL, P. B. (1940) Beiträge zur Ernährungsbiologie der Ascidien. *Pubbl. Staz. zool. Napoli,* **18,** 50–79.

VAN WEEL, P. B. (1949) On the physiology of the tropical fresh-water sponge, *Spongilla proliferanes* Annand. I. Ingestion, digestion and excretion. *Physiol. comp. oecol.* **1,** 110–26.

WELLS, G. P. and DALES, R. P. (1951) Spontaneous activity patterns in animal behaviour: The irrigation of the burrow in the polychaetes *Chaetopterus variopedatus* Renier and *Nereis diversicolor* O. F. Müller. *J. Mar. biol. Ass. U.K.* **29,** 661–80.

WERNER, B. (1951) Über die Bedeutung der Wasserstromerzeugung und Wasserstromfiltration für die Nahrungsaufnahme der ortsgebundenen Meeresschnecke *Crepidula fornicata* L. (Gastropoda Prosobranchia). *Zool. Anz.* **146,** 97–113.

WERNER, B. (1953a) Über den Nahrungserwerb der Calyptraeidae (Gastropoda Prosobranchia). Morphologie, Histologie und Funktion der am Narungserwerb beteiligten Organe. *Helgoländ. wiss. Meeresunters.* **4,** 260–315.

WERNER, B. (1953b) Ausbildungsstufen der Filtrationsmechanismen bei filtrierenden Prosobranchiern. *Verh. dtsch. zool.* 1952, pp. 529–46.

WERNER, B. (1955) Über die Anatomie, die Entwicklung und Biologie des Veligers und der Velichoncha von *Crepidula fornicata* L. (Gastropoda Prosobranchia). *Helgoländ. wiss. Meeresunters.* **5,** 169–217.

WERNER, B. (1959) Das Prinzip des endlosen Schleimfilters beim Nahrungserwerb wirbelloser Meerestiere. *Int. Rev. Hydrobiol.* **44,** 181–215.

WERNER, E. and WERNER, B. (1954) Über den Mechanismus des Nahrungserwerbs der Tunicaten, speciell der Ascidien. *Hegoländ. wiss Meeresunters,* **5,** 57–92.

WERNSTEDT, C. (1942) Studies on the food of *Macoma baltica* and *Cardium edule. Vidensk. Medd. dansk naturh. Foren. Kbh.* **106,** 241–52.

WESENBERG-LUND, C. (1943) *Biologie der Süsswasserinsekten.* Springer–Verlog.

WHITEHEAD, P. J. (1959) Feeding mechanism of *Tilapia nigra. Nature, Lond.* **184,** 1509–10.

WICKSTEAD, J. (1962) Food and feeding in pelagic copepods. *Proc. zool. Soc. Lond.* **139,** 545–55.

WICKSTEAD, J. H. and BONE, Q. (1959) Ecology of acraniate larvae. *Nature, Lond.* **184,** 1849–51.

WILLEMSEN, J. (1952) Quantities of water pumped by mussels (*Mytilus edulis*) and cockles (*Cardium edule*). *Arch. neérl. Zool.* **10,** 153–60.

WILLÉN, T. (1959) The phytoplankton of Görväln, a bay of Lake Mälaren. *Oikos,* **10,** 241–74.

Williams, P. M. (1961) Organic acids in Pacific Ocean waters. *Nature, Lond.* **189,** 219–20.

Wilska, A. (1953) Membrane structures in mucus. *Experientia,* **9,** 293.

Wilson, D. P. (1958) Some problems in larval ecology related to the localized distribution of bottom animals. Perspectives in Marine Biology. Ed. A. A. Buzzati-Traverso. *Int. Union biol. Sci. Ser. B.* **27,** 87–99.

Wilson, R. F. (1961) Measurements of organic carbon in sea water. *Limnol. Oceanogr.* **6,** 259–61.

Wimpenny, R. S. (1958) Carbon production in the sea at the Smith's Knoll Lightvessel. *Rapp. Cons. Explor. Mer,* **144,** 70–2.

Wintzell, J. (1918) *Bidrag till de skandinaviska Ophiuridernas biologi och fysiologi.* Uppsala. Diss.

Wolff, T. (1960) The hadal community, an introduction. *Deep-Sea Res.* **6,** 95–124.

Wood, E. J. F. (1953) Heterotrophic bacteria in marine environment of Eastern Australia. *Aust. J. Mar. Freshw. Res.* **4,** 160–200.

Wood, E. J. F. (1963) Heterotrophic micro-organisms in the oceans. *Ann. Rev. Oceanogr. Mar. Biol.* **1,** 197–222.

Wood, E. J. F. and Davis, P. S. (1956) Importance of smaller phytoplankton elements. *Nature, Lond.* **177,** 438.

Wright, J. C. (1959) Limnology of Canyon Ferry Reservoir. II. Phytoplankton standing crop and primary production. *Limnol. Oceanogr.* **4,** 235–45.

Yamaguchi, Z. (1931) Some notes on the physiology of *Styela clava. Sci. Rep. Tôhoku Imp. Univ.* **6,** 597–607.

Yasuda, F. (1960) The relationship of the gill structure and food habits of some coastal fishes in Japan. *Rec. Oceanogr. Works in Japan,* **5,** 139–52.

Yentsch, C. S. and Ryther, J. H. (1959a) Absorption curves of acetone extracts of deep water particulate matter. *Deep-Sea Res.* **6,** 72–4.

Yentsch, C. S. and Ryther, J. H. (1959b) Relative significance of the phytoplankton and nanoplankton in the waters of Vineyard Sound. *J. Cons. int. Explor. Mer,* **24,** 231–38.

Yentsch, C. S. and Scagel, R. F. (1958) Diurnal study of phytoplankton pigments, an *in situ* study in East Sound, Washington. *J. Mar. Res.* **17,** 567–83.

Yentsch, C. S. and Vaccaro, R. F. (1958) Phytoplankton nitrogen in the oceans. *Limnol. Oceanogr.* **3,** 443–8.

Yonge, C. M. (1926a) Structure and physiology of the organs of feeding and digestion in *Ostrea edulis. J. Mar. biol. Ass. U.K.* **14,** 295–386.

Yonge, C. M. (1926b) Ciliary feeding mechanisms in the thecosomatous pteropods. *Linn. Soc. J. Zool.* **36,** 417–29.

Yonge, C. M. (1927) Structure and function of the organs of feeding and digestion in the septibranchs, *Cuspidaria* and *Poromya. Phil. Trans. roy. Soc. Lond. B.* **216,** 221–63.

Yonge, C. M. (1928) Feeding mechanisms in the invertebrates. *Biol. Rev.* **3,** 21–76.

Yonge, C. M. (1931) Digestive processes in marine invertebrates and fishes. *J. Cons. int. Explor. Mer,* **6,** 175–212.

YONGE, C. M. (1932) Notes on feeding and digestion in *Pterocera* and *Vermetus,* with a discussion on the occurrence of the crystalline style in the Gastropoda. Great Barrier Reef Expedition 1928–29. *Sci. Rep.* **1,** 259–81.

YONGE, C. M. (1936) Mode of life, feeding, digestion and symbiosis with Zooxanthellae in the Tridacnidae. *Sci. Rep. Gr. Barrier Reef Exped. Brit. Mus.* (*Nat. Hist.*) **1,** 283–321.

YONGE, C. M. (1937a) Evolution and adaptation in the digestive system of the Metazoa. *Biol. Rev.* **12,** 87–115.

YONGE, C. M. (1937b) The biology of *Aporrhais pes-pelicani* (L.) and *A. serresiana* (Mich). *J. Mar. biol. Ass. U.K.* **21,** 687–704.

YONGE, C. M. (1938) Evolution of ciliary feeding in the Prosobranchia, with an account of feeding in *Capulus urgaricus. J. Mar. biol. Ass. U.K.* **22,** 453–68.

YONGE, C. M. (1939) The protobranchiate Mollusca: a functional interpretation of their structure and evolution. *Phil. Trans. roy. Soc. Lond. B.* **230,** 79–147.

YONGE, C. M. (1946a) On the habits and adaptations of *Aloidis* (*Corbula*) *gibba. J. Mar. biol. Ass. U.K.* **26,** 358–76.

YONGE, C. M. (1946b) On the habits of *Turritella communis* Risso. *J. Mar. biol. Ass. U.K.* **26,** 377–80.

YONGE, C. M. (1947) The pallial organs in the aspidobranch Gastropoda and their evolution throughout the Mollusca. *Phil. Trans. roy. Soc. Lond. B.* **232,** 443–518.

YONGE, C. M. (1948) Formation of siphons in Lamellibranchia. *Nature, Lond.* **161,** 198.

YONGE, C. M. (1950) On the structure and adaptations of the Tellinacea, deposit-feeding Eulamellibranchia. *Phil. Trans. roy. Soc. Lond. B.* **234,** 29–76.

YONGE, C. M. (1951a) Studies on Pacific coast mollusks. I. On the structure and adaptations of *Cryptomya californica* (Conrad). *Univ. Calif. Publ. Zool.* **55,** 395–400.

YONGE, C. M. (1951b) Studies on Pacific coast mollusks. IV. Observations on *Siliqua patula* Dixon and on evolution within the Solenidae. *Univ. Calif. Publ. Zool.* **55,** 421–38.

YONGE, C. M. (1951c) Studies on Pacific coast mollusks. V. Structure and adaptation in *Entodesma saxicola* (Baird) and *Mytilimeria nuttallii* Conrad with discussion on evolution within the family Lyonsiidae (Eulamellibranchia). *Univ. Calif. Publ. Zool.* **55,** 439–50.

YONGE, C. M. (1953) Mantle chambers and water circulation in the Tridacnidae (Mollusca). *Proc. zool. Soc. Lond.* **123,** 551–61.

YONGE, C. M. (1957) Mantle fusion in the Lamellibranchia. *Pubbl. Staz. zool. Napoli,* **29,** 151–71.

YONGE, C. M. (1960) *Oysters.* Collins, London.

YONGE, C. M. and ILES, E. J. (1939) On the mantle cavity, pedal glad, and evolution of mucous feeding in the Vermetidae. *Ann. Mag. nat. Hist.* **3,** 536–56.

ZEIN-ELDIN, Z. P. and MAY, B. Z. (1958) Improved N-ethylcarbazole determination of carbohydrates with emphasis on sea water samples. *Analyt. Chem.* **30,** 1935–41.

ZENKEVITCH, L. (1963) *Biolgy of the Seas of the U.S.S.R.* George Allen and Unwin Ltd., London.

ZEUTHEN, E. (1947) Body size and metabolic rate in the Animal Kingdom with special regard to the marine micro-fauna. *C. R. Lab. Carlsberg Sér. Chim.* **26,** 17–161.

ZOBELL, C. E. (1940) The effect of oxygen tension on the rate of oxidation of organic matter in sea water by bacteria. *J. Mar. Res.* **3,** 211–23.

ZOBELL, C. E. (1943) The effect of solid surfaces upon bacterial activity. *J. Bact.* **46,** 39–56.

ZOBELL, C. E. (1946) *Marine Microbiology*. Waltham, Mass. U.S.A.

ZOBELL, C. E. (1954) The occurrence of bacteria in the deep sea and their significance for animal life. *Int. Union biol. Sci. Sér. B.* **16,** 20–3.

ZOBELL, C. E. and ANDERSON, D. Q. (1936) Observations on the multiplication of bacteria in different volumes of stored sea water and the influence of oxygen tension and solid surface. *Biol. Bull.* **71,** 324–42.

ZOBELL, C. E. and FELTHAM, C. B. (1937–38) Bacteria as food for certain marine invertebrates. *J. Mar. Res.* **1,** 312–27.

ZOBELL, C. E. and FELTHAM, C. B. (1942) The bacterial flora of a marine mud flat as an ecological factor. *Ecology,* **23,** 69–78.

ZOBELL, C. E. and GRANT, C. W. (1943) Bacterial utilization of low concentrations of organic matter. *J. Bact.* **45,** 555–64.

ZOBELL, C. E. and LANDON, W. A. (1937) The bacterial nutrition of the California mussel. *Proc. Soc. exp. Biol. N.Y.* **36,** 607–9.

ZOBELL, C. E. and MORITA, R. Y. (1957–59) Deep-sea bacteria. *Galathea Rep.* **1,** 139–54.

AUTHOR INDEX

SUBJECT INDEX

OTHER TITLES IN THE ZOOLOGY DIVISION

General Editor: G. A. KERKUT

Vol. 1. RAVEN—*An Outline of Development Physiology*
Vol. 2. RAVEN—*Morphogenesis: The Analysis of Molluscan Development*
Vol. 3. SAVORY—*Instinctive Living*
Vol. 4. KERKUT—*Implications of Evolution*
Vol. 5. TARTAR—*The Biology of Stentor*
Vol. 6. JENKIN—*Animal Hormones—A Comparative Survey*
Vol. 7. CORLISS—*The Ciliated Protozoa*
Vol. 8. GEORGE—*The Brain as a Computer*
Vol. 9. ARTHUR—*Ticks and Disease*
Vol. 10. RAVEN—*Oogenesis*
Vol. 11. MANN—*Leeches (Hirudinea)*
Vol. 12. SLEIGH—*The Biology of Cilia and Flagella*
Vol. 13. PITELKA—*Electron-Microscopic Structure of Protozoa*
Vol. 14. FINGERMAN—*The Control of Chromatophores*
Vol. 15. LAVERACK—*The Physiology of Earthworms*
Vol. 16. HADZI—*The Evolution of the Metazoa*
Vol. 17. CLEMENTS—*The Physiology of Mosquitoes*
Vol. 18. RAYMONT—*Plankton and Productivity in the Oceans*
Vol. 19. POTTS and PARRY—*Osmotic and Ionic Regulation in Animals*
Vol. 20. GLASGOW—*The Distribution and Abundance of Tsetse*
Vol. 21. PANTELOURIS—*The Common Liver Fluke*
Vol. 22. VANDEL—*Biospeleology—The Biology of Cavernicolous Animals*
Vol. 23. MUNDAY—*Studies in Comparative Biochemistry*
Vol. 24. ROBINSON—*Genetics of the Norway Rat*
Vol. 25. NEEDHAM—*The Uniqueness of Biological Materials*
Vol. 26. BACCI—*Sex Determination*

OTHER DIVISIONS IN THE SERIES ON PURE AND APPLIED BIOLOGY

BIOCHEMISTRY

BOTANY

MODERN TRENDS IN PHYSIOLOGICAL SCIENCES

PLANT PHYSIOLOGY